C000170841

"Providing in-depth analyses of the paradoxi‹
routinely exploit nonhuman animals, the insig
Exploit Animals illuminate the critical import
with other animals to create a more compassionate socity for
for us, a win-win for all."

Marc Bekoff, PhD, co-author of *The Animals' Agenda: Freedom,*
Compassion, and Coexistence in the Human Age **and** *Unleashing Your*
Dog: A Field Guide to Giving Your Canine Companion the Best Life
Possible, **and Professor Emeritus of Ecology and Evolutionary**
Biology, University of Colorado, USA

"In this groundbreaking book, leading writers and animal advocates explore
what editors Kristof Dhont and Gordon Hodson rightly describe as 'one
of the most pressing contradictions in human behavior': our dual capacity
to love and harm animals. If we hope to halt climate change, advance social
justice, and build a healthy future, this book is essential reading."

Scott Plous, Professor of Psychology and Executive Director
of the Social Psychology Network, Wesleyan University, USA

"*Why We Love and Exploit Animals* is essential reading for anyone wishing to
improve their understanding of this vitally important yet little understood phenomenon – and to make choices that help create a better world for all beings."

Melanie Joy, PhD, author of *Why We Love Dogs, Eat Pigs, and Wear*
Cows: An Introduction to Carnism **and** *Powerarchy: Understanding the*
Psychology of Oppression for Social Transformation

"This volume offers new and unique interdisciplinary perspectives on an
issue, humans' relationships with other animals, that is of significant everyday
importance. The editors, both of whom are distinguished scholars in psychology, have brought leading international researchers and advocates together
in enlightening and constructive conversation. The chapters in this skillfully
edited volume are of consistently high quality. Individually and collectively,
the authors balance research and practice, complexity of issues with clarity of
presentation, and different and often competing perspectives to address provocative questions about the many facets of humans' relations with animals.
This book is of great value to researchers, who will be stimulated by the novel
issues it raises (e.g., about altruism, empathy, and identity), and advocates, who
will gain new insights into how people think about and treat animals. The
chapters in this volume will stimulate new inquiry and informed discussion
about a topic of great importance and interest today and in the future."

John F. Dovidio, Carl Iver Hovland Professor of Psychology
and Public Health, Yale University, USA

"Dhont and Hodson assemble an impressive list of leading researchers studying how we think, feel, and behave toward animals. This book is a must-read for anyone who wants to understand our paradoxical relationship with animals. Dhont and Hodson offer a compelling case for thinking seriously about how we both love and exploit animals – and how this paradox plays out both for individuals, but also for society as a whole."

Chris G. Sibley, Professor of Psychology,
University of Auckland, Australia

"The protection of animals has taken a front seat in the minds of the general public like never before. This book is for the seasoned animal activist, the academic, and anyone who has ever questioned the place of animals in society. Eminently readable, the authors help us to distil the changing landscape for animals, and our place within it. They take us on a journey of answering the hard questions we have all wondered about. A 'must-have' in your library."

Krista Hiddema, President of Happily Ever Esther
Farm Sanctuary, home of Esther the Wonder Pig, and Executive
Director of For the Greater Good, www.Kristahiddema.com

"Comprehensive, provocative, and exciting, this book provides timely insight into how we reconcile animal as man's best friend and dinner. Dhont and Hodson have gathered together world leaders in animal advocacy, and the science of understanding animal exploitation; together, they present research, case studies, and stories that explain how we can revere, use, and abuse our fellow creatures. Finally, this ground-breaking book augurs change towards a kinder, greener, planet for both human and animal."

Fiona Kate Barlow, Associate Professor of Psychology and
Director of the Centre for Research in Social Psychology,
University of Queensland, Australia

"Making legal progress for animals is impossible without grappling with the psychological and social mechanisms that allow people to simultaneously adore animals, while exploiting and harming them for food, fashion, experiments, and entertainment. Some of the insights in this stunning new book will delight you, some will discourage you, but they will all make you a better, smarter, and more evidence-driven animal advocate."

Camille Labchuck, Executive Director of Animal Justice

"Advocacy for nonhuman animals has long been separated from animal-related academic research, to the detriment of both fields. *Why We Love and Exploit Animals* bridges that separation and provides unique insights framed by social science and informed by leading advocates and thinkers. It's an important book for anyone interested in animal protection!"

Che Green, Founder and Executive Director of Faunalytics

"Dhont and Hodson's book is a very rich collection of chapters from the major scholars in the field. This book is very thought-provoking and a must read for everyone, vegetarians and non-vegetarians, interested in human-animal relations and animal advocacy!"

Julia Becker, Professor of Social Psychology, University of Osnabrück, Germany

ASKHAM BRYAN
COLLEGE
LEARNING RESOURCES

WHY WE LOVE AND EXPLOIT ANIMALS

This unique book brings together research and theorizing on human-animal relations, animal advocacy, and the factors underlying exploitative attitudes and behaviors towards animals.

Why do we both love and exploit animals? Assembling some of the world's leading academics and with insights and experiences gleaned from those on the front lines of animal advocacy, this pioneering collection breaks new ground, synthesizing scientific perspectives and empirical findings. The authors show the complexities and paradoxes in human-animal relations and reveal the factors shaping compassionate versus exploitative attitudes and behaviors towards animals. Exploring topical issues such as meat consumption, intensive farming, speciesism, and effective animal advocacy, this book demonstrates how we both value and devalue animals, how we can address animal suffering, and how our thinking about animals is connected to our thinking about human intergroup relations and the dehumanization of human groups.

This is essential reading for students, scholars, and professionals in the social and behavioral sciences interested in human-animal relations, and will also strongly appeal to members of animal rights organizations, animal rights advocates, policy makers, and charity workers.

Kristof Dhont, PhD, is a Senior Lecturer in Psychology at the University of Kent, UK, and Director of SHARKLab, studying human intergroup and human-animal relations. He investigates the psychological factors underpinning speciesism, racism, and sexism and serves as Associate Editor (*Group Processes & Intergroup Relations*) and Consulting Editor (*European Journal of Personality*).

Gordon Hodson, PhD, is a Professor of Psychology at Brock University, Canada. His research interests include prejudice, dehumanization and speciesism, ideology, and intergroup contact. He is an Editor-in-Chief (*European Review of Social Psychology*) and Associate Editor (*Group Processes & Intergroup Relations*). He is a Fellow of the Association for Psychological Science.

WHY WE LOVE AND EXPLOIT ANIMALS

Bridging Insights from Academia and Advocacy

Edited by Kristof Dhont and Gordon Hodson

Routledge
Taylor & Francis Group

LONDON AND NEW YORK

First published 2020
by Routledge
2 Park Square, Milton Park, Abingdon, Oxon OX14 4RN

and by Routledge
52 Vanderbilt Avenue, New York, NY 10017

Routledge is an imprint of the Taylor & Francis Group, an informa business

© 2020 selection and editorial matter, Kristof Dhont and Gordon Hodson;
individual chapters, the contributors

The right of Kristof Dhont and Gordon Hodson to be identified as the authors of the
editorial material, and of the authors for their individual chapters, has been asserted in
accordance with sections 77 and 78 of the Copyright, Designs and Patents Act 1988.

All rights reserved. No part of this book may be reprinted or reproduced or utilised
in any form or by any electronic, mechanical, or other means, now known or
hereafter invented, including photocopying and recording, or in any information
storage or retrieval system, without permission in writing from the publishers.

Trademark notice: Product or corporate names may be trademarks or registered trademarks,
and are used only for identification and explanation without intent to infringe.

British Library Cataloguing-in-Publication Data
A catalogue record for this book is available from the British Library

Library of Congress Cataloging-in-Publication Data
A catalog record has been requested for this book

ISBN: 978-0-8153-9664-2 (hbk)
ISBN: 978-0-8153-9665-9 (pbk)
ISBN: 978-1-351-18144-0 (ebk)

Typeset in Bembo
by Newgen Publishing UK

MIX
Paper from
responsible sources
FSC
www.fsc.org
FSC™ C013985

Printed in the United Kingdom
by Henry Ling Limited

To Ada, Nora, Elian, and Liene, for their love and the joy they bring to my life
every day, and to Frike, for her ceaseless support, patience, and love. (Kristof)

To the strong and independent women in my life who have supported and
inspired me: Myrtle, Holly, Tarah, and Eva. (Gordon)

CONTENTS

CONTRIBUTORS

Editors

Kristof Dhont, PhD, is a Senior Lecturer in Psychology at the University of Kent, UK. He is founder and director of SHARKLab, studying the role of personality and situational factors in human intergroup and human-animal relations. He investigates the psychological underpinnings and ideological roots of speciesism, racism, and sexism, and the moral psychology of eating and exploiting animals. He completed his PhD and a post-doctoral fellowship at Ghent University, Belgium, and he currently serves as Associate Editor for the journal *Group Processes & Intergroup Relations* and as Consulting Editor for the *European Journal of Personality*.

Gordon Hodson, PhD, is a Professor of Psychology at Brock University, Canada. His research interests include stereotyping, prejudice, and discrimination, with a focus on ideology, intergroup contact/friendship, and dehumanization. He co-edited *Advances in Intergroup Contact* (Hodson & Hewstone, 2013), and presently serves as Editor-in-Chief at the *European Review of Social Psychology* and Associate Editor at *Group Processes and Intergroup Relations*. He is a fellow of the Association for Psychological Science (APS), Society of Experimental Social Psychology (SESP), and the Society for Personality and Social Psychology (SPSP).

Authors

Carol J. Adams is the author of *The Sexual Politics of Meat* (1990), now in a 25th anniversary edition. It has been translated into German, Japanese, Chinese, Portuguese, Turkish, Portuguese, French, Spanish, and Korean. She is also the author of *Neither Man nor Beast* (1995/2018), *Burger* (2018) and the editor of five anthologies, including *Ecofeminism: Feminist Intersections with Other Animals and the Earth*, with Lori Gruen. With Virginia Messina, she has recently authored *Protest Kitchen*.

Catherine E. Amiot, PhD, is a Professor at the Department of Psychology, University of Québec in Montréal, Canada. Her research focuses on intergroup relations and self and identity processes, and in what ways these theoretical principles apply (or not) to human–animal relations in particular. Specifically, she seeks to understand how we can come to identify with animals as a broad and inclusive social group, and what social- and individual-level factors impede versus promote mutually beneficial human–animal relations.

Brock Bastian, PhD, is an Associate Professor in the Melbourne School of Psychological Sciences, The University of Melbourne, Australia. His research has touched on human animal relations from several different perspectives, seeking to understand not only how we relate to animals, but what that relationship can tell us about how we expand and retract our moral consideration for non-humans, and how we resolve moral dilemmas in everyday life.

Gene Baur is co-founder and president of Farm Sanctuary, and has been hailed as "the conscience of the food movement" by *TIME* magazine. Since 1986, he has campaigned to raise awareness about the abuses of animal agriculture. Gene was a pioneer in undercover investigations and instrumental in passing the first US laws to ban cruel factory farming practices. He has a master's degree in agricultural economics from Cornell University, USA, and is the author of two bestselling books.

Jon Bockman has held diverse leadership positions in nonprofit animal advocacy over the past decade. He served as Executive Director at Animal Charity Evaluators (ACE), a meta-charity that works to find and promote the most effective ways to help animals. He has overseen five rounds of charity evaluations, and has helped influence over $117 million in donations to ACE recommended charities. Prior to his career at ACE, Jon served as a director at a wildlife rehabilitation center, a humane investigator, a small vegan business owner, a member of the Institutional Animal Care and Use Committee at Northern Illinois University, USA, and founder of a farmed animal advocacy charity.

Matthew Calarco, PhD, is Professor of Philosophy at California State University, Fullerton, USA, where he teaches courses in Continental philosophy and animal and environmental philosophy. He is author of *Zoographies: The Question of the Animal from Heidegger to Derrida* (2008), *Thinking Through Animals* (2015) and co-editor with Peter Atterton of *Animal Philosophy* (2004). He is currently completing a book entitled *Altermobilities: Profaning the Streets.*

Thomas Davies is a PhD candidate at the University of Edinburgh, UK, supervised by Dr. Steve Loughnan and Dr. Adam Moore. Broadly speaking, his interests are in the social psychology of group relations – both intragroup and intergroup relations. His PhD thesis examines how challenging the notion of human supremacy over animals can have a positive impact on perceptions of various human groups.

Megan Earle is a PhD student in Social/Personality Psychology at Brock University, Canada. Her research interests include individual differences, prejudice, and discrimination, with a focus on ideology, intergroup contact, and the impact of media on social attitudes.

Susan T. Fiske, PhD, is Eugene Higgins Professor, Psychology and Public Affairs, Princeton University, NJ, USA, where she investigates cognitive stereotypes, emotional prejudices, and discriminatory tendencies, at cultural and interpersonal levels, with policy implications.

Brian Hare, PhD, is Professor of Evolutionary Anthropology, Psychology and Neuroscience at Duke University in North Carolina, USA and a core member of the Center for Cognitive Neuroscience. He received his PhD from Harvard University, founded the Hominoid Psychology Research Group while at the Max Planck Institute for Evolutionary Anthropology, and subsequently created the Duke Canine Cognition Center when arriving at Duke. He is fascinated by what animals can teach us about being human and using this new understanding to solve modern problems.

Nick Haslam, PhD, is Professor of Psychology at the University of Melbourne, Australia. His research interests include the psychology of intergroup relations, personality, and psychiatric classification. He has written extensively on dehumanization and related phenomena.

Harold Herzog, PhD, is Emeritus Professor of Psychology at Western Carolina University, USA. He has investigated psychological aspects of human-animal relationships for over 30 years. He is the author of *Some We Love, Some We Hate, Some We Eat: Why It Is So Hard To Think Straight About Animals.*

Christopher J. Holden, PhD, is an Assistant Professor of Psychology at Appalachian State University, USA. He is a personality psychologist who takes an evolutionary approach to understanding close relationships in humans. As many of us have emotional connections with our pets (i.e., a close relationship), he has expanded his research to include human-animal relationships.

Elise Holland, PhD, is an Honorary Research Fellow at the University of Melbourne and Senior Research and Evaluation Advisor at Our Watch, Australia's national foundation to prevent violence against women and their children. Her research focuses on the objectification and dehumanization of women in the media and in daily life.

Tobias Leenaert studied Germanic languages and comparative sciences of culture at the University of Ghent, Belgium. He is the author of *How to Create a Vegan World: A Pragmatic Approach*, and blogs about animal advocacy issues at www.veganstrategist.org. Tobias is a long-time speaker and strategist, and gives vegan advocacy trainings all over the world for the Center for Effective Vegan Advocacy (CEVA), which he co-founded with Dr. Melanie Joy. He is also co-founder of the Belgian vegetarian organization EVA, and of ProVeg International, a pro-vegan organization with the mission to reduce the global consumption of animals by 50 percent by the year 2040.

Ana C. Leite, PhD, is an Assistant Professor in social and organisational psychology at Durham University, UK. Ana completed her PhD in Psychology at the University of Porto, Portugal. Her research focuses on the factors underlying the social inclusion and exclusion of human and non-human animals. Given her professional background in consulting (Deloitte), she is also interested in uncovering and testing effective strategies for promoting well-being and equality at the workplace.

Steve Loughnan, PhD, is a Reader at the University of Edinburgh, UK. His interests focus on the human/animal divide. On one hand, this examines how people are seen as animals (dehumanization, objectification), and on the other how people think about and treat animals (consumption, anthropomorphism).

Jared Piazza, PhD, is a Lecturer of Psychology at Lancaster University, UK. He studies moral cognition, including the way we think morally about animals. His research has been published in journals such as *Appetite, Cognition*, and *Journal of Personality and Social Psychology*. He has ongoing collaborations with animal advocacy groups such as Greenpeace International and Beyond Carnism.

Jessica Pierce, PhD, is a bioethicist and writer. Her research covers a broad span of topics in bioethics, with special focus on animal ethics and environmental bioethics. Jessica is the author of nine books, including *The Animals' Agenda: Freedom, Compassion, and Coexistence in the Age of Humans* (with Marc Bekoff), *Run, Spot, Run: The Ethics of Keeping Pets* and *The Last Walk: Reflections on Our Pets at the Ends of Their Lives*. She holds a PhD from the University of Virginia and an MTS from Harvard Divinity School, USA.

Hank Rothgerber, PhD, is a Professor of Psychology at Bellarmine University, USA. He is an experimental social psychologist whose research focuses on the psychology of meat eating and its counterpart, vegetarianism. His work on meat eating adopts a cognitive dissonance perspective and seeks to identify and better understand the dissonance-reducing mechanisms by which individuals continue to eat meat in the face of increased criticism. In addition, over the last few years he has used quantitative methods to examine conscientious omnivores, semi-vegetarians, and questions of how vegetarians approach the dilemma of feeding their pets a meat-based diet.

Paul Rozin, PhD, received his PhD in Biology and Psychology from Harvard University, USA, under the sponsorship of Jean Mayer. He has been a member of the Psychology Department at the University of Pennsylvania for 55 years, where he is currently Professor of Psychology. His principal interest has been in how culture, in interaction with individual experience and biology, shapes the way humans deal with food and eating. Topics include the development of preferences and aversions, attitudes to meat, attitudes and beliefs about "natural," and conflicts with respect to food choice between pleasure, and health and moral concerns.

Matthew B. Ruby, PhD, is a Lecturer in Psychology at La Trobe University, Australia. His main research programs center on the psychology of food choice and eating behavior, in particular how people decide which (animal) foods are acceptable to eat and which are not, the tension many people feel between loving animals and loving meat, cultural differences in veganism/vegetarianism, and the factors that support and hinder people's transition to a plant-based diet.

Alina Salmen is a PhD candidate in Social Psychology at the University of Kent, UK. Her research focuses on the common ideological roots underlying exploitative tendencies towards women and animals, as well as the role of moral and individual difference factors in meat consumption and the exploitation of animals. More broadly, her research interests include political psychology, gender relations, and the dehumanization of women.

Verónica Sevillano, PhD, is an Assistant Professor in Social Psychology and Methodology Department at Universidad Autónoma de Madrid, Spain. Her research examines cognitive biases in environmental concern, and human-animal relations.

John Sorenson, PhD, is a Professor teaching courses on animals and human society in the Department of Sociology at Brock University, a leading Canadian institution in the rapidly-growing field of Critical Animal Studies. He has written and edited numerous books and articles on various aspects of human-animal relationships. His latest books are *Critical Animal Studies: Towards Trans-Species Social Justice* and *Dog's Best Friend? Rethinking Canid-Human Relations* (both co-edited with Atsuko Matsuoka).

Michelle Stratemeyer, PhD, is an Associate Lecturer at the University of Melbourne, Australia and a Research Fellow at the Centre for Ethical Leadership. Her research focuses on several areas in applied social and organizational psychology, including intimate partner violence, sexual objectification, harassment and bystander interventions, organizational diversity and inclusion, and leadership.

Vanessa Woods is a Research Scientist in Evolutionary Anthropology at Duke University, USA. She has worked in the Democratic Republic of Congo studying bonobos, and the People's Republic of Congo studying chimpanzees. Woods is also interested in science communication and has published popular articles and books about bonobos and dogs.

ACKNOWLEDGMENTS

This book would not have been possible without the assistance and support from others. We would like to thank our publisher, Routledge, particularly Eleanor Reedy for her support and encouragement throughout the entire process, and Alex Howard for his assistance.

We are extremely grateful to the many excellent scholars and authors who served as reviewers and provided thoughtful and constructive comments on earlier versions of the chapters. Special thanks go to Ulf Jürgens, Victoria Krings, and Alina Salmen for their editorial help with some of the chapters. Thanks to Brock University for their support to Gordon Hodson through the Brock University Chancellor's Chair for Research Excellence award. Finally, we are tremendously grateful to all authors in this book for their exceptional and inspiring contributions, for their willingness to revise their chapters in response to the feedback of the reviewers and our editorial feedback, and for making this book possible.

1

LOVING AND EXPLOITING ANIMALS

An introduction

Kristof Dhont and Gordon Hodson

In the summer of 2012, Steve and Derek, until then living a fairly ordinary life in Ontario (Canada), adopted a new pet animal. They named her Esther. Weighing four pounds and just a couple of weeks old, Esther instantly stole their hearts, and not much later, also the hearts of hundreds of thousands of others all over the world. It took Esther less than two years to become a social media star, with presently close to one and a half million followers on Facebook.

Esther is neither a dog nor a cat, but a pig, now weighing 650 pounds. Little did Steve and Derek know at that time what to expect from Esther. Not in the least bit did they expect that Esther would turn their lives upside down. For Steve and Derek, Esther was not considered a food animal but instead one of their pets, alongside their cats and dogs. This recategorization of pigs into a different group of animals caused a mental shift in their thinking about animals overall. By taking care of Esther they soon realized how clever and playful she is. Even more, they came to realize that Esther was not alone but rather that all pigs have unique personalities, sophisticated mental abilities, and needs. In 2014, Steve and Derek founded the Happily Ever Esther Farm Sanctuary (Ontario, Canada), where they take care of dozens of farm animals rescued from being slaughtered or other exploitative circumstances. They made a conscious choice to love and *not* exploit animals, and to dedicate their lives to caring compassionately for both companion and farm animals.

Although the number of farm sanctuaries is growing, Steve and Derek are vastly outnumbered by people who either go to great lengths to rationalize their involvement in the exploitation of animals or have never questioned their involvement. They represent an intriguing yet rare case in a world where we are socialized from the day we are born to live with the paradoxical societal norm of simultaneously loving and exploiting animals. Indeed, from an early age we have been taught that harming animals is morally wrong, and the majority of people find it emotionally disturbing to watch an animal suffering or being maltreated. Many people are genuinely captivated when observing a flock of starlings in flight or a 15-meter humpback whale jumping out of the water. We feel moved or emotionally lifted when our pet dog, cat, or backyard chickens greet us in the morning, and we enjoy watching wildlife documentaries (e.g., Planet Earth).

At the same time, we exploit animals in numerous different ways on a systemic level and on a massive scale, including for consumption, clothes, entertainment, sports, cosmetic tests, medical experiments, pet breeding, and warfare. Consider for instance the animal agriculture industry, which is responsible for the suffering of the largest number of animals inflicted by humans. In the USA alone, more than nine billion land animals are slaughtered for food every year (Sanders, 2018). Put differently, if you would start counting now, approximately 1000 animals have been killed by the time you were even able to finish reading this sentence (see also https://animalclock.org). Such numbers represent an enormous amount of animal suffering and are difficult to grasp. Now also think about how much energy, water, and land this industry requires to raise such gargantuan numbers of animals for the production of meat and other animal products. The negative environmental impact of animal agriculture is vast and now undeniable, being one of the key factors driving the loss of biodiversity, deforestation, water pollution and responsible for excessive water and energy use (e.g., Eshel, Shepon, Makov, & Milo, 2014; Godfray et al. 2018).

Animal agriculture is also one of the biggest contributors to climate change, with major papers recently published in the premier science outlets (Science; Nature; PNAS; Lancet) documenting how meat and dairy consumption account for more greenhouse gases than all forms of transportation combined (e.g., McKnight, 2014; Springman, Godfray, Rayner, & Scarborough, 2016; Tilman & Clark, 2015; Willet et al., 2019). To combat the climate crisis, monumental changes are needed, including the cutting back of red meat by more than 80 percent (Willet et al., 2019). But global demand for meat is at an all-time high. Worryingly, the actions we need to take are in direct conflict with the actions we actually take. We careen toward the destruction of our planet in large part due to our voracious appetites for the consumption of animals, while simultaneously professing to love animals. What can explain this great disconnect, arguably one of the most pressing contradictions in human behavior? How is it possible that we find it morally problematic to harm animals, yet also be directly or indirectly involved in harming animals on a daily basis? *Why do we love and exploit animals?* The answer to this question is not straightforward and must be explored from different angles and disciplines. This is precisely the purpose of this book.

Why we love and exploit animals is the central theme of the present book and runs as a thread through the entire package. This book consolidates the cross-disciplinary developments in a single and truly unique collection by bringing together (a) state-of-the-art contemporary research from social and behavioral sciences providing insights on the factors underpinning compassionate and exploitative attitudes and behaviors towards animals; (b) current views and hotly debated topics related to the ethics of eating, treating, and exploiting animals; and (c) accumulated insights from professionals and practitioners in the field of vegan and animal rights advocacy, along with the implications of research findings for their work and for social policy. We have assembled some of the world's leading academics, best-selling authors, and well-positioned animal advocates. All have influenced and advanced recent theoretical and practical developments in our understanding of people's thinking about and behavior towards animals, with many being pioneers in their field, with their work spanning decades. All have contributed cutting-edge research and/or have articulated revolutionary ideas.

By deliberately inviting authors from different backgrounds, academic disciplines, and outside of academia, we ensured that this diversity of perspectives would not result in one overarching or consistent message across chapters. Rather, you will discover different, at times clashing, ideas and conclusions from different authors. Bringing this variety of ideas together was purposefully done in order to facilitate critical thinking and challenge existing ideas.

We recognize that a substantial number of contributions have been written by authors from our own field, providing strong empirical bases in social, moral, and political psychology. Yet the book also has an explicit interdisciplinary scope, bridging psychology, sociology, and philosophy. On a broader level, the book uniquely bridges the gap between academics and practitioners/advocates by including contributions from leading authors and figures in the animal advocacy movement. In doing so, we emphasize the importance of connecting academics with advocates to increase the relevance and impact of both fields. Below, we provide an overview of the contributions to this book.

Contributions to this book

In Chapter 2, Brock Bastian and Catherine Amiot focus on the dynamic interplay between the extent to which people identify with animals on the one hand, and the extent to which people distance themselves from animals on the other. The intense psychological connections that people build with animals, but also the distance between humans and animals, represent key factors in shaping people's moral concerns and behaviors towards animals. As the authors argue, this dynamic shapes our own identities, the way we see ourselves, and, critically, the choices we make about both animal and human welfare issues.

The interconnection between attitudes towards animals and human outgroups is also of central importance in the next chapter. In Chapter 3, Kristof Dhont, Gordon Hodson, Ana Leite, and Alina Salmen specifically concentrate on the psychology of speciesism and address the question whether speciesism can be considered a type of prejudice and intergroup bias comparable to other types of prejudices and biases towards human groups (e.g., sexism and racism). They examine the common ideological belief systems underpinning different types of biases, highlighting the role of dominance desires and preferences for inequality in both human intergroup and human–animal relations. In discussing the parallels between different types of intergroup biases and biases towards animals, they consider the potential negative implications of such comparisons for animal advocacy and discuss the value of implementing experience- and behavior-based interventions in animal advocacy campaigns.

In Chapter 4, Jessica Pierce presents a thought-provoking discussion about the current state of animal welfare science, and how this research field systematically fails to seriously consider the interests of the animals in the treatment of captive animals. Indeed, drawing on her work with Marc Bekoff, Pierce highlights that even the field of animal welfare science typically puts the interest of humans first; this serves to legitimize animal exploitation within the different industries profiting from it (e.g., industrial farming, research laboratories, zoos) rather than meaningfully improving the lives of animals. A paradigm shift is needed, she argues, to transform the science of welfare into a science of well-being that incorporates new ethical principles, where animals are free from human captivity and from human-induced suffering.

In Chapter 5, Gordon Hodson, Kristof Dhont, and Megan Earle consider the impact of people's distorted thinking about animals, critically highlighting the implications of devaluing animals for how we think about other human groups. Specifically, relying on recent empirical research, the authors argue that considering animals to be of lower value than humans fuels the animalistic dehumanization of people belonging to another social group (e.g., ethnic minority group), which in turn leads to hostility and prejudice towards these other groups. They further expand this scope by considering the implications of people's beliefs about animals and meat consumption habits, for how people portray and treat vegans and vegetarians, who, as a social minority group, suffer prejudice, discrimination, and social isolation. The bottom line is that our thinking about animals is systematically connected to our thinking about human groups in ways that facilitate the exploitation and/or marginalization of both animals and human groups.

The idea that our thinking about animals is closely intertwined with our thinking about human outgroups is also manifested in the rich repertoire of animal metaphors that people use to describe other people. This is the central topic of Chapter 6, where Nick Haslam, Elise Holland, and Michelle Stratemeyer explore the psychology of animal metaphors and how such metaphors can be used to express praise or love for someone, but also (and even more frequently) to marginalize, dehumanize, or derogate others. Clearly, perceiving other people and social groups (ethnic or gender groups) in animalistic terms appears to be a widespread and often an automatic tendency in social perception, with potential detrimental consequences for how social groups are being treated. The dehumanization of people is also the focus in Chapter 7, where Vanessa Woods and Brian Hare focus on a specific type of dehumanization, namely the phenomenon of *simianization*, where people, especially Black people, are animalistically dehumanized by comparing them to great apes. Taking a historical perspective and citing sources dating back from the transatlantic slave trade, they convincingly illustrate the disturbing emotional relationship that humans have built with great apes. The close resemblance that we share with great apes has been a source of fascination, but at the same time it also elicits aversion and disgust as the result of apes being perceived as *too* humanlike. As the authors argue, precisely because of this aversive reaction, simianization has been a particularly powerful and popular form of dehumanization to denigrate and derogate other people.

Several chapters in this book specifically focus on meat consumption and people's distorted thinking about food animals relative to other types of animals. In Chapter 8, Jared Piazza addresses the question why it is so difficult to think ethically about our use of animals as food. He reviews empirical research documenting the disconnection between meat products and the animals killed for these products as well as the role of people's biased moral thinking about food animals and the rationalizations that sustain meat consumption. As Piazza argues, this disconnection allows meat eaters to love meat while at the same time being equally empathic as vegans in response to the suffering of farmed animals. Next, in Chapter 9 Christopher Holden and Harold Herzog zoom in on a specific yet powerful factor that distorts our thinking about animals, namely our perceptions of what is natural or unnatural. Specifically, the authors discuss how the belief that what is (perceived as) natural must be good acts unconsciously as a moral heuristic that biases people's moral judgements. Holden and Herzog apply this idea to the perception of animals and offer a newly developed scale (i.e., the Animal Naturalness Scale) to measure how people perceive the naturalness of animals in different situations and across categories, such as domesticated animals, animals in captivity or in the wild, and genetically modified animals. They find that people vary greatly in their evaluations of animal naturalness across different categories and that such perceptions are systematically associated with animal welfare concerns.

In Chapter 10, former executive director of Animal Charity Evaluators (ACE) Jon Bockman focuses on how to effectively reduce animal suffering. Relying on the principles of *effective altruism*, Bockman evaluates different domains of animal suffering (i.e., animals in the wild, animals in research, companion animals, and animals raised for food) in terms of scale, neglect, and tractability to learn how people can help the largest number of animals in the most effective way. Based on his careful analysis, Bockman reveals that helping farmed animals results in the biggest gains for animals overall. That is, farmed animal advocacy is identified as the most effective cause area. As discussed by Bockman, it is yet unclear, however, which are the most impactful interventions to help farmed animals. This highlights the need for more systematic research testing the effectiveness of different animal advocacy strategies.

The difficulties with developing successful strategies to change meat consumption behavior and tackle farm animal suffering become very apparent in the next two chapters, where the authors continue the psychological exploration of the paradox between being able to both eat animals and claim to love animals. Both the chapter of Steve Loughnan and Thomas Davies and that of Hank Rothgerber reveal that humans are psychologically well equipped with strong social and personal defence mechanisms that prevent them from feeling concerned about their meat consumption. In Chapter 11, Loughnan and Davies first map out the psychological factors at the level of the animals (eaten), the consumers (eaters), and the behavior (eating). In doing so, they also emphasize the importance of considering the broader social system that provides the context and incentives to preserve the status quo. In the next section, they move beyond the scope of animal exploitation and draw parallels with the other forms of exploitation, suggesting that animal exploitation is possibly "fundamental to other systems of oppression." In Chapter 12, Rothgerber elaborates on this idea and provides an accessible and timely overview of the empirical evidence on cognitive dissonance related to meat consumption. Unique for his chapter, however, is the consideration of the experiences of children. More specifically, Rothgerber addresses whether meat eating by children, as with adults, induces a negative emotional tension (i.e., dissonance) due to the awareness that meat consumption involves the killing and suffering of animals. The author also speculates about the different dissonance reduction strategies adopted by children and points out the differences and similarities with the strategies adopted by adults. More broadly, Rothgerber's discussion reveals that attitudes towards meat and animals among children is a largely unexplored area of research and highlights the potential effectiveness of focusing more on children both in research and in advocacy. Such a research approach would help us discover how our thinking about animals and meat is shaped from an early age. In terms of animal advocacy, targeting children in campaigns may prove to be essential for developing compassionate attitudes towards animals and to establish animal-friendly consumption behaviors, before the psychological rationalization infrastructure deeply entrenched in society takes over and makes it more difficult to change attitudes and behavior.

The societal context and many of the social factors that sustain and rationalize animal exploitation are further discussed in detail by John Sorenson. In Chapter 13, Sorenson takes a closer look at the role of the meat industry and its advertising partners in the creation of a distorted and euphemistic picture of reality of exploiting and slaughtering animals. The author illustrates how various rhetorical techniques are employed to conceal and misrepresent the truth about the industry behind animal product consumption, with "humane slaughter," "happy meat," and "free-range" chickens as the most widely-known examples of this practice. Sorenson also discusses the malicious narrative spread widely by the animal exploitation industry that portrays animal activists as aggressive extremists while portraying farmers as victims. Such techniques aim to criminalize animal activists and dismiss their concerns, while at the same time legitimating the exploitation and killing of animals.

Gene Baur, co-founder and president of Farm Sanctuary, further unpacks the many problematic aspects of industrial factory farming in Chapter 14. He draws on his decades of experience in animal activism and accumulated knowledge to sketch a clear picture of the many structural and psychological barriers that prevent people and society from becoming more animal friendly. He proposes that some traditional animal advocacy strategies may have been largely ineffective in bringing about change. During recent years, however, there is renewed optimism for societal and institutional reforms to occur. In particular, we witness the booming businesses of plant-based alternatives, the growth of mission-driven companies, and new advocacy organizations that focus on promoting plant-based food on all levels of society, targeting both consumers and institutions, and instilling new consumption habits and norms. In a similar vein, in Chapter 15, co-founder of the Center for Effective Vegan Advocacy (CEVA) and ProVeg International, Tobias Leenaert critically assesses the current state and effectiveness of the animal and vegan advocacy movement and argues convincingly that the movement urgently needs to adopt a more pragmatic approach to achieve the goal of a world free from animal suffering inflicted by humans. Leenaert proposes four ways in which this movement can become more pragmatic. This approach proposes a greater focus on (1) advocating for a reduction of animal products as an essential part of the road to the elimination of animal use; (2) using a wide range of arguments to motivate behavioral change, rather than relying on exclusively moral arguments; (3) investing in creating an environment that facilitates behavior change by developing accessible alternatives to animal products; and (4) being an inclusive movement that welcomes people who support the ideas of the animal rights movement (rather than only accepting people who already vegan). Leenaert's innovative and critical thinking about the movement provides ample ideas that can be put to the test by researchers.

In Chapter 16, Verónica Sevillano and Susan Fiske draw on theory and research in the field of social psychology to better understand how societal images of different animal species are constructed, and how such stereotypes are implicated in people's emotions and behavioral inclinations towards animals. This approach emphasizes the interplay between perceiving different animals in terms of warmth (*vs.* coldness) and competence (*vs.* incompetence), similar to the social perception of human groups. Applying this framework, the authors illustrate that their approach is helpful in understanding people's reactions when confronted with wildlife in human-animal conflict situations, thereby bridging the gap between social psychology and the field of conservation biology.

The next two chapters explore our moral thinking and judgments about animals and the consumption of animal products in a dialogue format. First, in a fictional debate between two Orthodox Jewish rabbis in Chapter 17, Paul Rozin and Matthew Ruby engage in a lively conversation critically exploring and challenging religious practices as prescribed by the Kosher dietary prohibitions on the one hand, and the ethical principles underpinning a vegetarian or vegan diet on the other. By comparing and contrasting these two diets the authors provide fundamental insights into the process of becoming vegetarian or vegan, which contributes to our understanding of moral change.

The second conversation, Chapter 18, is between Carol Adams and Matthew Calarco. This thought-provoking philosophical expedition explores the foundation for our ethical relations with animals. Adams and Calarco approach this topic from the feminist ethics-of-care tradition and discuss the differences and intersections of this approach with the writings of the French philosopher Jacques Derrida. Furthermore, they discuss how the feminist ethics of care deviates from the more standard philosophical approaches to animal ethics based on deontological or utilitarian principles, by emphasizing the importance of genuine feelings of affection or concern for the well-being of animals. In their conversation, they focus on themes such as the interdependency of human-animal relationships, the experience of suffering, and employing embodiment to emphasize putting care, emotions, and compassion, rather than reason and abstraction, on the forefront in the ethical treatment of animals.

In the final chapter, Chapter 19, as editors we provide a synthesis of the contributions to the present book. In doing so we highlight the reoccurring themes across chapters related to the question and problem of why people paradoxically love but exploit animals. Furthermore, we discuss the importance of connecting academics with practitioners to increase the relevance and impact of both fields, and highlight contemporary issues, shortcomings, and challenges presently faced by the field.

References

Eshel, G., Shepon, A., Makov, T., & Milo, R. (2014). Land, irrigation water, greenhouse gas, and reactive nitrogen burdens of meat, eggs, and dairy production in the United States. *PNAS*, 111, 11996–12001. http://dx.doi.org/10.1073/pnas.1402183111

Godfray, H. C. J., Aveyard, P., Garnett, T., Wall., J. W., Key, T. J., Lorimer, J.,... Jeb, S. A. (2018). Meat consumption, health, and the environment. *Science*, 361. eaam5324. doi:10.1126/science.aam5324

McKnight, T. (2014, Aug 4). Want to have a real impact on climate change? Then become a vegetarian. *The Guardian*. Retrieved from www.theguardian.com/commentisfree/2014/aug/04/climate-change-impact-vegetarian

Sanders, B. (2018, October 10). *Global animal slaughter statistics and charts*. Retrieved from https://faunalytics.org/global-animal-slaughter-statistics-and-charts/

Springmann, M., Godfray, H. C. J., Rayner, M., & Scarborough, P. (2016). Analysis and valuation of the health and climate change cobenefits of dietary change. *PNAS*, *113*, 4146–4151. www.pnas.org/cgi/doi/10.1073/pnas.1523119113

Tilman, D., & Clark, M. (2015). Global diets link environmental sustainability and human health. *Nature*, *515*, 518–524. doi:10.1038/nature13959

Willett, W., Rockström, J., Loken, B., Springmann, M., Lang, T., Vermeulen., S., ...Murray, C. J. L. (2019). Food in the Anthropocene: the EAT–Lancet Commission on healthy diets from sustainable food systems. *The Lancet*, *393*, 447–492. https://doi.org/10.1016/S0140-6736(18)31788-4

2

THE ANIMAL IN ME

Understanding what brings us closer and pushes us away from other animals

Brock Bastian and Catherine E. Amiot

Abstract

People have a conflicted relationship with animals, and therefore their own membership in the animal kingdom. In this chapter, we review research showing how our connection to animals can influence our attachment systems, how we construct and understand our social identities, and how this connection can be costly, sometimes forcing us to make hard choices between animal and human welfare. We often love and cherish animals, yet we also frequently use them for resources (e.g., meat) and can feel threatened when our superiority is questioned, motivating a desire to maintain psychological distance. Animals represent both a significant connection to our natural environment, but also serve as a reminder of our mortality. We discuss how research into human-animal relations impacts on many of the same psychological constructs relevant to human-human relations. Studying our relationship with animals therefore offers critical insights into broad psychological processes relevant for the human domain and also reveals that how we relate to animals can shape how we relate to other humans.

There is one basic fact about humans which is often overlooked – we are in fact part of the animal kingdom. Just like tigers, dogs, giraffes, and guinea pigs, we are properly considered an animal species. We eat, sleep, poo, procreate, fight and cooperate like other animals. Our behavior is driven by primal urges and latent instincts like any other animal. We also flourish in certain habitats, and perish in others. We face the risk of predators, and we hunt (or more frequently farm) our prey. Our biological systems are vulnerable to infection and disease, and our social systems are fundamental to our survival.

Take a moment to reflect on how you felt reading the first paragraph of this chapter. Did you feel a sense of belonging with animals? Or rather, did it make you feel awkward or uncanny? Like you wanted to distance yourself from animals? Perhaps your mind turned to examples of how humans are, in fact, quite different – and possibly superior – to other animals? What this highlights is that our relationship to other animals – in fact our relationship to our own membership in the animal kingdom – tends to be conflictual, inconsistent, irksome, and irrational (Herzog, 2011). As some have argued, being reminded of our animal nature can undermine psychological defenses which protect us from a fear of our own mortality (Greenberg, Solomon, & Pyszczynski, 1997).

In this chapter, we will explore the various ways in which the animal in all of us shapes how we think, feel, and behave, as well as what factors bring us closer or push us apart from other animals. For example, how meat eating motivates us to maintain distance from other animals (Bastian & Loughnan, 2017). Until recently, psychology has largely overlooked this aspect of what it means to be human and how our relationships with other animals share many similarities to how we relate to other humans (Amiot & Bastian, 2015). Studying how we relate to the animal within can shed light on our conflicted relationship with other animals, why we extend, but also sometimes limit, empathy toward other humans, and important individual differences in how we construct our social identities.

The animal in all of us

Whereas biologically, humans are indeed animals, how we relate to this part of our identity or to members of the broader category of animals, has only recently started to receive attention from researchers within the field of psychology (see Amiot & Bastian, 2015, 2017). Arising from this work has come the realization that our relationship to animals is both instinctual and complex, and that how we relate to animals can be understood by drawing on the psychology of how humans relate to each other. Paving the way for a psychology of human-animal relations, this field of research has sought to apply basic principles to how we interact with, and think about, other animals.

Our instinctual attraction to animals

If you are still finding the idea that humans are driven by primal urges, just like animals, a little confronting, then consider the idea that one of our instinctual drives may be to seek affiliation with animals. This has been referred to as the *biophilia hypothesis* which makes the case that humans appear to have an innate emotional affiliation with, and pay special attention to, other life forms, including animals (Kellert & Wilson, 1993; Wilson, 1984). One example of empirical evidence for this notion involves a study showing that children (aged 1–3) who had the option of interacting with live animals or interesting toys were more likely to choose the animals – they found them more interesting. They also talked more about the animals and asked more questions about them, compared to the toys (LoBue, Bloom Pickard, Sherman, Axford, & DeLoache, 2013).

This early and seemly inherent attraction to animals suggests that the human mind might have evolved to think differently about animals relative to inanimate objects, such as toys (Herzog, 2011). In fact, young animals appear to trigger our human tendency to care for human infants. This may be because they share many of the same features as human infants – for example, big eyes, large foreheads, and soft contours – and we respond to this by feeling attracted to young animals, wanting to protect and care for them, more so than older animals (Fridlund & MacDonald, 1998). It has been suggested that this innate drive to care for young animals may also have advantaged humans, as the domestication of animals has also been very important to our survival throughout history (Bradshaw & Paul, 2010).

Our connection to other animals hence exists at an instinctual level; we are interested and attracted to young animals and they may even exploit some of the same mechanisms designed to care for and look after other humans. As we will see however, this same connection can be understood at higher-order and more complex levels of analysis. In addition to these more evolutionary and built-in forces, the human-animal link is also shaped by relational (e.g., attachment) and social (e.g., socialization) factors.

Our psychological attachment to animals

Beyond the instinctual drives that direct our attention, interest, and affective engagement towards our fellow animals, our relationship to animals can also be understood using basic theories of human attachment. Perhaps one of the most fundamental insights into human psychology is that of attachment. With its beginnings in developmental theory and research, this work has revealed the importance of attachment figures for human development but also the various ways in which this can go wrong.

Attachment is defined as a deep and enduring emotional bond that connects one person to another across time and space (Ainsworth, 1973; Bowlby, 1969). The stability of this attachment has implications for social, emotional, and cognitive development and when disrupted this can lead to anxiety and maladjustment.

Various researchers have applied the theory of attachment to human-animal relations, observing that companion animals can serve as attachment figures for their owners (Zilcha-Mano, Mikulincer, & Shaver, 2011). Of course, humans also provide for the basic needs of their pets and in that sense both humans and animals serve as attachment figures for each other – it is just a little harder to ask animals about the nature of their emotional connection to humans and what it means to them! Nonetheless, research has applied indirect methods to investigate whether pets form psychologically meaningful attachments to their human owners.

One method, which has been used in developmental research, is referred to as the "strange situation." This involves placing a child in a room with secret mirrors and watching how they respond when they are separated from a parent but left with a stranger. Secure attachment is indicated when children continue to explore the room and interact with the stranger, even with their parent gone – they feel secure in the knowledge their parent will return, rather than anxious that they may not (Ainsworth, 1991). Insecure attachment becomes evident when they cry, become anxious, and fret about their parent leaving them alone. Secure attachment can also be observed in pet dogs who, when left in a room with a stranger, continue to play, explore, and interact with the stranger when their owner leaves the room. This is compared to dogs who do not feel securely attached, who, like some children, become anxious and inhibited when their owner is no longer present (Palmer & Custance, 2008).

While animals can become attached to people, people can also become attached to animals. Building directly on attachment theory, Zilcha-Mano et al. (2011) developed the Pet Attachment Questionnaire. This examines two forms of inse-cure attachment to pets; anxious and avoidant attachment. Insecure attachment to one's pet has been linked to the animal being kept indoors more frequently and more concerns for its wellbeing. Items from this questionnaire include: "I'm often worried about what I will do if something bad happens to my pet," "Sometimes I feel that I force my pet to show more commitment and desire to be close to me," "If I can't get my pet to show interest in me, I get upset or angry," "Signs of affect from my pet bolster my self-worth." As is clear, pets literally become attachment objects for people, and for some they feel anxious about how they would cope if their need for attachment is not reciprocated or is taken away.

Beyond being attachment objects for adults, secure attachment to pets is also psychologically healthy for a child's development. For instance, research shows that children who develop strong and secure attachment to a pet tend to have higher levels of self-confidence, are more resilient, and psychologically well-adjusted (Paul & Serpell, 1996). Just as secure human attachments are good for children, so are secure attachments to other animals.

Animals feed into our psychology, and our connection to other animals is important for our health. There is a noteworthy body of work which we will not review here, showing that pets can have a significant impact on the health and well-being of humans, quite apart from their role as attachment objects (see Amiot & Bastian, 2015). Yet, our sense of being connected to animals also goes beyond this more individual and relational aspect. Animals are not only of significance to our personal sense of self, they are also significant to our social sense of self.

Social identification with animals

Perhaps most relevant for this chapter is the extent to which people do, in fact, see themselves as part of the animal kingdom – that is, as animals. We know from a long history of research on social identification that people not only think of themselves as individuals, but also as members of social groups (Tajfel & Turner, 1986; Leach et al., 2008). The salience of particular group memberships can change, depending on the context, and this in turn can shape how people think, feel, and behave (Turner, Hogg, Oakes, Reicher, & Wetherell, 1987). For instance, at work

TABLE 2.1 Solidarity with animals scale

1. I feel a strong bond toward other animals
2. I feel solidarity toward animals
3. I feel close to other animals
4. I feel a strong connection to other animals
5. I feel committed toward animals

we might behave and respond to others like employees or co-workers, whereas at home we become members of a family, or perhaps a father, mother, or carer. These social roles are not only things that we do, they are social categories that we identify with, providing us with sense of how we fit into, and see ourselves identified with, the world around us.

Group memberships are also often nested within larger superordinate identities (Gaertner & Dovidio, 2000; Hornsey & Hogg, 2000). For instance, we might see ourselves as a member of a work team, perhaps competing with other work teams within the same organization, but when other people ask who we work for we identify with the organization as a whole. The same process may be evident when on a trip overseas and we are asked where we are from. Here, we might identify with our entire country – "I am from Australia" or "I am from Canada," yet at home we might see ourselves as Melbournians or Québécois and take pride in that identity when comparing ourselves to Sydneyites or Torontonians.

Recent research has shown that our superordinate identities can stretch all the way up to the entirety of humanity (McFarland, Webb, & Brown, 2012). Critically, the level at which people identify has implications for how they treat others. While teams may compete for resources within organizations, those same individuals may share resources when the comparative context leads them to see themselves as members of the same organization.

We wondered how far superordinate identities might go – could they stretch even beyond humanity? Although some previous research had examined whether people might feel psychologically connected to nature (Clayton, 2003), less work had directly examined the nature of our psychological connection to other animals.

To examine this, we constructed a scale designed to capture an especially relational dimension of identification – our Solidarity with Animals (Table 2.1; Amiot & Bastian, 2017). Previous work has examined solidarity as a dimension of social identification (Leach et al., 2008) which involves a person's psychological bond with, and commitment to, fellow in-group members. As such, it has a focus on how individuals relate to other group members, the concrete roles they occupy within this group, and whether they are motivated to act in ways that promote the interests of that group. In this sense, solidarity with animals involves a quite high level of commitment toward animals in general.

After developing a reliable scale to measure solidarity with animals – drawing on items developed in prior research on identification with human groups (e.g., Leach et al., 2008) – we found (in Study 2) that this aspect of identification was associated with a tendency to anthropomorphize animals (i.e., a tendency to see them as possessing more human-like mental capacities) and to feel anxious at the prospect that one's relationship with a pet may go wrong (i.e., anxious attachment), as a sign of high concern for this relationship. We also found (in Study 4a) that pet owners score higher on solidarity with animals – possibly because caring for an animal on a daily basis brings us closer to animals in general – and that the more pets a person owned, the more solidarity they felt towards animals in general. We also found (in Study 4b) that meat-eaters (versus vegetarians) scored lower on the solidarity with animal's measure and furthermore, as the quantity of meat eating went up, solidarity with animals went down; the more that people ate meat the less they felt identified with animals. These last results may be due to the fact that many people reduce or avoid eating meat altogether because of their concerns for the negative treatment of meat-animals (see Loughnan & Davies, this volume; see Piazza, this volume).

Together, these studies revealed that how people treat animals has implications for whether they feel identified with them. If we preference our own needs at the expense of theirs, such as is the case with meat eating, then we are less likely to feel identified with the broader category of animals. But could this also work the other way – the more identified we feel with animals as a large social category, the more likely we are to preference the needs of animals? Could this preference for animals even come at the cost of human interests? To provide some answers for these questions, we next tested whether people high on solidarity with animals might, to some extent, give preference to the needs of animals even when satisfying these needs conflicted with the needs of other humans.

To achieve this, in Study 7 of Amiot and Bastian (2017) we asked people to solve a series of moral dilemmas. We drew on a set of moral dilemmas that are commonly used to distinguish rational from emotional decision making – the trolley and footbridge dilemmas (Greene et al., 2001). In the trolley dilemma, people are told there is a trolley heading towards five workers on the tracks, and they will all be killed unless you throw a switch and redirect the trolley down an alternate route. The catch is, however, that there is one worker on those tracks who would inevitably be killed if you do. A majority of people tend to say they would sacrifice the life of one person to save five. In the footbridge dilemma, you are faced with the same scenario, except to avert the trolley from killing five workers, you would need to push a man over the edge of the bridge who would fall and stop the trolley in its tracks. In this scenario most people say they would not do this, because it feels more wrong to push someone to their death than to throw a switch.

In Study 7 we used these same dilemmas, except we asked people to imagine sacrificing the lives of dogs to save a human life. In the trolley dilemma, participants imagined that they were an innocent bystander watching a trolley speeding down some tracks where a person was working. They were informed that if they do nothing the worker would be killed, but if they throw the switch, the trolley will be diverted down a different track where there are ten dogs that would perish as a result. Across all participants 31 percent of people said they would prefer to do nothing and save the ten dogs. This percentage is not particularly small, which is quite surprising, given that people tend to value human lives over the lives of other animals (see Hodson, Dhont, & Earle, this volume). Yet, importantly, we found that people who felt strongly identified with animals were less likely to indicate they would throw the switch, through their inaction saving ten dogs but allowing the loss of one human life.

In a second footbridge dilemma, participants were given the same scenario, except that to save the human life they would need to push a kennel, containing ten dogs inside, over a bridge into the path of the trolley in order to save the human life. This time, 61 percent of people said they would do nothing, allowing the loss of human life. Again, however, it was people who felt strongly identified with animals that were less likely to sacrifice the dog and push the kennel over the side of the bridge.

Finally, using a different type of dilemma, we told participants to imagine that a ship was sinking and there was a life boat that could only fit five individuals. They were also told that there were only five individuals on the sinking ship. The catch was that there was also a dog, and if the dog took a place on the life boat then one human life would be lost. They were told that (a) they could choose to either throw the dog overboard, saving the human lives, (b) draw lots among the humans and throw the losing human overboard, therefore saving the dog's life, or (c) draw equal lots amongst the humans and the dog and throw the loser (be it a human or the dog) overboard. What did people choose? Again, it is surprising to find that across all participants each option was equally popular – about one third of people would throw the dog overboard, one third would save the dog and sacrifice a human, and one third would leave the lives of both the humans and the dogs to equal chance. As before, however, people who felt a strong sense of solidarity with animals were more likely to favor options that protected the dog's life at the potential, or actual, cost of human life.

Solidarity with animals, it seems, has some cost attached to it; feeling a strong psychological bond and commitment to other animals means that their needs are weighted more heavily in our decision making. When faced with the dilemma of where to allocate resources, or who to save, individuals who feel high solidarity with animals may turn their back on their fellow humans.

Critically, however, feeling committed to animals may also have benefits for our relationships with other humans. In Study 3 of Amiot and Bastian (2017) we gave people the solidarity with animals measure along with several other measures relating to prejudice. This revealed something very interesting. People who felt a strong connection to animals were also less likely to be prejudiced towards other human groups, were less likely to hold harsh attitudes towards others who disagreed with their perspective, and were less likely to seek to dominate other groups (for similar findings see Dhont, Hodson & Leite, 2016; also Dhont, Hodson, Leite, & Salmen, this volume). We also observed that people high on solidarity were more likely to be liberal and female. Although feeling very connected to animals means that animals' needs may come into conflict with those of other humans, when such a conflict is not apparent solidarity with animals may lead us to be more accepting and accommodating of others. As we will discuss later, our relationship to animals may play a central role in expanding our moral concern in general.

Similarity as a trigger for identification or threat?

One interesting question is the role that similarity plays in our tendency to feel identified with animals. Research on identification with human groups suggests that perceived similarity and homogeneity with fellow group members provides a basis to feel identified with others (Tajfel & Turner, 1986; Turner et al., 1987). Studies on human-animal relations have also found that perceiving similarities between animals and humans predicts greater empathy toward animals (Hills, 1995) and greater support for their rights (e.g., Kellert, 1980). Integrating these lines of work, we examined whether people also feel more committed to animals when they are reminded of their similarities to them? In Study 5 (of Amiot & Bastian, 2017) we showed some people pictures of animals with extraordinary human-like expressions (taken from the photographer Tim Flach), whereas we showed others pictures of animals showing more 'animal-like' expressions or acting in more 'animal-like' ways (i.e., animals in their natural habits, taking typical poses). We found that when people viewed the animals with human-like expressions they reported greater solidarity with animals. Study 6 replicated these findings, showing that solidarity helped to explain the link between perceived similarity of animals and the tendency to hold positive and supportive attitudes towards animals.

In a related series of studies (Amiot, Sukhanova, Greenaway, & Bastian, 2017) we observed that the more people perceived high human–animal similarities, the more likely they were to feel connected to animals, and the higher, in turn, this lowered their need to assert humans' superiority relative to animals. We measured connection and identification by drawing on our measure of solidarity with animals (see Table 2.1) as well as using a pictorial measure designed to capture the extent to which people saw animals as being close to them (e.g., Costello & Hodson, 2014), and as included within their own self-concept (i.e., a proxy for the animal within). Across the first two studies, we found that the tendency to perceive a lot of similarities between animals and humans predicted the tendency to identify with the group of animals. This sense of connection to and identification with animals then predicted a lower tendency to perceive humans as superior to animals.

We then replicated these findings in a third study (Amiot et al., 2017), by presenting people with pictures of animals that were objectively (i.e., phylogenetically) more versus less similar to humans. These were, respectively a: gorilla, white rhinoceros, common crane, iguana, flathead catfish, and May beetle. Following the picture of each animal, people were presented a set of questions that measured solidarity with this individual animal and its perceived status relative to humans. We found, as could be expected, that people felt more identified with the more phylo-genetically similar animals than with the less similar animals; they also perceived the more similar animals as closer to humans in terms of their social status. More interestingly, we found that when people had viewed the pictures of the more similar to human animals (i.e., gorilla, white rhinoceros, common crane) compared to the less similar animals (i.e., iguana, catfish, and beetle), this predicted higher solidarity with animals in general. This increased sense of solidarity then flowed on to a reduced general tendency to perceive humans as superior to animals. In this sense, these findings have concrete implications for activism and the presentation of effective messages; they imply that presenting pictures of animals that are more similar to humans can yield to more beneficial perceptions and evaluations, on the whole, of the status of diverse types of animals (so including animals who are both similar to humans and those who are less similar to humans).

As alluded to earlier, however, this effect of seeing animals as similar to humans can also work the other way, making people feel defensive or threatened. This is because humans have a fundamental fear of their own morality, and being reminded that we are just like animals breaks down the various defenses we have put in place to reduce this form of existential threat. For instance, when people are reminded of their own death they tend to become more strongly attached to their cultural and social worldviews. These ideological commitments are viewed as extending beyond life and death, and give people a sense of purpose that extends beyond their mere biological existence (Solomon, Greenberg, & Pyszczynski, 1991). These worldviews (sometimes explicitly) elevate us from animals by positing the existence of souls, spirits, or other uniquely human attributes that are not shared with other animals. When we are forced to face the more objective reality that there are far fewer, and less meaningful, differences between humans and animals than we might think, we can feel threatened and become defensive; we hence push back from other animals. This can be observed in studies showing that people often report feeling disgusted and desiring increased distance from animals after being reminded of their own mortality (Goldenberg et al., 2001), an effect that is even evident when people are asked to think about their own pets (Beatson, Loughnan, & Halloran, 2009).

Animals-as-us, versus us-as-animals

Similarity can be an important predictor of whether, and to what extent, people feel identified with animals. It can both promote a connection to animals, but under certain circumstances, we may also want to seek differentiation from animals, seeing them as less similar and ourselves as elevated beyond other animals' status. Yet, could the way we frame our similarity to animal's matter? For instance, seeing animals as intelligent and emotionally sensitive can make them seem more human-like. This may be quite different, in terms of how it impacts on us psychologically, to the first paragraph of this chapter which framed humans as animal-like. In both cases, humans and animals are framed as more similar to each other, but one has the implication of bringing the status of animals up to that of humans, and the other of lowering the status of humans down to that of other animals (see also Hodson et al., this volume). This of course assumes that such a status hierarchy exists – but one does not have to try very hard to see the very many examples which clearly suggest we see ourselves as higher in status than other animals.

One example of these different ways of framing similarity is illustrated in an advertising campaign within Australia which focused on the consumption of red meat and featured the Australian actor Sam Neil. This campaign promoted the idea that red meat was an important component of our evolutionary history, casting our consumption as instinctual behavior. Several television adverts show Neil providing an anthropological commentary on human meat consumption, and in one version dancing with an orangutan and discussing the benefits of red meat. This clearly framed humans as similar to other animals and meat eating as a behavior that is instinctual, habitual, and as shared with non-human animals through our common evolutionary past. Yet, through this increased identification with our animal nature, people increased their spending on red meat across Australia, from $8 billion in 2004/2005 to $9 billion in 2006/2007. The advert was also a finalist for an award from the Australian Market and Social Research Society (Peace, 2008).

The impact of this advertising strategy raised an important question. If viewing humans and animals as similar to each other theoretically should increase our concern for their welfare, presumably also decreasing speciesism and a tendency to exploit animals as a food resource, why did it do the opposite in this ad?

This made us wonder whether the way that our similarities to other animals are framed might have different implications for our relationship to them. We draw on a ground-breaking study by Costello and Hodson (2010) which had shown that emphasizing how animals are similar to humans increased the humanization of immigrants and produced more favorable attitudes towards them. This was compared to emphasizing either the human-animal divide or the various ways in which humans are similar to animals. In short, thinking about animals as "human-like" had made people more empathic and concerned about the plight of immigrants (by reducing the dehumanization of immigrants), whereas thinking about humans as "animal-like" had no such effect.

Building from this work we (Study 1; Bastian, Costello, Loughnan, & Hodson, 2012) had people write an essay about the similarities between humans and animals. We then grouped these essays according to whether people had focused on the ways in which animals are "human-like" or had focused on the ways in which humans are "animal-like." This revealed that people who wrote about animals as "human-like" expressed more moral inclusivity towards all animals – they were more likely to be concerned about their needs and rights – compared to those who wrote about humans as animal-like.

In Study 2 of Bastian and colleagues (2012), participants were exposed to one of two editorials discussing the similarities between humans and animals (from Costello & Hodson, 2010). In one version, animals were framed as being just like humans (animals-are-human-like) and were told that: "Animals are motivated to avoid pain and to seek pleasure, just like humans ... like humans, other animals possess the capacity to make choices, create their own destinies, and understand abstract concepts including cause and effect relationships." In the other version, humans were framed as being like animals (humans-are-animal-like) and were told that: "Humans are motivated to avoid pain and to seek pleasure, just like animals … like animals, much of human behavior is influenced by basic instincts such as hunger, lust, pain avoidance, and pleasure." Again, we found that people who had read about animals being like humans were more morally inclusive towards a range of animals (e.g., chimp, kangaroo, bat, snake, snail, cow).

We sought to replicate these same findings in Study 3, except this time experimentally allocating people to write essays either framing animals as like humans or framing humans as like animals. We also included a control condition. This again revealed that it was the animals-are-like-humans condition that increased moral concern for all animals. The humans-are-like-animals condition was similar to the control condition – it did not shift attitudes one way or the other. In this study, we also included a measure of speciesism (i.e., the use of animals for human ends) and, similar to Costello and Hodson (2010), a measure of moral concern for a range of human outgroups (e.g., immigrants). The findings were consistent: People in the animals-are-like-humans condition were less likely to endorse speciesism and expressed more moral concern towards human outgroups.

Together this research demonstrated that how we frame our similarities to other animals matters. It not only matters for how we think, feel, and act towards animals themselves, but it has implications for how we think, feel, and act towards other humans. This is a recurring theme of this work – that our relationships toward animals do not only matter for them, they also matter for us and feed into our psychology – including how we treat each other as humans.

Interestingly, our findings did not support the possibility that likening humans to animals might increase our mistreatment of them. This would have more straightforwardly explained why the Sam Neil red meat campaign worked so well. Yet, it may be that Australians watching these adverts never reflected on the harm that their meat consumption brought to some animals. Rather, they were simply responding to the message that eating meat is natural, normal, and necessary (see Piazza, this volume; Piazza et al., 2015). As detailed by Bastian and Loughnan (2017), a plethora of defenses have been developed both at the level of individuals and society, through which to keep the consumption of meat separated from reminders of its animal origins and the harm it brings. As we will see in the next section, when this well-maintained separation is challenged people can use a range of strategies through which to reduce discomfort and protect their self-image.

Is our relationship to animals shaped by resource concerns?

If humans can identify with animals and see them as part of the same superordinate category, much like we do with other human groups, then it also suggests that many of the same factors that tear human groups apart may also tear humans away from feeling connected to and identified with animals. A key theory in the literature on intergroup relations, referred to as realistic conflict theory (RCT; Sherif, 1966), proposes that as resources become limited or scarce, conflict over these resources can damage intergroup relations and lead to intergroup hostility. This would suggest that when resources are plentiful, our tendency to identify with animals should be stronger, but when resources become scarce we may retract to focus on more closely related ingroup members. This may mean that we are more likely to focus on our commitments to friends and family over more distant human relationships, and to humans in general over other animals.

Evidence for this comes from the observation of a positive link between people's economic affluence and their attitudes and concerns for animal welfare. In a largely multinational survey of attitudes towards animals using the Eurobarometer data, it was observed that people living in Scandinavian and eastern Mediterranean countries reported the greatest concern for animal welfare compared to people living in other European countries (European Commission, 2016). These results are consistent with the differences in economic stability and general affluence of these countries. Attending to animal welfare requires resources, and when resources are scarce they will be directed to those who are closer to us – our own ingroup members. Consistent with this is research showing that African Americans – a disadvantaged social group – reported being less attached to animals and having fewer pets compared to White Americans (Brown, 2002).

The influence of resources on human–animal relations has also been supported in experimental work. When participants were told that an insect was useful for human needs and represented a low threat to human resources, they expressed greater concern for its existence compared to when they were told that it was not useful for humans and represented a threat to our resources (Opotow, 1993). We like other animals (including insects), but when it comes down to us or them, as with our connection to other humans, we give preference to our own needs or the needs of those closest to us.

Does culture moderate our relationship with animals?

Another factor that can shape whether we feel identified with other animals versus detached from them is the culture in which we live. People living in Western cultures are often incensed when learning of the Korean practice of consuming dog meat, while Koreans wonder why Westerners are so socially isolated that they need to develop meaningful companionships with animals.

This example reveals two important aspects of culture that can shape how, and the extent to which, we feel close to versus disconnected from animals. The first involves cultural norms. There are culturally-situated prescriptions around how we should treat different animals, as with the example of dogs. In many countries horse meat is an acceptable part of the culinary landscape whereas in other countries horses are viewed strictly as pets, and eating them seems wrong and disgusting. In India cows are sacred, while in most other countries they are a primary source of protein. Pet keeping also varies across culture, and not only with regard to which animals are viewed as pets, but also how pets in general are viewed. In a study of pet keeping across 60 cultures (Gray & Young, 2011) it was revealed that significant cross-cultural variations emerged in pet keeping and in whether or not pets are elevated to the level of "friends" and "family members" (as they are often in the United States and other Western cultures). In fact, pets were treated this way in only five of the 60 cultures, demonstrating that elevating pets to these roles seems to be the exception rather than the rule. As noted by the authors, some cultures do not even have a word for pet!

Another way in which culture can shape our tendency to feel identified with animals has to do with basic elements of a culture. As noted above, it may be that Western individualism – as a tendency to be self-reliant rather than embedded with a larger social network – leads to the tendency to see pets as part of the family or the need to feel meaningfully connected to one's pet. This is because when human social relations are scarce, or we find ourselves lonely and isolated, we may seek solidarity and social connectedness with animals instead. Cultures influenced by Judeo-Christian thinking may also lead people to see animals as subordinate to humans – as we are their caretakers and occupy a stewardship role relative to them – whereas cultures influenced by Buddhism or Hinduism may see animals as more equal to humans, and quite literally as linked to humans through reincarnation.

In sum, this early work on how people relate to and feel connected to other animals shows that our thoughts, feelings, and even our very identities are tied up in our relationship with animals. It highlights that many of the same processes identified in how we relate to other humans can be observed in how we relate to other animals. By applying our understanding of relationships and intergroup dynamics to the domain of human-animal relations we are not only learning more about our relationship with animals in particular, but also more about our human psychological processes in general.

Seeking to separate ourselves from animals

So far, we have reviewed research showing the various avenues through which people may feel more connected to animals and how similarity plays an important role in our attitudes towards animals and identification with them. Yet, as the title of this book suggests, we sometimes exploit animals and this has implications for how we relate to them. At times we want to sever our connection to animals and this may not only have implications for how we treat animals, but may also have implications for how we see ourselves.

Exploiting animals and dissociation

One of the key ways in which animals are exploited is for their meat. A large majority of people globally eat meat, and as the wealth of nations continues to increase people are also able to afford to eat more meat. This has led to some genuinely horrific practices under the guise of factory farming (see also Sorenson, this volume). As the description of this practice suggests, animals are literally treated as commodities with little concern for their emotional or physical welfare. It is fair to say that most people do not find it easy to think about, let alone see for themselves, the brutal reality of the meat industry. This raises the important question of how, then, are people able to continue consuming meat apparently trouble free?

One way in which people keep discomfort attached to their meat-eating practices at bay is to maintain a clean disconnect between their meat-eating practices and reminders of animal harm. Bastian and Loughnan (2017) have recently developed a model detailing how people avoid or reduce discomfort associated with their meat consumption. This ranges from well-rehearsed psychological strategies that are efficiently brought online when reminders of harm are made salient (see Bratanova, Loughnan, & Bastian, 2011; Bastian, Loughnan, Haslam, & Radke, 2012; Loughnan, Bastian, & Haslam, 2014; Loughnan, Haslam, & Bastian, 2010) to a range of social and cultural institutions, practices, and norms that literally shield people from ever needing to think about where their meat comes from.

Of key interest for our purpose here is that how the exploitation of animals may impact on self-perception. Broadly, people do not like to think of themselves as immoral and as such find ways to shield their self-concept from the moral implications of their meat-eating practices. In one way or another, this tends to involve emphasizing the differences between humans and animals.

In a study examining this motivated process, Bastian, Loughnan and colleagues (2012) found that when people were reminded of the harm brought to meat-animals, they lowered their perception of the mental qualities those animals possessed. In Study 3 this was achieved by making one group of participants believe they were about to eat meat, while another group was made to believe they were about to eat green apples. Although all participants were meat-eaters, we found that it was those who thought they were about to consume meat that reduced their perception of animal's mental capacities – thereby making animals seem less similar to humans and reducing concern for their welfare. In fact, we found that it was via the denial of mental capacities to animals that people felt better about their future consumption of meat – it literally reduced their feelings of discomfort.

This process of distancing the self from the target of one's harmful actions has also been observed within human–human relations. When people cause harm to others one of the ways they morally disengage from their actions is by dehumanizing the victim of their actions (Bandura, 1999). By creating distance between their own human qualities and those of their victims, people obfuscate the immoral implications of their actions. Seeking to disconnect oneself from the 'animal within' allows people to justify their exploitation of animals. Yet, could this process have broader implications, also for human–human relations?

Does dissociating from animals make us worse people?

A somewhat contentious idea is referred to as "The Link" (e.g., Ascione & Arkow, 1999; Ascione & Shapiro, 2009) which suggests that people who cause harm to animals may also be more likely to cause harm to other humans. For instance, in one study college students who reported two or more incidents of animal abuse were also likely to report more previous criminal behaviors and were more likely to bully others (Schwartz, Fremouw, Schenk, & Ragatz, 2012). Another study focused on inmates in medium- and maximum-security prisons (Henderson, Hensley, & Tallichet, 2011), finding that four out of five inmates reported hitting animals and over one third reported having tried to shoot or kick animals. A smaller percentage reported drowning, choking, burning, or having sex with animals. Women surveyed in domestic violence shelters were nearly 11 times more likely to report their partner had hurt or killed pets, than a comparison group of women who had not experienced intimate violence (Ascione et al., 2007).

This link between animal violence and human violence is sometimes considered to be a particularly contentious issue in the study of human–animal interactions. Whether causing harm to animals increases the likelihood of causing harm to humans is unclear. It could be that latent tendencies towards antisocial and violent behavior are simply generalized across human and animal targets (for related but distinct arguments, see Dhont, Hodson, Leite, & Salmen, this volume). There is also some research showing that animal abuse is sometimes just as high in populations that do not have a history of criminal behavior against humans (Patterson-Kane & Piper, 2009).

The existence of this presumed "link" between human and animal abuse becomes even more ambiguous when we broaden out our conception of what counts as harmful behavior directed towards animals. Meat eating is a prime example. This normative, common, and socially condoned behavior clearly causes harm to animals, yet the indirect nature of this harm allows us to shield ourselves from its gruesome consequences. How might this impact on those who are at the coalface of meat production, however? There is almost no research examining the impact of these types of environments, except reports noting that unnecessary abuse of animals frequently takes place within abattoirs.

It would seem that "The Link" may be most evident when people cause harm to animals in ways that run against social standards, suggesting they have the potential to also break social standards around the treatment of humans. Yet, when harm becomes brutally apparent it runs against our own moral standards, and justifying harm or dissociating ourselves from it, may have subtle and as yet underexplored consequences.

Does connecting to the animal within make us better people?

Just as disconnecting from animals can bring out the worst in us, can maintaining a connection to other animals make us better people? If so, does it matter how this connection is framed? As noted above, research has found that likening animals to humans (i.e., seeing them as more human) as opposed to likening humans to animals (i.e., seeing ourselves as less human) increases moral concern, not only for animals, but also for a range of human outgroups (e.g., immigrants, stigmatized groups; Bastian, Costello et al., 2012; Costello & Hodson, 2010). This suggests that in order to become better people, it is not so much about seeing ourselves as animals, but rather elevating animals "up to" the status of humans. It is via this process that we also elevate our moral consideration for members of human outgroups; if animals are worthy of our consideration and concern then *ipso facto* so must all humans be.

This association between consideration for the rights and needs of animals, and a concomitant concern for the rights and needs of other humans has been demonstrated in a number of studies. For instance, Wagstaff (1991) showed participants photos of animals, such as a cow in an abattoir or piglets in a cramped cage, and asked them to rate how angry or upset they felt (as a measure of empathy). Participants who showed more empathy towards animals also tended to be more positive and sympathetic towards human beings and had fewer negative attitudes towards the poor. This same relationship has been observed in animal protection workers, who tend to report higher levels of empathy for humans compared to those who are not directly involved in animal protect efforts (Signal & Taylor, 2007). There is also evidence that developing positive connections with animals leads to the development of greater empathy towards other humans. For instance, university students who reported more intense and meaningful connection to a childhood pet also reported greater empathy towards humans (Paul & Serpell, 1993).

Conclusion

In this chapter we have reviewed the various ways in which we are connected to other animals, the implications of these connections, and the consequences of seeking disconnection. In short, we are animals, and engaging with this understanding of the self may be important in many ways. As we have reviewed, it may be that, through "elevating" our perceptions of animals to how we view humans, we become more inclusive and more pro-social toward both animals and humans. Through this process we connect with animals in a way that also elevates our concern for other humans. In contrast, when we "lower" our conception of humans to how we commonly view animals, this process may somehow lower our moral standards as well, and bring about more negative consequences, both for humans and for animals.

There is now a significant movement toward developing our understanding of human-animal relations, and this has revealed why this relationship is not only important for how we treat animals, but also for our own psychological processes and how we understand ourselves. Throughout history we have viewed ourselves as the caretakers of other animals, but we underestimate the influence that they also have in our lives. We see an exciting time ahead as we continue to develop a psychology of human-animal relations, and continue to develop a view of ourselves and our humanity that takes seriously the notion of connecting to the animal within.

References

Ainsworth, M. D. S. (1973). The development of mother-infant attachment. In B. M. Caldwell & H. N. Ricciuti (Eds.), *Review of Child Development Research* (Vol. 3, pp. 1–94). Chicago, IL: University of Chicago Press.

Ainsworth, M. D. S. (1991). Attachments and other affectional bonds across the life cycle. In C. M. Parkes, J. S. Hinde, & P. Marris (Eds.), *Attachment Across the Life Cycle* (pp. 33–51). New York: Routledge.

Amiot, C. E., & Bastian, B. (2015). Toward a psychology of human-animal relations. *Psychological Bulletin, 141*(1), 6–47. https://doi.org/10.1037/a0038147

Amiot, C. E., & Bastian, B. (2017). Solidarity with animals: Assessing a relevant dimension of social identification with animals. *PLoS ONE, 12*(1), e0168184. https://doi.org/10.1371/journal.pone.0168184

Amiot, C. E., Sukhanova, K., Greenaway, K. H., & Bastian, B. (2017). Does human-animal similarity lower the need to affirm humans' superiority relative to animals? A social psychological viewpoint. *Anthrozoös, 30*(3), 499–516. https://doi.org/10.1080/08927936.2017.1335117

Ascione, F. R., & Arkow, P. (1999). *Child Abuse, Domestic Violence, and Animal Abuse: Linking the Circles of Compassion for Prevention and Intervention.* West Lafayette, IN: Purdue University Press.

Ascione, F. R., & Shapiro, K. (2009). People and animals, kindness and cruelty: Research directions and policy implications. *Journal of Social Issues, 65*(3), 569–587.

Ascione, F. R., Weber, C. V., Thompson, T. M., Heath, J., Maruyama, M., & Hayashi, K. (2007). Battered pets and domestic violence: Animal abuse reported by women experiencing intimate violence and by nonabused women. *Violence Against Women, 13*(4), 354–373. https://doi.org/10.1177/1077801207299201

Bandura, A. (1999). Moral disengagement in the perpetration of inhumanities. *Personality and Social Psychology Review, 3*(3), 193–209.

Bastian, B., Costello, K., Loughnan, S., & Hodson, G. (2012). When closing the human-animal divide expands moral concern: The importance of framing. *Social Psychological and Personality Science, 3*(4), 421–429. https://doi.org/10.1177/1948550611425106

Bastian, B., & Loughnan, S. (2017). Resolving the meat-paradox: A motivational account of morally troublesome behavior and its maintenance. *Personality and Social Psychology Review, 21*(3), 278–299. https://doi.org/10.1177/1088868316647562

Bastian, B., Loughnan, S., Haslam, N., & Radke, H. R. M. (2012). Don't mind meat? The denial of mind to animals used for human consumption. *Personality and Social Psychology Bulletin, 38*(2), 247–256. https://doi.org/10.1177/0146167211424291

Beatson, R., Loughnan, S., & Halloran, M. (2009). Attitudes toward animals: The effect of priming thoughts of human-animal similarities and mortality salience on the evaluation of companion animals. *Society and Animals, 17*(1), 72–89. https://doi.org/10.1163/156853009X393774

Bowlby, J. (1969). *Attachment and loss.* New York: Basic Books.

Bradshaw, J. W. S., & Paul, E. S. (2010). Could empathy for animals have been an adaptation in the evolution of Homo sapiens? *Animal Welfare, 19*(2), 107–112.

Bratanova, B., Loughnan, S., & Bastian, B. (2011). The effect of categorization as food on the perceived moral standing of animals. *Appetite, 57*(1), 193–196. https://doi.org/10.1016/j.appet.2011.04.020

Brown, S. E. (2002). Ethnic variations in pet attachment among students at an American school of veterinary medicine. *Society & Animals, 10*(3), 249–266. https://doi.org/10.1163/156853002320770065

Clayton, S. D. (2003). Environmental identity. In S. D. Clayton & S. Opotow (Eds.), *Identity and the Natural Environment: The Psychological Significance of Nature* (pp. 45–65). Cambridge, MA: MIT Press.

Costello, K., & Hodson, G. (2010). Exploring the roots of dehumanization: The role of animal-human similarity in promoting immigrant humanization. *Group Processes & Intergroup Relations, 13*(1), 3–22. doi:10.1177/1368430209347725

Costello, K., & Hodson, G. (2014). Explaining dehumanization among children: The interspecies model of prejudice. *British Journal of Social Psychology, 53*, 175–197. DOI:10.1111/bjso.12016

Dhont, K., Hodson, G., & Leite, A. C. (2016). Common ideological roots of speciesism and generalized ethnic prejudice: The social dominance human–animal relations model (SD-HARM). *European Journal of Personality, 30*(6), 507–522.

Dhont, K., Hodson, K., Leite, A.C., & Salmen, A. (this volume). The psychology of speciesism. In K. Dhont and G. Hodson (Eds.), *Why We Love and Exploit Animals: Bridging Insights from Academia and Advocacy.* Abingdon: Routledge.

European Commission. (2016). Special Eurobarometer 442: "Attitudes of Europeans towards animal welfare. Brussels: European Commission.

Fridlund, A. J., & MacDonald, M. J. (1998). Approaches to goldie: A field study of human approach responses to canine juvenescence. *Anthrozoös, 11*(2), 95–100. doi:10.2752/089279398787000751

Goldenberg, J. L., Pyszczynski, T., Greenberg, J., Solomon, S., Kluck, B., & Cornwell, R. (2001). I am not an animal: Mortality salience, disgust, and the denial of human creatureliness. *Journal of Experimental Psychology: General, 130*(3), 427–435. https://doi.org/10.1037//0096-3445.130.3.427

Gaertner, S. L., & Dovidio, J. F. (2000). *Reducing Intergroup Bias: The Common Ingroup Identity Model.* New York: Psychology Press.

Gray, P. B., & Young, S. M. (2011). Human-Pet Dynamics in Cross-Cultural Perspective. *Anthrozoös, 24*(1), 17–30. doi:10.2752/175303711X12923300467285

Greene, J. D., Sommerville, R. B., Nystrom, L. E., Darley, J. M., & Cohen, J. D. (2001). An fMRI investigation of emotional engagement in moral judgment. *Science, 293*(5537), 2105–2108.

Greenberg, J., Solomon, S., & Pyszczynski, T. (1997). Terror management theory of self-esteem and cultural worldviews: Empirical assessments and conceptual refinements. In M. P. Zanna (Ed.), *Advances in Experimental Social Psychology* (Vol. 29, pp. 61–139). San Diego, CA: Academic Press.

Henderson, B. B., Hensley, C., & Tallichet, S. E. (2011). Childhood Animal Cruelty Methods and Their Link to Adult Interpersonal Violence. *Journal of Interpersonal Violence, 26*(11), 2211–2227. doi:10.1177/0886260510383038

Herzog, H. (2011). The impact of pets on human health and psychological well-being: Fact, fiction, or hypothesis? *Current Directions in Psychological Science, 20*(4), 236–239. https://doi.org/10.1177/0963721411415220

Hills, A. M. (1995). Empathy and belief in the mental experiences of animals. *Anthrozoös, 8*(3), 132–142. doi:10.2752/089279395787156347

Hodson, G., Dhont, K., & Earle, M. (this volume). Devaluing animals, "animalistic" humans, and people who protect animals. In K. Dhont and G. Hodson (Eds.), *Why We Love and Exploit Animals: Bridging Insights from Academia and Advocacy.* Abingdon: Routledge.

Hornsey, M. J., & Hogg, M. A. (2000). Assimilation and diversity: An integrative model of subgroup relations. *Personality and Social Psychology Review, 4*(2), 143–156. doi: 10.1207/S15327957PSPR0402_03

Kellert, S. R. (1980). American attitudes toward, and knowledge of animals: An update. *International Journal for the Study of Animal Problems, 1*(2), 87–119.

Kellert, S. R., & Wilson, E. O. (1993). *The Biophilia hypothesis.* Washington, DC: Island Press.

Leach, C. W., van Zomeren, M., Zebel, S., Vliek, M. L. W., Pennekamp, S. F., Doosje, B., …Spears, R. (2008). Group-level self-definition and self-investment: A hierarchical (multicomponent) model of in-group identification. *Journal of Personality and Social Psychology, 95*(1), 144–165. https://doi.org/10.1037/0022-3514.95.1.144

Lobue, V., Bloom Pickard, M., Sherman, K., Axford, C., & Deloache, J. S. (2013). Young children's interest in live animals. *British Journal of Developmental Psychology, 31*(1), 57–69. https://doi.org/10.1111/j.2044-835X.2012.02078.x

Loughnan, S., Bastian, B., & Haslam, N. (2014). The psychology of eating animals. *Current Directions in Psychological Science, 23*(2), 104–108. https://doi.org/10.1177/0963721414525781/

Loughnan, S., & Davies, T. (this volume). The meat paradox. In K. Dhont and G. Hodson (Eds.), *Why We Love and Exploit Animals: Bridging Insights from Academia and Advocacy.* Abingdon: Routledge.

Loughnan, S., Haslam, N., & Bastian, B. (2010). The role of meat consumption in the denial of moral status and mind to meat animals. *Appetite, 55*(1), 156–159. https://doi.org/10.1016/j.appet.2010.05.043

McFarland, S., Webb, M., & Brown, D. (2012). All humanity is my ingroup: A measure and studies of identification with all humanity. *Journal of Personality and Social Psychology, 103*(5), 830–853. https://doi.org/10.1037/a0028724

Opotow, S. (1993). Animals and the scope of justice. *Journal of Social Issues, 49*(1), 71–85.

Piazza, J. (this volume). Why people love animals yet continue to eat them. In K. Dhont and G. Hodson (Eds.), *Why We Love and Exploit Animals: Bridging Insights from Academia and Advocacy.* Abingdon: Routledge.

Palmer, R., & Custance, D. (2008). A counterbalanced version of Ainsworth's strange situation procedure reveals secure-base effects in dog-human relationships. *Applied Animal Behaviour Science, 109*(2), 306–319. doi:10.1016/j.applanim.2007.04.002

Patterson-Kane, E. G., & Piper, H. (2009). Animal abuse as a sentinel for human violence: A critique. *Journal of Social Issues, 65*(3), 589–614. doi:10.1111/j.1540-4560.2009.01615.x

Paul, E. S., & Serpell, J. A. (1993). Childhood pet keeping and humane attitudes in young adulthood. *Animal Welfare, 2*(4), 321–337.

Paul, E. S., & Serpell, J. A. (1996). Obtaining a new pet dog: Effects on middle childhood children and their families. *Applied Animal Behaviour Science, 47*(1), 17–29. doi:https://doi.org/10.1016/0168-1591(95)01007-6

Peace, A. (2008). Meat in the genes. *Anthropology Today, 24*(3), 5–10. doi:10.1111/j.1467-8322.2008.00583.x

Piazza, J., Ruby, M. B., Loughnan, S., Luong, M., Kulik, J., Watkins, H. M., & Seigerman, M. (2015). Rationalizing meat consumption. The 4Ns. *Appetite, 91*, 114–128. https://doi.org/10.1016/j.appet.2015.04.011

Schwartz, R. L., Fremouw, W., Schenk, A., & Ragatz, L. L. (2012). Psychological profile of male and female animal abusers. *Journal of Interpersonal Violence, 27*(5), 846–861. https://doi.org/10.1177/0886260511423254

Sherif, M. (1966). *In Common Predicament; Social Psychology of Intergroup Conflict and Cooperation.* Boston, MA: Houghton Mifflin.

Signal, T. D., & Taylor, N. (2007). Attitude to animals and empathy: Comparing animal protection and general community samples. *Anthrozoös, 20*(2), 125–130. doi:10.2752/175303707X207918

Solomon, S., Greenberg, J., & Pyszczynski, T. (1991). A terror management theory of social behavior: The psychological functions of self-esteem and cultural worldviews. In

M. E. P. Zanna (Ed.), *Advances in Experimental Social Psychology* (Vol. 24, pp. 93–159). San Diego, CA: Academic Press.

Sorenson, J. (this volume) Humane hypocrisies: Making killing acceptable. In K. Dhont and G. Hodson (Eds.), *Why We Love and Exploit Animals: Bridging Insights from Academia and Advocacy.* Abingdon: Routledge.

Tajfel, H., & Turner, J. C. (1986). The social identity theory of intergroup behavior. In S. Worchel & W. G. Austin (Eds.), *Psychology of Intergroup Relations* (pp. 7–24). Chicago, IL: Nelson-Hall.

Turner, J. C., Hogg, M. A., Oakes, P. J., Reicher, S. D., & Wetherell, M. (1987). *Rediscovering the Social Group: A Self-Categorization Theory.* Oxford: Blackwell.

Wagstaff, G. F. (1991). Attitudes toward animals and human beings. *The Journal of Social Psychology, 131*(4), 573–575.

Wilson, E. O. (1984). *Biophillia: The Human Bond with Other Species.* Cambridge, MA: Harvard University Press.

Zilcha-Mano, S., Mikulincer, M., & Shaver, P. R. (2011). An attachment perspective on human-pet relationships: Conceptualization and assessment of pet attachment orientations. *Journal of Research in Personality, 45*(4), 345–357. https://doi.org/10.1016/j.jrp.2011.04.001

3

THE PSYCHOLOGY OF SPECIESISM

Kristof Dhont, Gordon Hodson, Ana C. Leite, and Alina Salmen

Abstract

Despite being animals, humans distance themselves physically and mentally from (most) other animals and prioritize human interests. We exploit other animals to feed, clothe, and entertain ourselves, to name just a few animal exploitation practices. Such discrimination against other species, or *speciesism*, is the central focus of the present chapter. Drawing on recent scientific findings, we reveal the psychological connections between speciesism and prejudices such as racism and sexism. Those who support animal exploitation also tend to endorse sexist and racist views and rely on the belief in group dominance and human supremacy to justify systems of inequality and oppression. The common denominator is that the interests of disadvantaged groups like animals, women, and ethnic minorities, are considered subordinate to the interests and privileges of advantaged groups like humans in general, and white men in particular. Although recognizing this intersectionality is critical to the understanding of human-animal relations, explicitly referring to such parallels in animal advocacy campaigns can be easily misunderstood, and may be ineffective or even counterproductive. We see value in experience- and behavior-based interventions where people learn to connect psychologically with animals to change their animal-relevant beliefs, and more generally, to broaden the mind and challenge exploitative societal traditions.

What do they know – all these scholars, all these philosophers, all the leaders of the world – about such as you? They have convinced themselves that man, the worst transgressor of all the species, is the crown of creation. All other creatures were created merely to provide him with food, pelts, to be tormented, exterminated. In relation to them, all people are Nazis; for the animals it is an eternal Treblinka.
Isaac Bashevis Singer, 1968

On a mass scale and on a daily basis, animals[1] are being shot, gassed, suffocated, or electrocuted. This typically happens after living in miserable and abusive circumstances in extreme confinement. Animals are being mutilated through branding, dehorning, amputating tales or tongues, etc., and crammed into trucks for long distances often in severe weather conditions. In his writings, novelist and Nobel Prize laureate Isaac Bashevis Singer regularly compared the way people treat and slaughter non-human animals to the way Jews were treated and killed by the Nazis (I. B. Singer, 1968). Others have used metaphors and imagery referring to the period of slavery in America to illustrate standard procedures in modern factory farms (e.g., Plous, 2003; Spiegel, 1988). Yet purely in terms of numbers, if you consider the billions of animals being killed every year, the Holocaust comparison undersells the scope of the problem.

The systemic exploitation of other species by humans is often referred to as "speciesism," analogous to the discriminatory treatment of people based on gender (i.e., sexism) or race (i.e., racism) (Dhont, Hodson, & Leite, 2016; Horta, 2010; Ryder, 2006; Singer, 1975; Sorenson, 2016, this volume). From a psychological perspective, speciesism can be defined as the differential treatment (behavioral) or moral evaluation (attitudes and beliefs) of animals merely based on their species membership (see also Caviola, Everett, & Faber, 2019). Surprisingly, only recently has speciesism become a topic of empirical research in social and behavioral sciences, despite the long-standing debates and controversies surrounding this topic in philosophical and animal advocacy circles.

This chapter reviews recent empirical work on the psychology of speciesism and the proposed parallels with biases towards human groups such as racism and sexism. Is speciesism comparable to other types of prejudice? And if so, what are the common psychological factors driving and maintaining both speciesism and human intergroup biases? After addressing these questions, we then discuss the usefulness of such comparisons in animal advocacy. Indeed, comparing speciesism to racism or sexism in animal rights campaigns, and referring to past atrocities against minority groups to raise awareness about the treatment of animals, has often caused public commotion and anger towards animal advocates, rather than towards animal abusers. Finally, we discuss the political polarization of animal rights and reflect on possible solutions.

Generalized prejudice and speciesism

Gordon Allport (1954) famously observed that "If a person is anti-Jewish, he is likely to be anti-Catholic, anti-Negro, anti any out-group" (p. 68). This general tendency of holding negative attitudes towards a range of different outgroups is known as *generalized prejudice* (Akrami, Ekehammar, & Bergh, 2011; Hodson & Dhont, 2015; McFarland, 2010). Supporting this idea, empirical research across cultural contexts has shown that those who devalue people of one social group, for instance by endorsing racist views, also tend to devalue people of other groups, for instance by also endorsing sexist and homophobic views (Akrami et al., 2011; Hodson, MacInnis, & Busseri, 2017; Meeusen & Dhont, 2015; Zick et al., 2008). Recent work further refined the theorizing on generalized prejudice by demonstrating that the idea applies especially, and perhaps only, to prejudice towards marginalized (or low-status) groups, rather than *all* groups (Bergh, Akrami, Sidanius, & Sibley, 2016).

Roughly sketching their personality, people high on generalized prejudice tend to be less friendly, less compassionate, and less humble, not just toward members of other groups, but towards other people in general (Cichocka, Dhont, & Makwana, 2017; Hodson & Dhont, 2015; Sibley & Duckitt, 2008). They are also more conventional and closed-minded; not very open to innovative ideas or novel experiences. Such personality characteristics are further implicated in the type of ideological and political beliefs people endorse (Carney, Jost, Gosling, & Potter, 2008; Cichocka & Dhont, 2018; Hodson & Dhont, 2015; Hodson, Hogg, & MacInnis, 2009), which, as we will see, are relevant for our understanding of people's attitudes towards animals. People characterized by a greedy, interpersonally insensitive, and rude (as opposed to humble, compassionate, friendly, etc.) personality are, ideologically speaking, more acceptant of social inequality. Indeed, they strongly prefer a society with a strict hierarchical order where some social groups are dominating over others, as opposed to egalitarian intergroup relations. This ideological belief of endorsing inequality and group-based dominance has been termed social dominance orientation (SDO; Pratto, Sidanius, Stallworth, & Malle, 1994; Sidanius & Pratto, 1999). Furthermore, conventional people tend to resist societal change, and rather seek stability and order. They strongly value cultural and family traditions, are more compliant with laws and authorities, and tend to be intolerant towards individuals or groups violating or deviating from the cultural norms. This cluster of socially conservative ideological beliefs is known as right-wing authoritarianism (RWA; Altemeyer, 1996). Both SDO and RWA are strong predictors of generalized prejudice (Hodson & Dhont, 2015; Hodson et al., 2017; Sibley & Duckitt, 2008) and of support for parties on the right side of the political spectrum, and in the case of SDO, support for far-right parties (e.g., Van Assche, Van Hiel, Dhont, & Roets, 2019; Van Hiel, Cornelis, Roets, & De Clercq, 2007).

In general, relatively few people show explicit support for inequality between human social groups merely based on their group membership. Indeed, in principle, only few would say that certain social groups, like particular ethnic groups, or one gender or age group, are morally inferior to another group (see Kteily, Ho, & Sidanius, 2012; Sibley & Liu, 2010). Yet, comparatively, a substantially bigger number of people agree that an animal life is inherently of lesser value than a human life (Dhont & Hodson, 2014; Salmen & Dhont, in prep). Relative to other low-status groups, animals may represent the *quintessential low-status group* – targets of speciesism.

Illustrative of speciesism, in one of their studies, Caviola and his colleagues (2019) asked American participants to allocate $100 between a charity that focuses on helping humans and a charity that focuses on helping animals. For both charities, the researchers mentioned that an allocation of $10 would keep one individual free from pain and suffering for one day. Far from an equal split between both charities, an average of $68 went to helping humans versus $32 to helping animals. For these participants, the pain and suffering of a human being was worth more than twice as much as the pain and suffering of an animal. The authors further demonstrated that the differential allocation of money to a human or an animal cause was larger for those participants holding stronger speciesist beliefs, measured by asking participants to indicate their agreement with statements such as "*Morally, animals always count for less than humans.*"

At this point, some readers may object that the differential treatment or judgment of humans and animals are not based on species membership alone, thereby questioning the concept of speciesism altogether. Instead, one could argue that, compared to non-human animals, our higher cognitive abilities and capacity to suffer is the reason and also the justification for why people consider and treat the life of animals as less valuable. Such viewpoints not only fail to acknowledge the scientific observations of the rich cognitive and emotional lives of animals (Bekoff, 2007a; Bekoff & Pierce, 2009; de Waal, 2016), but can also not (fully) account for the devaluation of animals by humans (Caviola et al., 2019; Horta, 2010; P. Singer, 1975). Why else, other than their species membership, would those with severe mental disabilities, but not animals with high mental abilities like chimpanzees, be granted equal moral value just like all other humans? When the same participants of the study by Caviola and colleagues were asked to rate the intelligence levels and suffering capability of chimpanzees and those with severe mental disabilities, chimpanzees were rated as smarter, while the ratings of suffering capability did not significantly differ between these two groups. If anything, one would expect that the chimpanzees received the advantageous treatment if cognitive capacities matter, or no differences when considering the capacity to suffer. This is not what happened. With $100 at their disposal to allocate between the two groups, on average $72 went to helping those with severe mental disabilities leaving only $28 for helping the (smarter) chimpanzees. Speciesism is real, and the animals are paying the price.

Speciesism does not only manifest itself in the differential treatment and moral evaluation of humans and animals, but also in the way we treat and think morally about different animals. Consider for instance the horsemeat scandal that broke out in the UK in 2013, and then spread across Europe ("Q&A: Horsemeat scandal," 2013; Quinn, 2013). When the UK food safety authorities revealed that numerous beef products contained horsemeat, with some "beef" lasagnes even being 100 percent horsemeat, this caused public rage and fury among plenty of British consumers – not because of the incorrect labelling or fraudulent practice, but because of the cultural taboo against eating horses in the UK. Britons are effortlessly willing to devour pigs and cows, as their Sunday roast drowned in gravy, or as a meat pie with a crust of mashed potato, yet they are disgusted by the idea of eating horses. The lack of moral objections to eating cows or pigs indicates that there is no other reason for the horrified reactions over horsemeat than the fact that horses are not considered food animals in the UK ("Why are the British revolted by the idea of horsemeat?", 2013). In contrast, many French people living in the UK did not understand what the fuss was all about and, at worst, experienced this issue as an inconvenience for not knowing the ingredients. While the horsemeat scandal temporarily increased the sales of vegetarian meat alternatives in the UK, it also sparked a rise in sales of approximately 15 percent for traditional horse butchers in France (Chrisafis, 2013)! It seems that the (bad) publicity and chatter about horsemeat had ignited renewed interest in traditional horsemeat dishes in France and reminded French people about the availability of this "low-fat, low-cholesterol" alternative to beef and pork.

The moral concern for animals is thus largely based on criteria that, in principle, would be considered irrelevant when judging whether or not certain decisions and behaviors are morally acceptable. Indeed, our social and cultural relation with specific animal species, the function of the animal category for humans (e.g., edible or not), the way they look (e.g., cute, ugly, similar to humans), and how we portray or perceive their behavior (e.g., as dangerous or threatening), all feed into the differential concern and treatment of animals (Bratanova, Loughnan, & Bastian, 2011; Herzog, 2010; Joy, 2010; see also Piazza, this volume; Woods & Hare, this volume). In a recent online survey study, we presented a list of twenty animals to American participants and asked them to select those animals that they feel morally obligated to show concern for (Leite, Dhont, & Hodson, 2019). We included several *companion animals* like dogs, cats, and horses, several animals typically considered as *food* such as cows and pigs, but also ducks and chickens, and thus a range of food animals that are very different from each other from a purely biological perspective. Furthermore, participants also rated a range of *wild animals*; some *more appealing* and of *higher status*, such as chimps, bears, and dolphins, as compared to *less appealing* wild animals like snakes, frogs, and crocodiles. As expected, participants selected animals belonging to the same categories about equally frequently. Yet, as illustrated in Figure 3.1, there were huge differences between the categories of animals. A vast majority of the respondents felt morally obliged to show concern

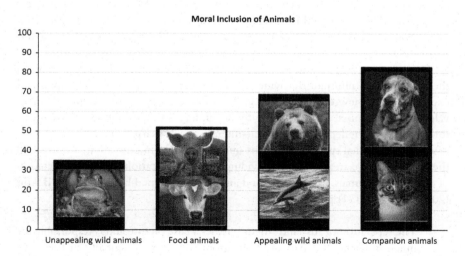

FIGURE 3.1   Mean percentage of moral inclusion of different animal categories (based on Leite, Dhont, & Hodson, 2019).

for the welfare and interests of companion animals (cats, dogs, and horses), clearly more than for any other animal category. Most participants also indicated that they care about appealing wild animals (e.g., dolphins, chimps), yet the proportion of participants feeling moral concern dropped significantly to just over 50 percent when it concerned food animals (e.g., chickens, pigs, cows), with a further drop for unappealing wild animals (e.g., snake, snail). These findings demonstrate that judgments about the need to morally care for an animal vary widely between animal species, and are directly linked to whether or not the animal is typically exploited or ignored by people.

A similar discrepancy between different animals was observed in the study by Caviola and colleagues (2019) discussed earlier, when participants were asked to allocate $100 between a charity helping dogs and a charity helping pigs. On average, participants allocated more than double the amount to help dogs ($69) than to help pigs ($31). In our study (Leite et al., 2019) participants cared most about dogs, of all animals, with 90 percent indicating that they feel morally obliged to care about dogs. If the detection of horse meat already leads to public calls for a largescale boycott of affected supermarkets, can you imagine what would happen if food safety authorities were ever to find traces of dog meat in beef products? Thinking of chewing on a steak of Labrador or gulping down Beagle stew not only feels disturbing and physically repulsive for most people, it also violates our moral intuitions (see also Joy, 2010). Dogs are our buddies, part of the family, and who would eat their friends or family? The special social relation we have with dogs, rooted in Western cultural traditions, makes Westerners perceive and think about dogs differently than how we perceive and think about pigs and cows. Despite the absence of fundamental differences in their emotional and cognitive capacities, we value the life of a dog much higher than the life of a pig or a cow. Yet, if you were born and raised in Vietnam or China, there is a chance that you would be slurping tastefully from a bowl of dog meat soup every now and then, and if born in India, consider the cow a sacred being and never eat beef.

This selection of scientific findings and the example of the horsemeat scandal should suffice to illustrate the widespread prevalence of speciesist principles and expressions. We view ourselves, humans, as inherently superior to animals and we value different animal species differently depending on their functional or cultural role for humans, reflecting human self-interests. Such sharp demarcation between us and other animal species underpins our prejudicial thinking about animals and the biased way we treat animals, providing moral justification for the exploitation of all kinds of animals, particularly food animals. Renowned biologist Marc Bekoff (2007b, pp. 170–171) puts it this way:

> The separation of "us" and "them" creates a false picture and is responsible for much suffering. It is part of the in-group/out-group mentality that leads to human oppression of the weak by the strong as in ethnic, religious, political, and social conflicts.

Arguably, in many respects, how we discriminate between human and non-human animals resembles the way that low-status human groups are being discriminated against. Such insights suggest that the concept of generalized prejudice is not limited to prejudices towards human groups but can be extended to prejudice towards animals. The parallels or interconnectedness between speciesism and prejudices towards human groups have often been discussed by influential scholars like Peter Singer and Carol Adams. Moreover, some animal advocacy posters allude to these similarities, for instance those showing a picture of a chained animal next to a picture of a chained human slave. Yet, only recently have these parallels been empirically investigated in psychological research.

Common ideological roots of speciesism and prejudice towards human groups

Is speciesism a type of prejudice comparable to traditional types of prejudice like homophobia or racism, and can it be integrated in the construct of generalized prejudice?

The central idea of generalized prejudice is that someone who is highly prejudiced towards Group X is highly prejudiced towards Groups Y and Z (Allport, 1954; Akrami et al., 2011; Meeusen & Dhont, 2015). Hence, a first, relatively easy way of testing whether the concept can be broadened to include animals as another outgroup (and target of prejudice), is to investigate whether those who express negative and prejudiced views toward human outgroups (e.g., ethnic and religious groups) also endorse exploitative attitudes toward animals (speciesist attitudes). We tested this idea in a number of different samples, first in a Canadian sample of students and subsequently in a series of studies conducted in Belgium, the UK, and the USA using student and more heterogeneous samples of adults (Dhont, Hodson, Costello, & MacInnis, 2014; Dhont et al., 2016). Confirming the associations across all samples, those who expressed more negative attitudes against ethnic and racial minority groups also expressed greater support for a range of practices of animal exploitation such as hunting, factory farming, animal testing, whaling, and using animals for human entertainment. Following up on this research line, several other studies further showed that stronger endorsement of gender-based prejudice (i.e., sexism) and sexuality-based prejudice (i.e., homophobia) is positively related to greater speciesism (Caviola et al., 2019; Salmen & Dhont, in prep). Taken together, the first conclusion from these studies is that speciesism is a type of prejudice, reliably related to prejudicial attitudes towards a range of human outgroups. These findings point to the psychological similarities between these different types of prejudices and suggest they all share a common, generalized component. But *why* is this the case?

In the next step in this research line, we addressed the question of *why* prejudicial tendencies towards human outgroups and animals are connected. What is the psychological key ingredient underpinning both speciesism and prejudice towards human outgroups? The Social Dominance Human-Animal Relations Model (SD-HARM; Dhont et al., 2016) proposes that the desire for group-based dominance and social inequality, that is SDO, represents the unifying ideological motive constituting the common core of both speciesism and prejudice towards human groups. According to Social Dominance Theory, social dominators (i.e., those higher on SDO) want to preserve hierarchical social structures and further enhance the dominant position of high-status and advantaged groups (Sidanius & Pratto, 1999). To this end, they deploy a range of discriminatory belief systems like racism and sexism to rationalize and morally legitimize the implementation of institutionalized discriminatory policies. Hence, as a robust predictor of generalized prejudice, SDO operates in a wide-ranging fashion, targeting marginalized and low-status groups "regardless of the manner in which these groups are defined" (Sidanius & Pratto, 1999, p. 61; see also Kteily et al., 2012).

Given this generalized tendency for group-based dominance and inequality, the impact of SDO may extend well beyond the realm of human intergroup relations and be implicated in human-animal relations. Indeed, research showed that those higher on SDO tend to see greater differences between animals and humans and more strongly believe that humans are superior to animals (Costello & Hodson, 2010, 2014; Dhont & Hodson, 2014; Hodson, Dhont, & Earle, this volume). Furthermore, across all our samples, those higher on SDO not only showed higher levels of ethnic prejudice but also more strongly endorsed speciesist attitudes (Dhont et al., 2014, 2016). Theoretically, much like how social dominators adhere to legitimizing ideologies such as racism and sexism to justify the oppression of low-status human groups, they also endorse human supremacy and speciesist beliefs to defend the exploitation and consumption of animals (Dhont & Hodson, 2014; Hyers, 2006).

Critically, we also formally tested whether the significant association between ethnic prejudice and speciesism can be explained by their joint association with SDO, as proposed by the SD-HARM model (Dhont et al., 2016). In statistical terms, this means that the correlation between ethnic prejudice and speciesism would drastically decrease or disappear (i.e., become statistically non-significant) after accounting for SDO. This was confirmed in each of our studies: when SDO was modelled as the common ideological core underpinning both ethnic prejudice and speciesism, the residual relation between ethnic prejudice and speciesism approached zero and was nonsignificant. Caviola and colleagues (2019) replicated this pattern of results and further showed that SDO also accounted for the associations between speciesism and homophobia.

In testing the critical role of SDO in these relations, we also investigated whether other ideological variables such as RWA and left-right political orientation would explain the associations between ethnic prejudice and speciesism. We found no evidence for this, which rules out several alternative explanations for the observed associations and emphasizes the unique role of *group-based dominance motives* in connecting prejudicial tendencies in human intergroup and human-animal relations. Yet, interestingly, political conservatism also predicted speciesism even after accounting for SDO and RWA. This finding may suggest that beyond group-dominance motives and the preservation of cultural traditions, political conservatives may oppose animal rights (and endorse speciesist practices) possibly because support for animal rights, just like pro-environmentalism, is perceived as a left-wing topic. We come back to this ideological divide between the left and the right later in this chapter.

In conclusion, by revealing the key role of SDO, SD-HARM offers a unifying and parsimonious theoretical account for the connections between attitudes towards animals and attitudes towards ethnic outgroups. Our studies empirically substantiate the long-held views of scholars in philosophy and animal advocates who argued that common or interconnected belief systems sustain the exploitation of both animals and human groups (e.g., Adams, 2015; Adams & Gruen, 2014; Ryder, 2006; Singer, 1975; Spiegel, 1988; see also Dhont, Hodson, Loughnan, & Amiot, 2019; Hodson & Costello, 2012; Plous, 2003). Desires for power and the might of one group over another motivate oppressive and discriminatory views and practices towards human and non-human outgroups.

Extensions of SD-HARM

Part of the appeal of SD-HARM is that the model relies on a single principle, with SDO as the main protagonist. Yet, we acknowledge wholeheartedly that a variety of psychological and contextual factors, along with historical and economic factors, play a role in how speciesism develops, and need to be considered to fully understand why speciesism is deeply entrenched in society and in people's way of life. In testing and confirming SD-HARM, our studies focused on ethnic intergroup attitudes and support for animal exploitation across a range of different animals and practices. Hence, this approach restricts the conclusions in at least two ways, leaving room to extend the model. First, the findings tell us little about the ideological or psychological drivers of why different animals are being perceived and treated differently from other animals within and between cultural contexts. As we have seen earlier, eating horse or dog meat are common practices in some countries, but met with revulsion and outrage in other countries. Along similar lines, bullfights in Spain and whaling practices in Norway and Iceland are defended passionately and vehemently by a sizable number of citizens living in those countries, but are seen as inexcusable brutalities to many others. These examples highlight the systematic differences between cultures in support for specific types of animal exploitation, with particular animals being loved in one culture but exploited in others. Cultural norms and traditions often dictate what is acceptable and inacceptable. Yet people also differ from each other in the extent to which they value traditions and cultural norms. Hence, while SDO, as a generalized tendency, reliably predicts greater support for many types of animal exploitation within and across cultures, cherishing specific practices from one's own culture likely reflects ideological values of social conformity and traditions. This idea is consistent with the finding that RWA, as a typical indicator of social-cultural conservatism, correlates robustly with meat consumption (Dhont & Hodson, 2014; Hodson et al., this volume). The more socially conservative people are, the more meat they eat. Extending this idea to other practices of animal exploitation, we suspect that social-cultural conservatives (e.g., those higher in RWA) will show greater support for specific practices of animal exploitation typical for their own cultural region, but not necessarily for foreign cultural practices.

As an initial test of this idea, we surveyed an online sample of about 250 American adults and asked them about their views regarding three different practices of animal exploitation (Dhont, Leite, Hodson, & Milfont, in prep). Two of these practices, bullfighting and whaling, are not part of the cultural traditions in the USA, whereas the third practice, rodeos, happens widely over the USA, and for many Americans is closely tied to national or regional traditions. We also measured their levels of SDO and RWA. Consistent with our line of reasoning, although SDO was related to stronger support for all three practices of animal exploitation, RWA only predicted support for rodeos, and did not significantly predict support for whaling or bullfighting, after statically accounting for SDO. These findings are a first step in extending the SD-HARM model and show that in addition to group-dominance motives, desires to preserve cultural traditions and resistance to cultural change can be a powerful ideological motive to defend practices of animal exploitation. Yet a critical test of these ideas involves the cross-cultural comparison of people's attitudes.

Secondly, the SD-HARM studies only investigated the associations between speciesism and ethnic prejudice, but no other types of outgroup prejudice like sexuality-based prejudice, sexism, or ageism. As we have seen, at least one published study demonstrated that SD-HARM also applies to the association between speciesism and homophobia (Caviola et al., 2019, Study 3). Noteworthy however, the findings of Caviola and colleagues also suggested that SDO accounts for only part of the association between speciesism and sexism, indicating that other factors likely play a role in this relation. Indeed, the associations between sexism and speciesism may be more complex because of the multifaceted nature of sexism (Glick & Fiske, 1996, 2001). Consider for instance the use of aggressive and derogatory metaphors by calling women "bitches" or "cows" (Haslam, Holland, & Stratemeyer, this volume), as witnessed during the 2016 US presidential campaign when Donald Trump fans were wearing t-shirts and badges screaming "*Trump That Bitch!*". The explicit association of women with animals in these metaphors is meant in an intimidating and pejorative way, reflecting a hostile type of sexism. Moreover, such views inherently imply the inferior status of animals relative to humans, and thus a belief in human dominance over animals (see also Adams, 2015; MacKinnon, 2004). By putting women in this lower "animal status," not only dominance motives in human intergroup relations (i.e., SDO) but also human supremacy beliefs serve to sustain and legitimate hostile sexist views of male dominance over women (Salmen & Dhont, in prep).

At the same time, plenty of media commercials depict women (but seldom men) frivolously swirling in a natural environment, such as washing their hair under a glorious waterfall in a crystal blue lagoon. Such images portray, ostensibly harmlessly, women as being closely connected to nature and animals in ways that are meant to communicate a kind, positive message. We also often hear that women are more "in tune" with nature, or on other occasions, that they have better natural, "maternal" instincts. Yet despite being perceived as subjectively benevolent, connecting women to nature nevertheless sustains the traditional view of women as fragile and in need of protection and control. In other words, it implies a paternalistic view of gender roles, as also expressed in patronizing animal metaphors when calling women "kittens" or "pets" (see Haslam et al., this volume). Just like hostile sexism, albeit in a more subtle way, this benevolent form of sexism (Glick & Fiske, 1996, 2001; Hopkins-Doyle, Sutton, Douglas, & Calogero, 2019) reflects how people think about both women (relative to men) and animals (relative to humans). Based on this theorizing, Salmen and Dhont (in prep) proposed that beliefs in human dominance over animals and nature underpin both hostile and benevolent sexist ideologies and serve as a justification for the lower status of women and animals in society, hence lying at the core of sexism and speciesism.

Putting these ideas to the test, Salmen and Dhont (in prep) conducted several online survey studies totalling more than 1500 adults from the USA and the UK (approximately 55 percent female). Across all studies, the results confirmed that hostile and benevolent sexists hold stronger speciesist views, providing further evidence that speciesism can be considered a type of prejudice linked to other types of prejudice. Critically, the results also consistently demonstrated that human supremacy belief was significantly related to both hostile and benevolent sexism, above and beyond the effects of gender. When further testing the role of dominance desires in human intergroup relations by simultaneously investigating SDO and human supremacy beliefs, the findings revealed that both SDO and human supremacy beliefs uniquely predicted sexism. In other words, consistent with the SD-HARM model, speciesism and different sexism dimensions are indeed meaningfully related and rooted in common ideological belief systems, with SDO of crucial importance. Yet these findings further extend SD-HARM by highlighting that also people's perception of a hierarchical divide between humans and animals guides how they perceive and treat women relative to men. Indeed, those with stronger desires for human dominance over animals tend to endorse hostile and benevolent sexist views, which, in turn, are implicated in harmful behaviours such as downplaying or justifying sexual violence and restricting women's autonomy (Abrams, Viki, Masser, & Bohner, 2003; Petterson & Sutton, 2018; Salmen & Dhont, in prep).

To conclude, the way we think about animals has implications for the way we think about human groups. And the other way around, what we consider permissible in the (mal)treatment of humans, translates and often aggravates what we consider permissible in the (mal)treatment of animals. Systems of oppression and exploitation of humans and animals not only show important outward similarities, they also connect psychologically, deep down in the human mind, in how we think categorically about the world, in how we shape our social constructions about human-animal and human intergroup relations, and in how we create intractable ingroup-outgroup divisions. Yet the challenge remains: What can we do about it?

Implications for animal advocacy

How helpful or convincing is it to compare the struggle for animal rights with other social justice issues in animal advocacy? Establishing empirical evidence for the connections between different forms of oppressions does not in itself demonstrate that communicating about these connections in animal advocacy is an effective strategy. Unfortunately, as far as we are aware, no empirical research has addressed this question yet. We can merely speculate about the possible consequences of alluding to the Holocaust, slavery, or to racism, sexism, and the like as a means of reducing speciesism or the consumption of animal products. However, we can foresee reasons for being cautious about such approaches.

From the viewpoint of a vegan or animal rights activist, the slogans like "racism = speciesism = sexism" make complete sense. Such observations can also stimulate interesting theoretical debates within the animal advocacy movement and lead to a deeper understanding of different social issues. Yet, within this movement, most people accept the underlying morally motivated presumption that the life and suffering of an animal is of equal value as (or similar enough to) the life and suffering of a human. As we have seen, however, the majority of people disagree with this basic idea and value human life more than the life of an animal, especially farm animals. For them, such slogans are likely perceived as unreasonable, silly, or antagonistic, and therefore further ignored or resisted. Using such comparisons can, therefore, lead to unintended, harmful consequences that work against the cause of animal advocacy.

One possible reaction of a typical meat-eating lay audience is that they may not only feel that they are being judged and criticized for eating meat, but also think that they are being called racist or sexist. Rather than raising awareness about animal suffering, such interpretation of the message will likely upset and offend the audience. Indeed, as we can reasonably assume that the vast majority of an audience of ordinary people frankly rejects racism, (unintentionally) accusing them of racism will create unwanted negativity or even plain hostility between the messenger and the receiver. This turns people off and leads to disapproval of the animal advocacy organization or activist.

A second possible interpretation of the message is that it is considered an attempt to downplay the severity of the atrocious historical events or trivialize other social justice issues. Needless to say, depreciating extremely sensitive issues by mentioning the Holocaust, for instance, will be met with outrage and resistance. Between 2003 and 2006, the animal rights organization People for the Ethical Treatment of Animals (PETA) travelled around the world with exhibits entitled "The Holocaust On Your Plate" and "Animal Liberation Project: We Are All Animals." The first one displayed enormous campaign boards with headers such as "To animals, all people are Nazis," that juxtaposed photos showing rows of chickens in battery cages with photos of rows of prisoners in a Nazi concentration camp lying in wooden bunks. The other exhibition displayed massive panels with photos of a range of practices of animal exploitation (circus animals, animals in factory farms) next to photos of a range of historical examples of violence and exploitation of marginalized groups including slavery and lynching scenes. In both cases, various journalists and a number of leaders from the Jewish and black communities and civil rights organizations sharply denounced the exhibits as disgraceful and morally appalling (Kim, 2011). In both cases, the moral outcry from mainstream media and minority community members led to public apologies from PETA's president. Germany's highest court even banned PETA's Holocaust campaign after the Central Council of Jews in Germany had sued PETA. Holocaust survivor and president of the Central Council, Paul Spiegel, described the campaign as "the most disgusting abuse of the memory of the Holocaust in recent years." Similarly, African Americans and survivors of racial violence in the USA found it utterly repulsive that PETA had used other people's suffering for their "own" cause. Rather than making people rethink the treatment and value of animals, Jewish and Black people, among others, these groups felt that PETA dehumanized them by comparing their group to animals.

To be clear, in the absence of empirical data, it is impossible to tell how effective PETA's campaigns were in increasing awareness about animal suffering. Clearly such campaigns can cause counter-productive outcomes, perhaps leading to long-term grievances towards animal advocates within certain communities. We know from psychological research that such negative cycles are hard to break. People who feel dehumanized by a certain group will, in turn, also show a strong hostile counter-reaction and reciprocal dehumanization toward that group (Hodson et al., this volume; Kteily, Hodson, & Bruneau, 2016). We doubt that this was an intended feature of PETA's campaign strategies. In sum, relying on the parallels between speciesism, racism, sexism, etc. in animal advocacy can be counterproductive, particularly when communicated in ways that could offend your audience.

What is the relevance, therefore, of revealing the connections between different types of prejudices? The discovery that common ideological beliefs underpin different types of biases also implies that implementing interventions that change precisely those common factors responsible for a wide range of prejudices may have a far-reaching impact on a range of intergroup biases. Our findings suggest that techniques aiming at attenuating people's fundamental preferences for group-based dominance and inequality in human intergroup and human-animal relations (i.e., SDO and human supremacy beliefs) have this potential. But again, there is a catch.

When explicitly emphasizing principles of social equality and advocating for social justice and change, the topics of animal rights and veganism seem to belong inherently in the left-wing domain. It is common knowledge, that the majority of vegetarians or animal advocates consider themselves liberal or left-wing, as also demonstrated by a recent Gallup poll conducted in the USA (Reinhart, 2018). Among the 5% of Americans calling themselves vegetarians, there is a staggering ideological divide of 11% vegetarians among American liberals compared to only 2% vegetarians among conservatives. Only focusing on values that are particularly important for left-wing adherents may further increase the political polarization on this topic, or even encourage conservatives to eat more meat and take pride in it, as it is likely seen as a conservative thing to do. Hence, framing the case for animal rights only in terms of equality values (or egalitarianism) and social change/ justice values likely turns off conservatives, given that these values are either not important to them or contradict their values (see Jost, 2017; Jost, Glaser, Kruglanski, & Sulloway, 2003). By being more mindful of the values of people across the political spectrum, and especially those of conservatives, there is potential to reach a broader audience. Indeed, even though political liberals are slightly more sensitive to suffering and harm than conservatives, the principle of "do no harm" appeals to people on both sides of the ideological spectrum (Graham, Haidt, & Nosek, 2009). Future research can test the impact of different moral framing techniques in animal advocacy messages.

Another method to side-step the ideological polarization of animal rights is to avoid the explicit use of ethically loaded arguments or messages altogether. Many omnivores, no matter their ideological beliefs, care about animals (to some extent), but continue paradoxically and happily with exploiting them, for instance by eating meat. The psychological flexibility of our moral reasoning and behavior is mind-boggling, as discussed in detail in other chapters of this book (e.g., Loughnan & Davies, this volume; Piazza, this volume), meaning that various motivational, social, and external obstacles prevent people from stopping or reducing their meat consumption. Therefore, interventions that do not actively try to convince people about what is right or wrong, but rather aim to change aspects of people's behavior, might prove particularly promising (see also Leenaert, this volume). Indeed, research on cognitive dissonance (Festinger, 1957, 1962) suggests that after people engage in the desired behavior, they may bring their attitudes in line with their behavior (Olson & Stone, 2005), whereas attempts to change people's attitudes to bring about behavior change is often met with resistance (Knowles & Linn, 2004; Wegener, Petty, Smoak, & Fabrigar, 2004). Here we will focus on one such technique: creating opportunities for positive and meaningful interactions between members of different groups, including those between humans and farm animals.

An extensive body of research has long established that favorable contact between people of different ethnic or religious backgrounds or with different sexual orientations reduces mutual prejudice and improves intergroup relations (see Hodson & Hewstone, 2013; Pettigrew & Tropp, 2011). Yet, positive intergroup contact also has wider implications that generalize beyond the immediate intergroup context, as illustrated by what is known as the secondary transfer effect of intergroup contact (Pettigrew, 2009; Tausch et al., 2010). Research conducted on representative samples from divided communities demonstrated for instance that cross-community contact between Catholics and Protestants in Northern Ireland improves attitudes not only towards the other religious community but also towards racial minorities uninvolved in the contact situation (Tausch et al., 2010). Similar observations have been reported for contact between Greek and Turkish Cypriots leading to better attitudes toward mainland Greeks and Turks, and between Black and White American students resulting in improved attitudes towards Hispanics. By stimulating outgroup empathizing and taking the perspective of the opposing group, positive intergroup contact not only leads to a better mutual understanding, but also generates an open and flexible way of thinking about intergroup relations and about the world (Hodson, Crisp, Meleady, & Earle, 2018; Meleady, Crisp, Dhont, Hopthrow, & Turner, in press). In doing so, it can change general beliefs about intergroup hierarchy and inequality (Dhont, Van Hiel, & Hewstone, 2014). For instance, Belgian high school students going on a one-week trip to Morocco where they socialized with Moroccan students and their family not only showed lower anti-Moroccan prejudice after the school trip, but also lower levels of SDO (Dhont et al., 2014). In another study, conducted in a university housing setting in the USA, first-year students were randomly allocated to share a room with either someone from the same race or someone from a different race at the beginning of the semester (Shook, Hopkins, & Koech, 2016). At the end of the semester, those living in interracial rooms, but not those in same-race rooms, showed a decrease in SDO as well as more favorable attitudes towards both the other racial group (the roommates' group) and toward other groups uninvolved in the housing setting, such as Muslims. These findings thus show that by decreasing the levels of SDO, intergroup contact reduces the ideological underpinnings of a range of intergroup biases, and a core component of speciesism.

The ideas and findings from intergroup contact research can be further developed and applied to human-animal relations. The possible applications are countless, yet largely unexplored. A first series of new research questions could focus on the generalized effects of contact with human outgroups. Can positive contact with human outgroups reduce speciesist attitudes by lowering SDO levels and stimulating cognitive flexibility? Under what conditions does the secondary transfer effect of contact generalize to improved attitudes towards animals? A second research line can delve into the effects of interactions of animal advocates or vegans with meat eaters. Can positive contact with vegetarians and vegans not only reduce prejudice towards vegans and vegetarians but also facilitate changes in dietary habits and reduce the consumption of animal products? What is the role of social networks (e.g., peer groups and parents) in smoothening or hindering attempts to go vegan? And what are the types of interactions that really put off meat eaters? For instance, recent research reveals that those relatively more conservative (vs liberal) are more likely to lapse back to eating meat after attempting to stop, in part because they experience inadequate social support for their endeavors (Hodson & Earle, 2018). Such findings suggest that positive and supportive social interactions with other people, who are pro-animal in nature, can help people achieve their own goals to exploit animals less.

Finally, a third research line can focus on interspecies contact. Plenty of anecdotal evidence tells us that personal contact with animals has strong potential to form a long-lasting interspecies friendship bond, not only with companion animals (Auger & Amiot, 2017) but also with farm animals (see Baur, this volume) and wild animals like bonobos (see Woods & Hare, this volume). Such cross-species friendships engender empathic ties with the animal that likely generalize towards the entire species in ways that increase opposition to the exploitation of the species. Along with the affectionate bond, befriending farm animals can create greater awareness of their cognitive abilities and sentience, leading to the rejection of harming or killing them (e.g., for meat production). Regular visits to farm sanctuaries provide this possibility, and intervention studies could test the impact of such visits. Of course, getting people to visit farm sanctuaries in the first place and letting them build a connection with animals would be a first challenge to overcome. Schools and youth or community organizations could play a meaningful role here to make this happen. Where direct interspecies contact is difficult to establish, other techniques that rely on the power of human imagination and let people take the perspective of animals may be effective. This can be achieved through the use of virtual reality, media channels and storytelling, techniques that are worth to investigate in empirical research.

Conclusion

Humans live conflicted lives with regard to animals, both loving them (especially as pets or exotic wild animals) but also exploiting them (particularly for food and clothing). Much of this ambivalence remains largely unresolved because of the speciesist attitudes that run hard and deep throughout culture and across time. Animals, relative to humans, are undervalued and thus afforded less consideration and concern. In conjunction with the many rationalizations about "naturalness" and human supremacy, the psychological ambivalence is something that most people learn to live with and not typically resolve. As we have outlined in this chapter, human-animal relations are intimately intertwined with human-human interactions and the overarching justifications for inequality, with one form (e.g., animal exploitation) propping up the other (e.g., sexism, racism) and vice versa. Although we urge caution in implementing interventions that explicitly highlight the similarities between speciesism and prejudice, we recognize that their intersectionality is critical to address and confront. Given the power of intergroup contact, we encourage that interventions to change beliefs or moral thinking about human-animal relations are paired with contact-based experiences that broaden the mind and challenge societal traditions and assumptions.

Note

1 Throughout the chapter, we use "animals" to refer to non-human animals for the sake of brevity, nonetheless recognizing that humans are animals.

References

Abrams, D., Viki, G. T., Masser, B., & Bohner, G. (2003). Perceptions of stranger and acquaintance rape: The role of benevolent and hostile sexism in victim blame and rape proclivity. *Journal of Personality and Social Psychology, 84*(1), 111–125. doi: 10.1037/0022-3514.84.1.111

Adams, C. J. (2015). *The Sexual Politics of Meat: A Feminist-Vegetarian Critical Theory* (25th anniversary edition). New York: Bloomsbury Academic.

Adams, C. J., & Gruen, L. (2014). *Ecofeminism: Feminist Intersections with Other Animals and the Earth.* New York: Bloomsbury.

Akrami, N., Ekehammar, B., & Bergh, R. (2011). Generalized prejudice: Common and specific components. *Psychological Science, 22*(1), 57–59. doi: 10.1177/0956797610390384

Allport, G. W. (1954). *The Nature of Prejudice.* Reading, MA: Addison-Wesley.

Altemeyer, B. (1996). *The Authoritarian Specter.* Cambridge, MA: Harvard University Press.

Auger, B., & Amiot, C. (2017). Testing and extending the pets as ambassadors hypothesis: The role of contact with pets and recategorization processes in predicting positive attitudes toward animals. *Human-Animal Interaction Bulletin, 5*(1), 1–25.

BBC (2013, January 18). Why are the British revolted by the idea of horsemeat? Retrieved from www.bbc.co.uk/news/magazine-21043368

Bekoff, M. (2007a). *The Emotional Lives of Animals: A Leading Scientist Explores Animal Joy, Sorrow, and Empathy – and Why They Matter.* Novato, CA: New World Library.

Bekoff, M. (2007b). *Animals Matter: A Biologist Explains Why We Should Treat Animals with Compassion and Respect.* Boston, MA: Shambhala.

Bekoff, M., & Pierce, J. (2009). *Wild Justice: The Moral Lives of Animals.* Chicago, IL: Chicago University Press.

Bergh, R., Akrami, N., Sidanius, J., & Sibley, C. G. (2016). Is group membership necessary for understanding generalized prejudice? A re-evaluation of why prejudices are interrelated. *Journal of Personality and Social Psychology, 111,* 367–395. doi: 10.1037/pspi0000064

Bratanova, B., Loughnan, S., & Bastian, B. (2011). The effect of categorization as food on the perceived moral standing of animals. *Appetite, 57,* 193–196. doi: 10.1016/j.appet.2011.04.020

Carney, D. R., Jost, J. T., Gosling, S. D., & Potter, J. (2008). The secret lives of liberals and conservatives: Personality profiles, interaction styles, and the things they leave behind. *Political Psychology, 29,* 807–840. doi: 10.1111/j.1467-9221.2008.00668.x

Caviola, L., Everett, J. C., & Faber, N. (2019). The moral standing of animals: Towards a psychology of speciesism. *Journal of Personality and Social Psychology, 116,* 1011–1029. doi:10.1037/pspp0000182

Chrisafis, A. (2013, February 21). Horsemeat scandal triggers 15% rise in sales for France's equine butchers. *The Guardian.* Retrieved from: www.theguardian.com/uk/2013/feb/21/horsemeat-scandal-rise-sales-france-butcher

Cichocka, A., & Dhont, K. (2018). The personality bases of political ideology and behavior. In V. Zeigler-Hill & T. K. Shackelford (Eds.). *SAGE Handbook of Personality and Individual Differences: Volume III: Applications of Personality and Individual Differences* (pp. 323–352). London: SAGE.

Cichocka, A., Dhont, K., & Makwana, A. P. (2017). On self-love and outgroup hate: Opposite effects of narcissism on prejudice via social dominance orientation and right-wing authoritarianism. *European Journal of Personality, 31,* 366–384. doi: 10.1002/per.2114

Costello, K., & Hodson, G. (2010). Exploring the roots of dehumanization: The role of animal-human similarity in promoting immigrant humanization. *Group Processes & Intergroup Relations, 13,* 3–22. doi:10.1177/1368430209347725

Costello, K., & Hodson, G. (2014). Explaining dehumanization among children: The interspecies model of prejudice. *British Journal of Social Psychology, 53,* 175–197. doi:10.1111/bjso.12016

de Waal, F. (2016). *Are We Smart Enough to Know How Smart Animals Are?* New York: W. W. Norton & Company.

Dhont, K., & Hodson, G. (2014). Why do right-wing adherents engage in more animal exploitation and meat consumption? *Personality and Individual Differences, 64,* 12–17. doi: 10.1016/j.paid.2014.02.002

Dhont, K., Hodson, G., Costello, K., & MacInnis, C. C. (2014). Social dominance orientation connects prejudicial human-human and human-animal relations. *Personality and Individual Differences, 61–62*, 105–108. doi: 10.1016/j.paid.2013.12.020

Dhont, K., Hodson, G., & Leite, A. C. (2016). Common ideological roots of speciesism and generalized ethnic prejudice: The social dominance human-animal relations model (SD-HARM). *European Journal of Personality, 30*(6), 507–522. https://doi.org/10.1002/per.2069

Dhont, K., Hodson, G., Loughnan, S., & Amiot, C. E. (2019). Rethinking human-animal relations: The critical role of social psychology. *Group Processes & Intergroup Relations, 22*(6), 769–784. doi: 10.1177/1368430219864455

Dhont, K., Leite, A. C., Hodson, G., & Milfont, T. (in prep). *A Cross-Cultural Investigation of the Ideological Roots Underpinning Support for Traditions of Animal Exploitation.* Manuscript in preparation.

Dhont, K., Van Hiel, A., & Hewstone, M. (2014). Changing the ideological roots of prejudice: Longitudinal effects of ethnic intergroup contact on social dominance orientation. *Group Processes & Intergroup Relations, 17*, 27–44. doi: 10.1177/1368430213497064

Festinger, L. (1957). *A Theory of Cognitive Dissonance.* Stanford, CA: Stanford University Press.

Festinger, L. (1962). Cognitive dissonance. *Scientific American, 207*(4), 93–107. doi:10.1038/scientificamerican1062-93.

Glick, P., & Fiske, S. T. (1996). The ambivalent sexism inventory: Differentiating hostile and benevolent sexism. *Journal of Personality and Social Psychology, 70*, 491–512. doi: 10.1037/0022-3514.70.3.491

Glick, P., & Fiske, S. T. (2001). An ambivalent alliance: Hostile and benevolent sexism as complementary justifications for gender inequality. *American Psychologist, 56*, 109–118. doi: 10.1037/0003-066X.56.2.109

Graham, J., Haidt, J., & Nosek, B. A. (2009). Liberals and conservatives rely on different sets of moral foundations. *Journal of Personality and Social Psychology, 96*, 1029–1046.

Haslam, N., Holland, E., & Stratemeyer, M. (this volume). Kittens, pigs, rats, and apes: The psychology of animal metaphors. In K. Dhont & G. Hodson (Eds.), *Why We Love and Exploit Animals: Bridging Insights from Academia and Advocacy.* Abingdon: Routledge.

Herzog, H. (2010). *Some We Love, Some We Hate, Some We Eat: Why It's So Hard to Think Straight About Animals.* New York: HarperCollins Publishers.

Hodson, G., & Costello, K. (2012, December 12). The human cost of devaluing animals. *New Scientist,* 2895, 34–35. Retrieved from: www.newscientist.com/article/mg21628950.400-the-link-between-devaluing-animals-and-discrimination/

Hodson, G., Crisp, R. J., Meleady, R., & Earle, M. (2018). Intergroup contact as an agent of cognitive liberalization. *Perspectives on Psychological Science, 13*, 523–548. doi: 10.1177/1745691617752324

Hodson, G., & Dhont, K. (2015). The person-based nature of prejudice: Individual difference predictors of intergroup negativity. *European Review of Social Psychology, 26,* 1–42. doi: 10.1080/10463283.2015.1070018

Hodson, G., Dhont, K., & Earle, M. (this volume). Devaluing animals, "animalistic" humans, and people who protect animals. In K. Dhont & G. Hodson (Eds.), *Why We Love and Exploit Animals: Bridging Insights from Academia and Advocacy.* Abingdon: Routledge.

Hodson, G., & Earle, M. (2018). Conservatism predicts lapses from vegetarian/vegan diets to meat consumption (through lower social justice concerns and social support). *Appetite, 120*, 75–81. doi: 10.1016/j.appet.2017.08.027

Hodson, G., & Hewstone, M. (2013). *Advances in Intergroup Contact.* London: Psychology Press.

Hodson, G., Hogg, S. M., & MacInnis, C. C. (2009). The role of "dark personalities" (narcissism, Machiavellianism, psychopathy), Big Five personality factors, and ideology in explaining prejudice. *Journal of Research in Personality, 43,* 686–690. doi: 10.1016/j.jrp.2009.02.005

Hodson, G., MacInnis, C. C., & Busseri, M. A. (2017). Bowing and kicking: Rediscovering the fundamental link between generalized authoritarianism and generalized prejudice. *Personality and Individual Differences, 104,* 243–251. doi: 10.1016/j.paid.2016.08.018

Hopkins-Doyle, A., Sutton, R. M., Douglas, K. M., & Calogero, R. M. (2019). Flattering to deceive: Why people misunderstand benevolent sexism. *Journal of Personality and Social Psychology, 116,* 167–192. doi: 10.1037/pspa0000135

Horta, O. (2010). What is speciesism? *Journal of Agricultural and Environmental Ethics, 23,* 243–266. doi: 10.1007/s10806-009-9205-2

Hyers, L. (2006). Myths used to legitimize the exploitation of animals: An application of social dominance theory. *Anthrozoos, 19,* 194–210. doi: 10.2752/089279306785415538

Jost, J. T. (2017). Ideological asymmetries and the essence of political psychology. *Political Psychology, 38,* 167–208. doi: 10.1111/pops.12407

Jost, J. T., Glaser, J., Kruglanski, A. W., & Sulloway, F. J. (2003). Political conservatism as motivated social cognition. *Psychological Bulletin, 129,* 339–375. doi: 10.1037/0033-2909.129.3.339

Joy, M. (2010). *Why We Love Dogs, Eat Pigs, and Wear Cows: An Introduction to Carnism.* San Francisco, CA: Conari Press.

Kim, C. J. (2011) Moral extensionism or racist exploitation? The use of Holocaust and slavery analogies in the animal liberation movement. *New Political Science, 33,* 311–333. doi: 10.1080/07393148.2011.592021

Knowles E. S., & Linn J. A. (Eds.) (2004). *Resistance and Persuasion.* Mahwah, NJ: Lawrence Erlbaum Associates.

Kteily, N., Ho, A. K., & Sidanius, J. (2012). Hierarchy in the mind: The predictive power of social dominance orientation across social contexts and domains. *Journal of Experimental Social Psychology, 48,* 543–549.

Kteily, N., Hodson, G., & Bruneau, E. (2016). They see us as less than human: Meta-dehumanization predicts intergroup conflict via reciprocal dehumanization. *Journal of Personality and Social Psychology, 110,* 343–370. doi: 10.1037/pspa0000044

Leite, A. C., Dhont, K., & Hodson, G. (2019). Longitudinal effects of human supremacy beliefs and vegetarian threat on moral exclusion (*vs.* inclusion) of animals. *European Journal of Social Psychology, 49,* 179–189. https://doi.org/10.1002/ejsp.2497

Loughnan S., & Davies, T. (this volume). The meat paradox. In K. Dhont & G. Hodson (Eds.). *Why We Love and Exploit Animals: Bridging Insights from Academia and Advocacy.* Abingdon: Routledge.

MacKinnon, C. A. (2004). Of mice and men: A feminist fragment on animal rights. In C. R. Sunstein & M. C. Nussbaum (Eds.), *Animal Rights: Current Debates and New Directions* (pp. 263–276). New York: Oxford University Press.

McFarland, S. (2010). Authoritarianism, social dominance, and other roots of generalized prejudice. *Political Psychology, 31*(3), 453–477. doi: 10.1111/j.1467-9221.2010.00765.x

Meeusen, C., & Dhont, K. (2015). Parent-child similarity in common and specific components of prejudice: The role of ideological attitudes and political discussion. *European Journal of Personality, 29,* 585–598. doi: 10.1002/per.2011

Meleady, R., Crisp, R., Dhont, K., Hopthrow, T., & Turner, R. (in press). Intergroup contact, social dominance and environmental concern: A test of the cognitive-liberalization hypothesis. *Journal of Personality and Social Psychology.* doi:10.1037/pspi0000196

Olson, J. M., & Stone, J. (2005). The influence of behavior on attitudes. In D. Albarracín, B. T. Johnson, & M. P. Zanna (Eds.), *The Handbook of Attitudes* (pp. 223–271). Mahwah, NJ: Lawrence Erlbaum Associates.

Pettigrew, T. F. (2009). Secondary transfer effect of contact: Do intergroup contact effects spread to noncontacted outgroups? *Social Psychology, 40,* 55–65. doi: 10.1027/1864-9335.40.2.55

Pettigrew, T. F., & Tropp, L. R. (2011). *When Groups Meet: The Dynamics of Intergroup Contact.* New York: Psychology Press.

Piazza, J. (this volume). Why people love animals yet continue to eat them. In K. Dhont & G. Hodson (Eds.). *Why We Love and Exploit Animals: Bridging Insights from Academia and Advocacy.* Abingdon: Routledge.

Plous, S. (1993). Psychological mechanisms in the human use of animals. *Journal of Social Issues, 49,* 11–52. doi: 10.1111/j.1540–4560.1993.tb00907.x

Plous, S. (2003). Is there such a thing as prejudice toward animals? In S. Plous (Ed.), *Understanding Prejudice and Discrimination* (pp. 509–528). New York: McGraw-Hill.

Pratto, F., Sidanius, J., Stallworth, L. M., & Malle, B. F. (1994). Social dominance orientation: A personality variable predicting social and political attitudes. *Journal of Personality and Social Psychology, 67,* 741–763. doi: 10.1037/0022-3514.67.4.741

Q&A: Horsemeat scandal (2013, April 10). *BBC.* Retrieved from www.bbc.co.uk/news/uk-21335872

Quinn, B. (2013, January 16). Horsemeat discovered in burgers sold by four British supermarkets. *The Guardian.* Retrieved from www.theguardian.com

Reinhart, R. J. (2018, August 1). Snapshot: Few Americans vegetarian or vegan. *Gallup.* Retrieved from: https://news.gallup.com/poll/238328/snapshot-few-americans-vegetarian-vegan.aspx

Ryder, R. D. (2006). Speciesism in the Laboratory. In P. Singer (Ed.), *In Defense of Animals: The Second Wave* (pp. 87–103). Malden, MA: Blackwell Publishing.

Salmen, A., & Dhont, K. (in prep). The psychological pillars of sexism and speciesism. Manuscript submitted for publication.

Shook, N. J., Hopkins, P. D., & Koech, J. M. (2016). The effect of intergroup contact on secondary group attitudes and social dominance orientation. *Group Processes & Intergroup Relations, 19,* 328–342. doi: 10.1177/1368430215572266

Sibley, C. G., & Duckitt, J. (2008). Personality and prejudice: A meta-analysis and theoretical review. *Personality and Social Psychology Review, 12,* 248–279. doi: 10.1177/1088868308319226

Sibley, C. G., & Liu, J. H. (2010). Social dominance orientation: Testing a global individual difference perspective. *Political Psychology, 31,* 175–207. doi: 10.1111/j.1467-9221.2009.00748.x

Sidanius, J., & Pratto, F. (1999). *Social Dominance: An Intergroup Theory of Social Hierarchy and Oppression.* Cambridge: Cambridge University Press.

Singer, I. B. (1968). *The Séance and Other Stories.* New York: Farrar Straus Giroux.

Singer, P. (1975). *Animal Liberation: A New Ethics for Our Treatment of Animals.* New York: HarperCollins.

Sorenson, J. (2016). *Constructing Ecoterrorism: Capitalism, Speciesism and Animal Rights.* Nova Scotia, Canada: Fernwood Publishing.

Sorenson, J. (this volume). Humane Hypocrisies: Making killing acceptable. In K. Dhont & G. Hodson (Eds.), *Why We Love and Exploit Animals: Bridging Insights from Academia and Advocacy.* Abingdon: Routledge.

Spiegel, M. (1988). *The Dreaded Comparison: Human and Animal Slavery.* New York: Mirror Books.

Tausch, N., Hewstone, M., Kenworthy, J. B., Psaltis, C., Schmid, K., Popan, J.... Hughes, J. (2010). Secondary transfer effects of intergroup contact: Alternative accounts and underlying processes. *Journal of Personality and Social Psychology*, *99*, 282–302. doi: 10.1037/a0018553

Van Assche, J., Van Hiel, A., Dhont, K., & Roets, A. (2019). Broadening the individual differences lens on party support and voting behavior: Cynicism and prejudice as relevant attitudes referring to modern-day political alignments. *European Journal of Social Psychology*, *49*, 190–199. doi:10.1002/ejsp.2377

Van Hiel, A., Cornelis, I., Roets, A., & De Clercq, B. (2007). A comparison of various authoritarianism scales in Belgian Flanders. *European Journal of Personality*, *21*, 149–168. doi: 10.1002/per.617

Wegener, D. T., Petty, R. E., Smoak, N. D., & Fabrigar, L. R. (2004). Multiple routes to resisting attitude change. In E. S. Knowles & J. A. Linn (Eds.), *Resistance and Persuasion* (pp. 13–38). Mahwah, NJ: Lawrence Erlbaum.

Woods, V. & Hare, B. (this volume). Uncanny valley of the apes. In K. Dhont & G. Hodson (Eds.), *Why We Love and Exploit Animals: Bridging Insights from Academia and Advocacy*. Abingdon: Routledge.

Zick, A., Wolf, C., Küpper, B., Davidov, E., Schmidt, P., & Heitmeyer, W. (2008). The syndrome of group-focused enmity: The interrelation of prejudices tested with multiple cross-sectional and panel data. *Journal of Social Issues*, *64*, 363–383. doi: 10.1111/j.1540-4560.2008.00566.x

4

PUTTING THE "FREE" BACK IN FREEDOM

The failure and future of animal welfare science

Jessica Pierce

Abstract

Our collective understanding of the cognitive and emotional capacities of animals has exploded over the past several decades, yet the expanding body of scientific knowledge has not translated into noticeable improvements in the treatment of captive animals. This chapter explores one explanation for this "knowledge translation gap": the science of what animals think and feel has largely been co-opted by industry and funneled into the narrow field of animal welfare science. Welfare science claims to improve the lives of animals by figuring out what they want and how they feel under certain captive conditions. But these incremental improvements do little for animals, and in fact serve to further entrench and legitimize the exploitation of animals in human industry. Moving forward, this chapter argues that the "Five Freedoms," a classic formulation of the welfare science agenda, could be destabilized. Detached from their status quo roots, the Five Freedoms could be an aspirational tool in shaping a more appropriate moral response to human-imposed suffering.

Our collective understanding of the cognitive and emotional capacities of animals has exploded over the past several decades. If you were to draw a curve representing growth in the scientific understanding of what animals think, know, and feel, you would see a sharp upward trend beginning in the 1960s. Since the first scattering of articles on chimpanzee intelligence, social behavior, and personality, the daily newsfeed is now cram-packed with headlines representing the latest revelations into who animals really are. In any given week, we might learn that individual fishes have unique personalities (Balcombe, 2016), that chickens have empathy (Marino, 2017), that bees can learn to use tools (Alem et al., 2016), that a sea lion can remember a trick she learned ten years ago (Kastak & Schusterman, 2002), and so much more.

With the evolution in our knowledge about animals we might expect also to have seen a shift in ethical attitudes toward them. It seems logical that the accumulation of scientific knowledge about how much like us animals are in their basic capacities for happiness, fear, attachment, loss, and pain and suffering (see, for example, Bekoff, 2000, 2007; De Waal, 2017; Panksepp, 1998; Safina, 2015) would be reflected in a pattern of increasing sensitivity and a subsequent decline in practices that we know to cause physical and emotional suffering. Yet the cultural shift has not yet occurred; human behavior seems increasingly out of sync with science.

For me and for many other people working in animal advocacy, the failure of science to produce an evolution in ethical values and commitments has been a bitter disappointment. An early feeling of optimism has given way to frustration, even alarm, about what is happening around the globe. Despite an increasingly sophisticated set of data points on animal cognition and emotion, the deeper picture has somehow failed to come into focus. We continue to inflict suffering on millions upon millions of living creatures, all of whom we manipulate and exploit in the name of human comfort and scientific "progress" as if they were mere insensate objects, as if their feelings and thoughts and suffering were of no real importance – even as centuries-long traditions of moral thought and moral common sense bulge and distort to counteract the increasing internal contradictions.

By many measures, animals are objectively worse off than ever before. Despite the extensive database on the cognitive and emotional capacities of pigs (Marino & Colvin, 2015), cows (Marino & Allen, 2017), and chickens (Marino, 2017), meat production and consumption continue to rise globally. Although captivity is known to cause profound emotional trauma to large mammals such as orcas and elephants, entertainment venues continue to keep these animals on display and crowds of people continue to flock to zoos and aquaria (NAVS, 2018; Whale and Dolphin Conservation, 2018). Although numerous non-animal alternatives to animal experimentation are readily available, the numbers of monkeys, sheep, guinea pigs, hamsters, rats, and mice used in research and testing continue to grow (Bekoff & Pierce, 2017; for US figures, see USDA/APHIS, 2017). More and more exotic animals such as geckos, sloths, and fennec foxes are being marketed and purchased as pets, despite the profound unsuitability of the human home environment to these animals' biological needs (Grant, Montrose & Wills, 2017). Problems are not limited to animals used in human industries but extend also to billions of wild animals who are losing the freedom to thrive, much less survive, within increasingly unstable and degraded ecosystems.

Where has all the science gone?

So, why, if we know so much more about the thinking and feeling capacities of animals, has the situation of animals not improved apace? Why has the science not led to a parallel evolution in moral commitments and behavior? The interplay between science, values, and practices is infinitely complex, so I do not offer the following remarks as an attempt to provide any kind of comprehensive explanation for why knowing more about animals has not altered our patterns of exploitation. Instead, I would like to highlight one small piece of the puzzle: the burgeoning field of animal welfare science. Welfare science has been able to offer incremental "welfare" improvements for animals, while at the same time undercutting the possibilities for more revolutionary and meaningful reform. It has created a singularly unlucky vortex for animals but may also (ironically) offer a path forward.

Animal welfare science: Some background

The so-called "Five Freedoms" will be familiar to many people working in animal advocacy. The Five Freedoms originated in the early 1960s as part of a British government study informally known as the Brambell Report (Brambell, 1965). The study was a response to public outcry over the abusive treatment of animals within increasingly intensive, high-density agricultural systems. After looking carefully at the conditions under which farm animals were raised for food – in which animals were exposed to levels of cruelty that offended a basic human sense of decency – the Brambell commission thought it prudent to establish some minimal moral guidelines for the industry. The Brambell commission thus specified that animals should have the "freedom to stand up, lie down, turn around, groom themselves and stretch their limbs." With these 14 words, the commission established what might be considered the first "animal welfare" standards, introduced the concept of animal welfare into the lexicon of animal industries and into the public consciousness, and set in motion a research agenda now known as animal welfare science. Let us be content, for the moment, to call it animal welfare "science"; we'll see, though, that it is as much about setting a values agenda than it is about gathering scientific data and formulating theories.

The initial Brambell welfare standards were expanded in 1992 by the Farm Animal Welfare Council into their current and now very familiar formulation as the Five Freedoms, which serve as a cornerstone for animal welfare science. In the language of the FAWC, the Five Freedoms state that all animals should have:

1. **Freedom from hunger and thirst**, by ready access to water and a diet to maintain health and vigour.
2. **Freedom from discomfort**, by providing an appropriate environment.
3. **Freedom from pain, injury and disease**, by prevention or rapid diagnosis and treatment.
4. **Freedom to express normal behaviour**, by providing sufficient space, proper facilities and appropriate company of the animal's own kind.
5. **Freedom from fear and distress**, by ensuring conditions and treatment, which avoid mental suffering (Farm Animal Welfare Council, 2002. Numbering and bold text have been added to improve readability).

The Five Freedoms are now invoked not only in relationship to farmed animals but are used to guide welfare ideals for animals in research laboratories, and zoos and aquariums. The freedoms appear in nearly every book about animal welfare, can be found on nearly every website dedicated to food-animal or lab-animal or zoo-animal welfare, form the basis of many animal-welfare auditing programs, and are taught to many students entering fields of animal husbandry. Indeed, they are so ubiquitous that they have become a kind of shorthand for human ethical responsibility to animals.

The questions welfare researchers have explored reflect the basic formula of the Five Freedoms: What do animals want to eat and when? What kind of shelter do they prefer? Is a particular experience causing them pain or distress? Welfare researchers have even tried to go beyond the relatively easy problems presented by the first four freedoms and have tackled the fifth and most difficult freedom, allowing animals to exercise normal species-specific behavior – sometimes exceedingly challenging within captive settings. Researchers explore, for example, what kind of industrial-scale housing system allows hens to perform the greatest repertoire of hen-like behaviors, such as pecking, scratching, and dust bathing (see Bergman et al., 2017, as a representative study).

What started to emerge during the research and writing of the Brambell Report is that what animals want and need is not always clear. Even something as seemingly obvious as providing appropriate housing turns out to be complicated, because what we might assume animals would want might not actually *be* what they want. Figuring out what animals want involves trying to gain access to the animals' subjective experiences. Like the broader science of ethology, welfare science has sought to reveal the inner workings of the animal mind through observing and interpreting outward behaviors. Yet unlike ethology, the purpose of welfare-focused science is not a greater understanding of who animals are, per se, but a clearer sense of how animals might be more efficiently and "humanely" exploited by humans.

There is huge variety in the methods welfare science uses to "look inside" the minds and hearts of animals. One of the most basic welfare experiments, often called *preference testing*, simply offers animals a choice between two alternatives to determine which the animals prefer. Preference might be measured by giving access to two options and measuring how much time the average animal spends in each one. Take, for example, one of the initial concerns that motivated the Brambell commission: the practice of confining egg-laying hens to so called "battery cages," which are long, stacked rows of very small and dense enclosures in which chickens are confined either as singletons (so that individual egg-laying productivity can be tracked) or, more commonly, in groups of 5–10 birds. The space allotment for each chicken averages about 67 square inches, a little less than the size of an 8" x 11" piece of paper (Humane Society of the United States, 2018). Battery cages at that time were typically constructed of wire mesh ("chicken wire") to which the anatomy of a chicken's foot is particularly ill-suited. Battery-housed hens suffered from painful foot deformities, broken bones, lameness, and open sores. The Brambell Report recommended that egg producers use a heavier metal mesh instead, because the chicken wire was so obviously problematic. But there was no empirical evidence that one type of flooring was better than another as far as the hens were concerned. So, researchers decided to "ask" the hens by giving them a choice of flooring. The hens preferred the chicken wire, most of the time, though they did not show a strong preference (Hughes & Black, 1973).

Another early battery-cage study by Marian Dawkins used preference testing to assess how much space Ross Ranger hens "need." Using hens housed four to a cage, she gave the groups a choice between smaller cages (247 square inches) and larger (988 square inches). She then recorded how long it took the hens to move into each cage and determined that because they moved more quickly into the larger cages, they preferred these (Dawkins, 1983).

Incidentally, the early chicken research offers insight into of the most serious scientific problems with animal welfare science – a problem which continues to dog welfare research to this day: it relies on an averaging of behavioral responses. If 40 chickens are tested to see whether they prefer to eat earthworms or crickets, and 30 of the chickens choose earthworms, the results of the preference test will be that "the Chicken prefers earthworms." Thereafter, earthworms may be all that chickens ever get, which is a shame for those 10 chickens who do not particularly like earthworms. (Of course, in real life, battery-caged chickens never get to enjoy either crickets or earthworms, but spend their entire lives eating industrial mush.) What animal welfare science fails to recognize is that there is no such thing as "the Chicken," or "the Elephant," or "the Rat." This averaging out is also a problem ethically, because it means that many individuals are not actually getting what they want or need.

Researchers also developed tests to determine how strongly animals prefer one choice over another, by making different choices have different "costs" and seeing whether animals are willing to "pay." For example, animals might be exposed to a loud, frightening sound after choosing their preferred option, or they might be given an electric shock. The researchers can then see how much it takes to dissuade an animal from her preferred choice. In a study of whether broiler chickens preferred plastic or mesh flooring, the experimenters added an "aversive." They illuminated the wire flooring at 800 lux. In this iteration, the chickens changed their preference from mesh to plastic flooring (Zulkifli & Khatijah, 1998). We might pause here for a moment to note that these "welfare" experiments are inherently unethical because they impose unnecessary suffering on the animals involved.

Researchers have also used preference testing to assess how much animals like and desire improvements to their environment. For example, hens will consistently choose "furnished" accommodations, which have perches and nesting boxes, over a barren environment. Barren environments are known to compromise the welfare of chickens and other animals because they lack complexity, stifle sensory engagement with the world, and lead to boredom and other negative affective states (see, for example, Tahamtani, 2015). Yet furnished accommodations are more expensive and require more space, so researchers also need to determine the optimal price point for welfare improvements of this sort: how much does laying productivity improve, if hens are given furnished cages and does the increased laying productivity cover the cost of the welfare improvements?

More nuanced kinds of preference testing also have been developed, often drawing on techniques from the fields of economics and consumer behavior. For instance, you can see whether there is inelastic demand for a given "commodity" by testing whether an animal still shows a preference when he is asked to work harder for it, where work might involve pulling a weight or pressing a lever repeatedly. To give one example, this type of preference testing has been used to study motivation in a cichlid fish called the Mozambique tilapia. Researchers applied a so-called push-door paradigm, in which the fishes had to push a door open with their snout to access food, a social partner, or additional space. How hard the fishes were willing to push was taken as a rough measure of how motivated they were for a given reward (Galhardo et al., 2011).

Another iteration of preference testing is called *cognitive bias testing*, which seeks to gain access to an animal's feelings by observing how he or she makes decisions (Mendl et al., 2009). The underlying idea is that emotions influence the decision-making process, a fact that has been well established within human and animal psychology. Running this in reverse, you can "see" an animal's mood state reflected in the kinds of decisions he or she makes. As an illustration, consider a recent study in which a cognitive bias model was used to explore which of two types of handling laboratory mice found more distressing, the standard "tail handling" (which, as you might guess, involves picking a mouse up by her tail) and cupped handling, where a mouse is held in a cupped human hand. Researchers compared the performance of tail-handled and cupped mice in an eight-arm radial maze. As hypothesized, mice performed less well in the maze after tail-handling, because tail-handling is likely more distressing and cognitive processing tends to be less accurate when an animal is experiencing a negative affective state (Novak et al., 2015).

Because cognitive bias testing often requires extensive training of the animals being tested, researchers have also looked for quicker ways to assesses animals' affective states. One of these short cut approaches is to look at what are called *attentional biases*. The idea is that animals who are feeling anxious will spend more time attending to perceived threats than animals who are feeling calm. An obvious way to test this is to experimentally induce states of anxiety in animals. Consider, for example, a 2016 study of attention bias in sheep. Using pharmaceutical compounds, researchers manipulated sheep anxiety levels and then exposed the sheep to a threatening dog. The sheep were given 1-methyl-chlorophenylpiperazine to induce high levels of anxiety and given diazepam to chill them out. The ramped-up sheep did, indeed, show increased levels of "vigilance" (which is welfare-speak for "fear") in response to the dog. Although the researchers claim that their study has the potential to "improve animal welfare protocols," it is hard to ignore the welfare compromises experienced by these particular sheep (Lee et al., 2016).

As should by now be evident to the reader, studies not only ask animals what they like the best, but also what causes them the most stress or pain. "Preference" research is, in fact, very often actually "aversion" research, focusing on things animals will distinctly not like but that we are going to do to them anyway. Research studies have measured animals' relative aversion to noise, vibration, and heat in various combinations. For example, the bibliography of the New South Wales Animal Research Review Panel (2007) provides an extensive database on how to make rats miserable through a rainbow of different sensory assaults. Scientists have also explored how chronic exposure to polluted air causes mice to display depressive-like responses and impairments in spatial learning and memory (Fonken et al., 2011), which euthanasia gases are most distressing to the mice who are being killed (Makowska, 2009), and have shown that "cattle subjected to prolonged painful treatment, such as having excessive numbers of students palpate them in artificial insemination class," become very stressed and "absolutely refused to re-enter the race [or chute] where this occurred" (Grandin & Shivley, 2015, p. 1241). Scientists have studied how much stress animals experience when exposed to overcrowding, small cages, incorrect social groupings or distribution of group members, nearness of predators, and all manner of other horrors. In most of these studies, the aversiveness of given activity or stimuli has already been established through painful and often deadly experiments on animals, often "replicated" multiple times to test their scientific rigor. The information to be gleaned by researchers is just how serious a welfare issue something is and, consequently, whether it will compromise commercial productivity or experimental results.

Where is the "free" in the five freedoms?

Welfare science has developed, ostensibly, to elucidate how animals feel and what they want so that we humans can "improve their welfare." On its surface, welfare science looks like a great benefit to animals worldwide. Who wouldn't want better welfare for animals? Indeed, many of the articles within the welfare science literature include reference to "ethics" and "doing right" and offer anodyne claims about humans needing to be good stewards of animals. Yet as we can see from looking at the kinds of questions welfare science asks, the kind of answers it seeks, and the methods it uses to find data points, the whole enterprise seeks to bolster the status quo, at every turn reinforcing a particular way of seeing, relating to, and (de)valuing animals. Welfare science is about what animals feel *within the context of the status quo* and takes for granted that "humane use" of animals is morally acceptable. It never openly acknowledges its most audacious ethical claims: (1) that nonhuman animals constitute a legitimate and appropriate resource for human experimentation, research, and industrial production and consumption (borrowing language from Peggs & Smart, 2017, p. 185), and (2) that inflicting suffering on millions, even billions, of thinking, feeling, desiring creatures is normal, acceptable, beyond reproach, and indeed essential to "progress." As much as welfare scientists may seek to approach their studies objectively and from a rigorously evidence-based stance, the entire enterprise is suffused with value assumptions about who animals are in relation to humans, and how human needs and desires take unquestioned precedence over the needs and desires of the animals themselves.

Indeed, welfare science has been instrumental in reinforcing the assumption that animals are here for the taking and that depriving them of the freedom to live their own lives is no problem, so long as we do it "scientifically and humanely" and follow "welfare guidelines." Going back to Brambell and the Five Freedoms, we can see this narrow, instrumental focus taking shape. Recall that the Brambell Report was written in response to concerns about the suffering of animals within the increasingly large-scale industrial farms that were replacing traditional methods of husbandry during the middle of the twentieth century. The practices that were under scrutiny were those used to raise laying hens, broiler chickens, pigs, and veal calves. At issue was not the question of whether to continue developing intensive food-production systems, but how to make modifications to these systems that would offset the worst suffering of animals and salve the public's conscience.

Did the crafters of the Five Freedoms recognize the fundamental paradox of their welfare formulation? How can an animal in an abattoir or battery cage be free? Being fed and housed by your captor is not freedom; it is simply what your captor does to keep you alive. (I have been unable to find any explanation as to why the language of "freedom" was chosen in the original Brambell report.) The Five Freedoms are not really concerned with freedom per se, but rather with keeping animals under conditions of such profound deprivation that no honest person could possibly describe them as free. And this is entirely consistent with the development of the concept of animal welfare. Welfare science aims to improve the lives of animals within situations in which human activities impose "welfare burdens." Welfare science may ask animals what they want, but the animals have been offered a very narrow range of questions. Welfare science might "ask" captive mink on a fur farm: Would you like a slightly taller cage so you can stretch your body to full length? Yet while waiting for the "scientific" answer to this scintillating question, let's step out of the status quo, step out of the sphere of "normal science," step back from blind dedication to the "better welfare." From this outside vantage point, we can see the cruelty and pointlessness of the question, as far as mink themselves are concerned. Based on the science collected thus far, and on moral common sense, the mink would surely prefer not to spend their lives in cages at all. Captivity imposes daily suffering, and it deprives mink of the opportunity to live their own lives, on their own terms, in their own natural ecological niche for which they have evolved. It deprives them of any semblance of a normal mink life. Instead, they subsist in an artificial and very stressful captive environment, while waiting for their children and friends and then themselves to be slaughtered and then have their skins removed.

Welfare science, and the Five Freedoms, fails to mention the most basic things that animals want and need: *freedom from captivity, from human exploitation, and freedom from human-imposed suffering.* If you look through the entire corpus of books and articles on welfare science, you will find no discussion of Freedom with a capital "F." You will find little to nothing that challenges the status quo, or that raises serious moral objections to the various ways in which we routinely compromise the lives and well-being of our fellow creatures for our own pleasure and profit. This needs to change.

The animals speak: "Animal welfare science is not for us!"

The fact that welfare assessments are "science-based" gives them a patina of accept-ability. Yet it must be kept in the forefront of our attention that welfare science serves a particular group of stakeholders, and these stakeholders are not animals. Take, as a quick case in point, a few representative entries from recent issues of some of the leading animal welfare science journals:

> Effects of Various Cleaning Agents on the Performance of Mice in Behavioral Assays of Anxiety (*Journal of the American Society for Laboratory Animal Science*).

> Effect of Environmental Enrichment and Herbal Compounds-Supplemented Diet on Pig Carcass, Meat Quality Traits, and Consumers' Acceptability and Preference (*Animals*).

> Using mildly electrified grids to impose costs on resource access: A poten-tial tool for assessing motivation in laboratory mice (*Applied Animal Behavior Science*).

> Pre-slaughter cattle welfare indicators for use in commercial abattoirs with voluntary monitoring systems: A systematic review (*Meat Science*).

> Effect of Chronic Social Stress on Prenatal Transfer of Antitetanus Immunity in Captive Breeding Rhesus Macaques (*Macaca mulatta*) (*Journal of the American Society for Laboratory Animal Science*).

And this is how it goes. Article after article details how to assess and monitor suffering under aversive conditions. This welfare-research endeavor sits comfortably within a broader scientific paradigm in which nonhuman animals are assigned an instrumental rather than an intrinsic value. As Kay Peggs and Barry Smart suggest, the "humane treatment" and "welfare improvement" approach to nonhuman animal suffering "leaves this instrumental relation undisturbed" (Peggs & Smart, 2017, p. 188). Animal suffering is seen as a necessary and justified cost of doing science, even when the science is supposedly being pursued to help animals. This is a bizarre moral contortion, made even more twisted by the fact that research on animal cognition and emotion continually reaffirms that humans and nonhuman animals share a basic anatomical and physiological blueprint and feel essentially the same things and suffer in comparable ways.

Getting back to our original question of why the science of animal cognition and emotion has not saved animals to the extent that it might, we can now for-mulate at least a partial explanation: the science of animal cognition and emotion has been largely framed and conducted within a scientific and moral paradigm that accords animals only instrumental value and that condones animal suffering as an accepted cost of progress.

Welfare science claims to improve the lives of animals *within* industrial agri-culture, laboratory research, zoos, circuses and aquariums, and many other venues within which animals are currently being exploited. It may question certain outrageous practices within these industries – castrating animals without the benefit of anesthesia, for example, or keeping pregnant sows in tiny gestation crates – but the larger cultural and ethical commitments to using and abusing animals are left intact. Indeed, the commitment to abusing animals becomes even further entrenched because those within industry can feel good about their work and consumers buying animal products or consuming goods which involve cruel treatment of animals in one way or another can feel good about "good welfare practices." They can buy their cage-free eggs and humanely raised grass-fed beef and slip past the uncomfortable fact that animals have suffered.

Simply put, the appeal to "animal welfare" has become a powerful form of moral justification for continuing human exploitation of animals. Welfare is a dirty little lie we tell ourselves about how kindly we treat animals, about how much we care when, in fact, we are quite willing to be savages. When you see the word "welfare" – and the word "humane," which has been similarly co-opted by industry – you can bet that something unpleasant is being done to animals.

Incrementalism: A moral cul-de-sac?

Welfare scientists might say, "we are actually doing animals more good than animal activists because we are providing research that will allow us to implement actual, practical solutions to current animal welfare problems." For example, Temple Grandin's work (Grandin & Deesing, 2008) on slaughterhouse design has been broadly celebrated as a boon for animals.[1] In one of her most eulogized achievements, Grandin hypothesized that being led down a narrow chute toward the killing floor is stressful for cattle. She provided the data to prove that cattle are scared (by testing physiological markers of stress such as increased levels of cortisol) and proposed that the architects of slaughterhouses redesign the chutes to be curved rather than straight. Making this change results in a small but measurable reduction in stress, since the cattle do not see what lies ahead of them. And a reduction in the terror felt by cattle about to be shot in the head with a captive bolt gun is called, by Grandin and her acolytes, a welfare improvement. "These cattle have better lives," they proclaim.

Similarly, scientists have been able to offer welfare improvements for caged mink on fur farms. In addition to researching preferred cage height, scientists have "operationally defined and validated" measures of boredom ("a negative state caused by under-stimulation") in mink, and have proven that environmental enrichment can reduce this negative state (Meagher, Campbell, & Mason, 2017; Meagher & Mason, 2012). This kind of research, if implemented into changes in mink cage size and the addition of toys and other objects of interest into the cages, could provide a welfare improvement. But things are still pretty awful for the mink. It is hard to argue with making their miserable lives a bit better, but should we be satisfied with miserable lives?

These small, incremental improvements may help animals by easing their suffering. But ultimately this kind of work does more harm than good, because it perpetuates the status quo. As Peggs and Smart (2017, p. 186) suggest in their challenge to the "humane use" approach to scientific research, "rather than arguing against the use of nonhuman animals the emphasis is placed upon organizing laboratories more efficiently and showing more care and consideration towards nonhuman animals who nevertheless continue to be exposed to pain and suffering in the course of scientific research." The "humane use" and "welfare" agenda not only fails to challenge the speciesist assumption that imposing suffering on millions of sentient creatures is acceptable and normal, it actually makes it seem okay (see also Sorenson, this volume). It keeps us and, more importantly, animals, trapped in a moral cul de sac, going around in circles rather than making progress forward (for other critiques of "welfarism," see Francione, 2007 and Francione & Charlton, 2015; see also Donaldson & Kymlicka, 2011).

Welfare science has taught us an extremely important lesson: We can objectively measure and observe animal suffering and, on the reverse, animal well-being. But rather than respond to suffering with small welfare improvements such as a few inches of additional space for battery caged hens or fur-farmed mink or a curved rather than a straight chute up to the slaughterhouse floor for cattle – all of which are ultimately self-serving for humans – we need instead to diligently put science into the service of animals themselves.

Welfare science: Could it work *for* animals?

As awful as much of the welfare science is – as much as it serves to reinforce violent and oppressive behaviors toward animals – it is nonetheless a scientific endeavor that could and perhaps even does help animals. It must be admitted that welfare science has refined and advanced our methods of "asking" animals how they are doing. We keep building a database of what animals find aversive, which sorts of enclosures or procedures cause trauma, what interventions seem to blunt suffering, and more broadly, why being held captive itself a profound moral problem for animals. We are getting to know the preferences of animals better and better, learning how to translate observations of behavior into "feeling correlates," and, perhaps most importantly, building a huge knowledge base about animals as sentient beings who experience a range of positive and negative emotions very much like our own.

Furthermore, animal activists have been able to use data provided by welfare scientists to effectively shape and promote their work on behalf of animals. For example, much of the research into captive elephant welfare in zoos falls into the category of welfare science and has provided the kind of objective information that helps activists argue that elephants should not be kept in zoos at all. We know, for example, that stereotypic behaviors are the second most common behavior observed in zoo elephants (second only to feeding) (Greco et al., 2016). A stereo- typic behavior is an abnormal, repetitive behavior such as pacing, swaying, or nodding. Stereotypies are not observed in wild elephants (or any other wild animal for that matter) and appear to be a "side-effect" of captivity. It is generally agreed that stereotypies are a behavioral manifestation of profound psychological distress and, when observed, are indicative of serious welfare compromise (for example, Mason, 1991). Now activists can argue, quite convincingly, that elephants do not belong in zoos and can put pressure on zoos to stop exhibiting elephants. The Detroit Zoo decided to permanently stop exhibiting elephants based on concerns highlighted by welfare science research and made accessible to the public through awareness-raising campaigns by animal advocacy organizations. In similar fashion, wide dissemination of information about the cognitive and emotional lives of orca whales, particularly in the popular documentary *Blackfish*, turned public opinion very strongly against the incarceration and exploitation of orcas by SeaWorld and led SeaWorld to grudgingly agree to stop captive breeding of orcas. SeaWorld has yet to stop holding orcas captive and forcing the whales to perform for audiences.

Welfare science has helped crystalize why the public can broadly affirm a "love" for animals – public opinion polls consistently find that most people have positive feelings toward animals and are made uncomfortable by the notion of animal suffering – while at the same time eagerly participating in industries that rely on the exploitation of animals. A recent Gallup Poll, for example, found that a third of Americans believe animals should have the same rights as people and about two thirds expressed concern about the treatment of animals in research, agriculture and entertainment venues (Gallup, 2015). Yet only about 3% of Americans report following a vegan diet, leaving us to wonder what exactly "loving" animals means to people.[2] Welfarism makes it easy to feel okay about exploitation, because we are told that animals were treated "humanely" and that they had "good welfare." Yet increased public awareness of the suffering inflicted by "humane" practices, and an increased caginess about the "good welfare" claims of industry, may begin to unsettle the meaning of these terms and reduce their power to salve the public conscience.

An inherent tension has dogged the science of welfare, particularly as it has drawn data, methods, and inspiration from the broader scientific fields of ethology and comparative psychology and evolutionary anthropology. Welfare science is predicated on the concept of evolutionary continuity – the fact that nonhuman animals have subjective experiences very like our own. Although many in the field downplay these very same subjective feelings when it comes drawing moral conclusions from their work, the evidence for continuity continues to build and a strict scientific and moral dichotomy between humans and other animals becomes harder and harder to defend. The science of how animals think and feel challenges the assumption that humans are qualitatively different from and more intrinsically valuable than other creatures; it challenges human exceptionalism, speciesism, and anthropocentrism. The tension is building, and a shift in paradigm may be in the works. Instead of legitimizing the status quo, a transformed welfare science could be used to destabilize and transform it.

One way to make this transformation of value systems and institutions happen is through education, and particularly through what Peggs and Smart (2017) call a critical pedagogy of nonhuman animal suffering. Within our educational system, they argue that we have the potential to "address [animal] suffering effectively" by helping students unseat "the ontological and epistemological assumptions, hege-monic cultural and ethical values, and systemic mechanisms that cause the suffering of nonhuman animals and maintain established scientific research and teaching practices" (p. 185). If we could catch students in undergraduate, graduate, and vet-erinary programs and "promote a critical engagement with the ethics and pol-itics of suffering to which nonhuman animals are exposed," we could move them beyond welfarist concerns "to a position that releases nonhuman animals from their abject subordinate role" (p. 194). Indeed, we can move back further still, and pro-vide (indeed, require) that young children receive humane education in schools, to develop the basic skills of empathy for nonhuman animals (Weil, 2004).

Putting the "free" back in freedom

Abundant scientific research supports the idea that animals need to feel in control of their own lives and choices and that they suffer from the loss of self-determination. Early work on learned helplessness in animals, which forms the basis of our current understanding of human depression, demonstrated that animals who are exposed to repeated and inescapable "stressors" (which is welfare-speak for "aversive" experiences) suffer a kind of mental collapse, where efforts to escape the aversive stimuli give way to despair. (Seligman's barbaric experiments on dogs and rats were foundational to his "learned helplessness" paradigm, as for example in Seligman 1972 and Seligman & Beagley 1975). Captivity, social isolation, and chronic exposure to stress – the fate of millions upon millions of animals – lead to measurable physio-logical changes in the brain, including the loss of neural plasticity (Salvanes et al., 2013). The loss of freedom also manifests in observable behaviors. Captive animals often display abnormal behavior patterns (the stereotypies mentioned above in relation to elephants) – for example, a polar bear pacing back and forth in a zoo enclosure or a rat ruining her teeth by obsessively biting the bars of her cage. These animals are telling us in no uncertain terms that the conditions in which they live are driving them mad.

The bottom line is that even "good welfare" cannot be good enough. Welfarist science puts human needs first, and tries to accommodate animals within the "human needs first" framework. Marc Bekoff and I offer an alternative to welfarist science, which we call the *science of animal well-being*. A science of well-being pays attention to every *individual* animal and commits to radically improved freedoms for animals – especially freedom from human captivity and exploitation and from human-induced suffering. Well-being broadens the question of "what animals want and need" beyond the welfare box, and tries to understand animal preferences from the animals' point of view.

The Five Freedoms were forward-thinking for their time and could remain a crucial moral guide for our interactions with animals – if we begin to take Freedom with a capital "F" seriously. The science of what animals feel and want can and should be diligently used in the service of animals. What animals really want and need is not more welfare science, but a more radical shift in how scientists and others think about who animals are. What we're learning about animal feelings and how animals cope with challenging situations can help increase our empathy and decrease our willingness to impose suffering. After all, it does not take rocket science, or even animal welfare science, to know that animals would prefer to live their own lives, in their own environments, on their own terms. Just like us, animals value their Freedom.

Acknowledgment

This chapter is based on my collaborative work with Marc Bekoff, particularly our 2017 book *The Animals' Agenda: Freedom, Compassion, and Coexistence in the Human Age* (Beacon Press). Many thanks to Marc for his contributions to the evolution of my own thinking about animals, science, and ethics.

Notes

1 Although Grandin enjoys celebrity status among many cattle ranches, carnivores, and welfarists, she has her share of critics. Marc Bekoff has been vocal in challenging Grandin's work (for example: "My Beef with Temple Grandin," www.psychologytoday.com/us/ blog/animal-emotions/201304/my-beef-temple-grandin-seemingly-humane-isnt-enough), as has James McWilliams (see "Why the Meat Industry Loves Temple Grandin," http://dailypitchfork.org/?p=536), and John Sorenson (this volume). Incidentally, Grandin has been challenged not only by animal advocates but also by people who question her claim that animals think like autistic people (e.g., Vallortigara et al., 2008).

2 Statistics on veganism in the USA vary, with some sources placing the number of vegans lower, at 0.5% or as high as 6%. The 3% statistic cited here is from https://news.gallup. com/poll/238328/snapshot-few-americans-vegetarian-vegan.aspx.

References

Alem S., Perry C.J., Zhu X., Loukola O.J., Ingraham T., Søvik E., et al. (2016) Associative mechanisms allow for social learning and cultural transmission of string pulling in an insect. *PLoS Biology*, 14(10), e1002564. https://doi.org/10.1371/journal.pbio.1002564.

Balcombe, J. (2016). *What a Fish Knows: The Inner Lives of Our Underwater Cousins.* New York: Scientific American / Farrar, Straus and Giroux.

Bekoff, M. (2007). *The Emotional Lives of Animals: A Leading Scientist Explores Animal Joy.* Novato, CA: New World Library.

Bekoff, M. (2000). Animal Emotions: Exploring Passionate Natures. *Bioscience*, 50, 861–870.

Bekoff, M., & Pierce, J. (2017). *The Animals' Agenda: Freedom, Compassion, and Coexistence in the Age of Humans.* Boston, MA: Beacon Press.

Bergman, S., Schwarzer, A., Wilutzky, K. Louton, H., Bachmeier, J. Schmidt, P. ...Rauch, E. (2017). Behavior as welfare indicator for the rearing of broilers in an enriched husbandry environment—A field study. *Journal of Veterinary Behavior: Clinical Applications and Research*, 19, 90–101.

Brambell, R. (1965). *Report of the technical committee to enquire into the welfare of animals kept under intensive livestock husbandry systems.* London: Her Majesty's Stationary Office.

Dawkins, M. S. (1983). Cage size and flooring preferences in litter-reared and cage-reared hens. *British Poultry Science, 24*, 177–182.

De Waal, F. (2017). *Are We Smart Enough to Know How Smart Animals Are?* New York: W. W. Norton & Company.

Díez-León, M., Quinton, M., & Mason, G. (2017). How tall should a mink cage be? Using animals' preferences for different ceiling heights to improve cage design. *Applied Animal Behaviour Science, 192,* 24–34.

Donaldson, S., & Kymlicka, W. (2011) *Zoopolis: A Political Theory of Animal Rights.* Oxford: Oxford University Press.

Farm Animal Welfare Council (2002). Farm animal welfare in Great Britain: Past, present and future. https://assets.publishing.service.gov.uk/government/uploads/system/uploads/attachment_data/file/319292/Farm_Animal_Welfare_in_Great_Britain_-_Past__Present_and_Future.pdf.

Fonken, L., Xu, X., Weil, Z., Chen, G., Sun, Q., Rajagopalan, S., & Nelson, R. (2011). Air pollution impairs cognition, provokes depressive-like behaviors and alters hippocampal cytokine expression and morphology. *Molecular Psychiatry, 16,* 987–973. http://doi.org/10.1038/mp.2011.76

Francione, G. (2007). The four problems of animal welfare: In a nutshell. www.abolitionistapproach.com/the-four-problems-of-animal-welfare-in-a-nutshell/

Francione, G., & Charlton, A. (2015) *Animal Rights: The Abolitionist Approach.* Exempla Press.

Fraser, D. & Nicol, C.J. (2011). "Preference and motivation research." In *Animal Welfare*, edited by Michael Appleby et al., 183–99. 2nd ed. Cambridge: CABI.

Galhardo, L., Almeida, O., & Oliveira, Rui. (2011). Measuring motivation in a cichlid fish: An adaptation of the push-door paradigm. *Applied Animal Behaviour Science, 130,* 60–70. 10.1016/j.applanim.2010.12.008.

Gallup (2015). In U.S., more say animals should have same rights as people. https://news.gallup.com/poll/183275/say-animals-rights-people.aspx.

Grandin, T., & Deesing, M. (2008). *Humane Livestock Handling.* North Adams, MA: Storey.

Grandin, T., & Shivley, C. (2015). How farm animals react and perceive stressful situations such as handling, restraint, and transport. *Animals, 5,* 1233–1251. http://doi.org/10.3390/ani5040409

Grant, R. A., Montrose, V. T., & Wills, A. P. (2017). ExNOTic: Should we be keeping exotic pets? *Animals,* 7(6), 47. http://doi.org/10.3390/ani7060047

Greco, B. J., Meehan C. L., Hogan J. N., Leighty K. A., Mellen J., & Mason G. J, (2016). The days and nights of zoo elephants: Using epidemiology to better understand stereotypic behavior of African elephants (*Loxodonta africana*) and Asian elephants (*Elephas maximus*) in North American zoos. *PLoS ONE,* 11, e0144276. https://doi.org/10.1371/journal.pone.0144276

Hughes, B. O., & Black, A. J. (1973). The preference of domestic hens for different types of battery cage floor. *British Poultry Science, 14,* 615–19.

Humane Society of the United States (2018). Cage-free versus battery-cage eggs. www.humanesociety.org/issues/confinement_farm/facts/cage-free_vs_battery-cage.html.

Lee, C., Verbeek, E., Doyle, R., & Bateson, M. (2016). Attention bias to threat indicates anxiety differences in sheep. *Biology Letters, 12,* 20150977. http://dx.doi.org/10.1098/rsbl.2015.0977

Kastak, C. R., & Schusterman, R. J. (2002). Long-term memory for concepts in a California sea lion (*Zalophus californianus*). *Animal Cognition, 5,* 225–232.

Marino, L. (2017). Thinking chickens: A review of cognition, emotion, and behavior in the domestic chicken. *Animal Cognition, 20,* 127–147. http://doi.org/10.1007/s10071-016-1064-4

Marino, L., & Allen, K. (2017). The psychology of cows. *Animal Behavior and Cognition, 4,* 474–498. https://dx.doi.org/10.26451/abc.04.04.06.2017

Marino, L. & Colvin, C.M. (2015). Thinking pigs: A comparative review of cognition, emotion, and personality in *Sus domesticus*. *International Journal of Comparative Psychology, 28*. https://escholarship.org/uc/item/8sx4s79c

Makowska, J., Vickers, L., Mancell, J., & Weary, D. (2009). Evaluating methods of gas euthanasia for laboratory mice. *Applied Animal Behaviour Science, 121*, 230–235.

Mason, G. (1991). Stereotypies: A critical review. *Animal Behaviour, 41*, 1015–37.

Meagher, R. & Mason, G. J. (2012). Environmental enrichment reduces signs of boredom in caged mink. *PLOS One, 7*, e49180. https://doi.org/10.1371/journal.pone.0049180

Meagher, R. K., Mason, G. J., & Campbell, D. (2017). Boredom-like states in mink and their behavioural correlates: A replicate study. *Applied Animal Behaviour Science, 197*, 112–119,

Mench, J., & Swanson, J. (2000). "Developing science-based animal welfare guidelines." Paper presented at University of California Poultry Symposium and Egg Processing Workshop. Modesto, CA, November 7. Available at http://animalscience.ucdavis.edu/avian/mench.pdf.\

Mendl, M., Burman, O., Parker, R. & Paul, E. (2009). Cognitive bias as an indicator of animal emotion and welfare: Emerging evidence and underlying mechanisms. *Applied Animal Behaviour Science, 118*, 161–181.

National Anti-Vivesection Society (2018). Elephants in captivity in the United States. www.navs.org/what-we-do/keep-you-informed/legal-arena/wildlife/elephants-captivity-united-states/#.W1sry9JKiUk.

New South Wales Animal Research Review Panel (2007). *Guidelines for the housing of rats in scientific institutions*. www.animalethics.org.au/__data/assets/pdf_file/0014/222512/housing-rats-scientific-institutions.pdf.

Novak, J., Bailoo, J. D., Melotti, L., Rommen, J., & Würbel, H. (2015). An exploration-based cognitive bias test for mice: Effects of handling method and stereotypic behaviour. *PLoS ONE, 10*, e0130718. http://doi.org/10.1371/journal.pone.0130718

Panksepp, J. (1998). *Affective Neuroscience*. New York: Oxford University Press.

Peggs, K., & Smart, B. (2017). Nonhuman animal suffering: Critical pedagogy and practical animal ethics. *Society & Animals, 25*, 181–198.

Safina, C. (2015). *Beyond Words: How Animals Think and Feel*. New York: Henry Holt and Company.

Salvanes, A. G. V., Moberg, O., Ebbesson, L. O. E., Nilsen, T. O., Jensen, K. H., & Braithwaite, V. A. (2013). Environmental enrichment promotes neural plasticity and cognitive ability in fish. *Proceedings of the Royal Society B: Biological Sciences, 280*, 20131331. http://doi.org/10.1098/rspb.2013.1331

Seligman, M. E. P. (1972). Learned helplessness. *Annual Review of Medicine, 23*, 407–412. doi:10.1146/annurev.me.23.020172.002203

Seligman, M. E., & Beagley, G. (1975). Learned helplessness in the rat. *Journal of Comparative and Physiological Psychology, 88*, 534–541. http://dx.doi.org/10.1037/h0076430

Sorenson, J. (this volume). Humane hypocrisies: Making killing acceptable. In K. Dhont & G. Hodson (Eds.), *Why We Love and Exploit Animals: Bridging Insights from Academia and Advocacy*. Abingdon: Routledge.

Tahamtani, F. M., Nordgreen, J., Nordquist, R. E., & Janczak, A. M. (2015). Early life in a barren environment adversely affects spatial cognition in laying hens (*Gallus gallus domesticus*). *Frontiers in Veterinary Science, 2*. http://doi.org/10.3389/fvets.2015.00003

United States Department of Agriculture/Animal and Plant Health Inspection Service (2017). Annual report animal usage by fiscal year. www.aphis.usda.gov/animal_welfare/downloads/reports/Annual-Report-Animal-Usage-by-FY2016.pdf.

Vallortigara, G., Snyder, A., Kaplan, G., Bateson, P., Clayton, N. S., & Rogers, L. J. (2008) Are animals autistic savants? *PLoS Biology, 6*, e42. https://doi.org/10.1371/journal.pbio.0060042

Weil, Z. (2004). *The Power and Promise of Humane Education.* Gabriola Island, Canada: New Society Publishers.

Whale and Dolphin Conservation (2018). Fate of captive orcas. http://us.whales.org/wdc-in-action/fate-of-captive-orcas.

Zulkifli, I. A., & Khatijah, S. (1998). The relationships between cage floor preferences and performance in broiler chickens. *Asian-Australasian Journal of Animal Sciences, 11,* 234–38.

5

DEVALUING ANIMALS, "ANIMALISTIC" HUMANS, AND PEOPLE WHO PROTECT ANIMALS

Gordon Hodson, Kristof Dhont, and Megan Earle

Abstract

The central premise of the present chapter is that humans routinely undervalue animals relative to themselves. This devaluing has implications not only for animals, in terms of welfare and exploitation, but also for humans. For instance, devaluing animals increases the social value of representing other social groups as animal-like, thus denying these human groups the protections otherwise afforded to humans (and one's own group). But there are also implications for those who protect animals or, at minimum, refuse to engage in the exploitation of animals. Recent research demonstrates that among many meat eaters, vegans and vegetarians are relatively disliked and viewed as threatening. This is particularly the case for vegans and vegetarians who cite animal justice (vs health or environmental concerns) for their renunciation of meat. Overall the research record increasingly shows that our thinking about animals is intimately and systematically linked to our thinking about other human groups in ways that entrench dominance over animals and those mentally associated with animals. The implications of these associations are explored.

My food is not that of man; I do not destroy the lamb and the kid to glut my appetite; acorns and berries afford me sufficient nourishment. My companion will be of the same nature as myself, and will be content with the same fare…the sun will shine on us as on man, and will ripen our food. The picture I present to you is peaceful and human, and you must feel that you could deny it only in the wantonness of power and cruelty.

Mary Shelley, 1818/2016, pp. 128–129

One could be forgiven for not recognizing the passage above as that from Dr. Frankenstein's "monster," a being created in a laboratory and comprised of both human and animal corpses.[1] In the 200 years since the publication of that book, this creature has arguably become one of the most reviled and hated villains of literature and film, gripping society with both fascination and horror. Of note, this monster was both made *of* animal parts (and thus uncivilized and not human) and was disinclined to eat or exploit animals[2]. As a non-human vegan, therefore, the monster is an intriguing character, being animal-like in composition while yearning for an enlightened existence that harms no animals (affording animals more rights than humans are generally willing to grant). Physically he is *not human enough*, yet mentally he is *too human* given his desperation to be accepted as human at all costs. The Frankenstein monster is thus an apt metaphor for the theme of this chapter, recognizing the complexity of our attitudes toward animals, animal-like humans, and humans who protect the rights of animals. Our central premise is that humans generally devalue animals and place a premium on the value of humans – except those deemed animal-like or who eschew the exploitation of animals – and that there are consequences of these valuation processes for animals and humans alike.

Devaluing animals relative to humans

Throughout history humans have generally placed themselves at the top of a hierarchy of animals (the so-called "Chain of Being," see Brandt & Reyna, 2011).[3] This conceptualization places a premium on humanity relative to all other species, allowing humans to feel entitled to use or exploit virtually any animal or land space for their survival or even mere pleasure. This premium is so omnipresent that any reader who is unconvinced by this point would be unlikely to become convinced by arguments or evidence we could raise here. But it is worth noting some of the more obvious examples that illustrate this point. Consider the notions of liberty and freedom: the vast majority of animals are denied personhood and the ensuing rights that are afforded to persons. Indeed, there exists fierce pushback against the notion of granting animals *person status*, even for animals widely considered "human-like" (e.g., chimpanzees) (Rosenblatt, 2017). In contrast, completely abstract and lifeless entities such as corporations are granted rights and personhood, enshrined by the US Supreme Court (Totenberg, 2014). Among objectors, granting some animals personhood would, it is feared, open the floodgates to include a multitude of other animals, robbing humans of their unique status and privilege. Such action would also take away the social value to be derived from representing human outgroups as animalistic and "lesser" (see Hodson, MacInnis, & Costello, 2014). To those with privilege and power, the devaluing of animals represents a central plank in a broader social structure that justifies the status quo and is thus deemed indispensable. Its removal could seriously jeopardize ideologies that justify inequality as an acceptable practice in the abstract.

Consider also the routine use of animals for research purposes. In the medical domain, much of this research is of such questionable value as to be considered mean-spirited. For example, Volkswagen was exposed for forcing monkeys to squat in cramped air-tight boxes while breathing car emissions for hours at a time (Connelly, 2018). Comparable procedures have been used by humans in committing extreme behaviors such as suicide or genocide (e.g., the Holocaust), yet are simply considered the cost of doing business when animals are the victims. Even worse, the majority of medical research studies on animals are ineffective at finding cures for human diseases (for an analysis on pharmaceutical testing see Kramer & Greek, 2018). These outdated methods of discovery are inefficient and slow to change given, in large part, the expendable nature of creatures that are so devalued relative to humans. Therefore, devaluing animals translates into harm not only for animals but for humans. Yet the public continues to assume that animal research aimed at curing human diseases is efficacious; the public would be much less supportive if they knew the true ineffectiveness of such research (Joffe, Bara, Anton, & Nobis, 2016). Without such knowledge, animal suffering for the benefit of humans to many is considered a cost of medicine; saving a human life at the cost of countless animal lives serves as its own justification. The devaluing of animals, and the premium afforded to humans, is so deeply engrained that essentially any potential benefit to humans is pursued at the cost of animal welfare and life.

Much closer to home we see ample evidence that, within the discipline of psychology, animals are regarded as lower beings than humans. Among the psychological community and related sciences, animals have been denied basic characteristics or qualities such as intelligence, personality, mind, and even pain perception (de Waal, 2009, 2016; Singer, 1975; Weiss, 2017). It should come as no surprise, therefore, that humans *likened* to animals have similarly been psychologically robbed of these characteristics in the minds of those seeking to rationalize the status quo and/or deny rights to these targets (e.g., Hodson et al., 2014), a topic we turn to next. In closing this present section, suffice it to say that there is considerable evidence that animals, relative to humans, are dramatically devalued by humans. Indeed, we are presently witnessing our planet's so-called "Sixth Mass Extinction," with human populations *growing* rapidly while levels of non-human species and lifeforms *drop* precipitously and dangerously (see Ceballos, Ehrlich, & Dirzo, 2017). To some extent these patterns reflect not only human dominance but our priorities. Thinking about animals as holding lesser value is unfortunately used to justify the mistreatment of animals and, as we next review, the mistreatment of human groups associated with animals.

Thinking about animals shapes thinking about "animal-like" humans

To this point we have focused on how humans tend to devalue animals relative to themselves. Next we seek to highlight some of our own research demonstrating how devaluing animals creates problems for people mentally represented as being animal-like, a phenomenon known as dehumanization. As a general process, dehumanization is operationalized as "*the perception and/or belief that another person (or group) is relatively less human than the self (or ingroup)*" (Hodson et al., 2014, p. 87, italics in original). Under this broader umbrella, there are different ways to dehumanize others, such as thinking of others as animal-like, as machine-like, as being devoid of "mind" and experience (see Haslam & Loughnan, 2014)[4]. In the present chapter we focus on *animalistic dehumanization*, a process whereby other humans are denied qualities that are generally deemed uniquely human and thus unrepresented in animals, or whereby humans are metaphorically likened to animals (see Loughnan, Haslam, & Kashima, 2009). Critically, this is a relative process, comparing the other to the self or one's own group; it is rare for a human group to be entirely denied humanness. For example, prejudice-prone people oppose helping immigrants in need when such groups pose a supposed realistic (i.e., tangible) threat to the host nation. Such persons also oppose help to immigrants who supposedly pose symbolic (i.e., non-tangible) threats, such as speaking a different language or supporting a different religion – but in order to rationalize the rejection of a group not posing a realistic threat the group is also dehumanized (Costello & Hodson, 2011). Representing the other group as relatively less human facilitates the decision to offer them less help, because help is generally reserved for those most human.

Interspecies model of prejudice (IMP)

In thinking about or conceptualizing others as animal-like, and in particular to denigrate the "other" by doing so, necessarily draws a link between thinking about human outgroups and thinking about animals. In short, how we think about animals shapes our thinking about groups we consider animal-like. To put it simply, it would be no insult, or of no strategic social advantage, to label another human as animal-like if animals themselves were not devalued in the first place. The Interspecies Model of Prejudice (IMP; Hodson et al., 2014) proposes the following: greater perceived divide between humans and animals, that is, seeing animals as different from and inferior to humans, fuels the tendency to think about human outgroups as less human than our ingroup (the group to which we belong). In turn, this animalistic dehumanization of the outgroup fuels or facilitates a host of biases toward that group (e.g., prejudice, stereotypes, discrimination, failure to help). Put together, this process takes the following form: human-animal divide → outgroup animalistic dehumanization → outgroup biases. The implications of this model are considerable: human-human prejudices (e.g., racism) may find some of their origins in devaluing animals (i.e., the human-animal divide).[5]

This proposal was first tested by Costello and Hodson (2010). In the first study, a sample of predominantly White Canadians self-reported: (a) the extent to which they perceived a human-animal divide; (b) the extent to which they considered immigrants less human than Canadians (using subtle measures whereby participants rated the extent to which each group is characterized by personality or emotional tendencies that have previously been shown to be considered particularly "human"); and (c) their attitudes toward immigrants. The IMP model was supported; the more participants distanced animals from humans, the greater their tendency to think about immigrants as relatively less human, which in turn predicted less favorable attitudes toward immigrants (see also Costello & Hodson, 2014a, 2014b). In a second study, participants' perceptions of a human-animal divide were experimentally manipulated. More specifically, instead of measuring naturally occurring human-animal divide beliefs, participants were randomly assigned to read scientific-appearing articles on either the similarities or the differences between humans and animals. The rationale underpinning the study is that if one can reduce the human-animal divide perception, particularly in ways that "elevate" animals up to the level of humans, this should rob people of the ability or tendency to dehumanize a human outgroup. After all, when revaluing animals to put them on par with humans, there is little social value in then referring to the outgroup as animal-like; the "insult" has lost its sting. This prediction was supported – increasing the psychological value of animals reduced the animalistic dehumanization of immigrants, which in turn lessened prejudice toward that outgroup. Central to the theme of this chapter, therefore, interventions to change thinking about human-animal relations can have profound impact on human-human relations, opening up a whole new raft of potential prejudice interventions.

Presumably, these psychological linkages between humans and animals take root in the early formative years of development, and preliminary evidence supports this contention. Costello and Hodson (2014a) provided the first known demonstration that White children as young as 6–10 years animalistically dehumanize Black children. Using measures adapted from researchers assessing racial prejudices in children, Costello and Hodson asked these children to put cards, marked with words, into boxes that were fronted with a face of a White or Black child. These cards contained traits and emotions typically considered unique to humans (e.g., sympathy; embarrassment; open to experience) or shared with animals (e.g., happiness; fear; agreeable). Across both studies, young White children showed a striking tendency to attribute the more human qualities to White children, and to relatively deny such qualities to Black children (see Figure 5.1). Children, it appears, learn at an early age the social value in representing the outgroup as less human (and hence more animal-like). The children also employed a board that contained sliding images of animals and humans, both of which could be moved, allowing the children to express their perception of the human-animal divide. The more that these children considered animals as separate from and inferior to humans the more they dehumanized (and expressed negative attitudes toward) Black children (see Figure 5.2). Moreover, children seeing humans and animals as more different (*vs.* similar) also scored lower on a cognitive task relevant to thinking about the overlap between group categories generally. In children, therefore, more advanced cognitive abilities might be required to understand that targets that look different on the surface, such as humans and animals, may nonetheless be similar.

Interestingly, as shown in Figure 5.2, the more that children devalued animals, the more that their parents did the same, and the greater the parents' racial prejudice and endorsement of inequality between human social groups (i.e., social

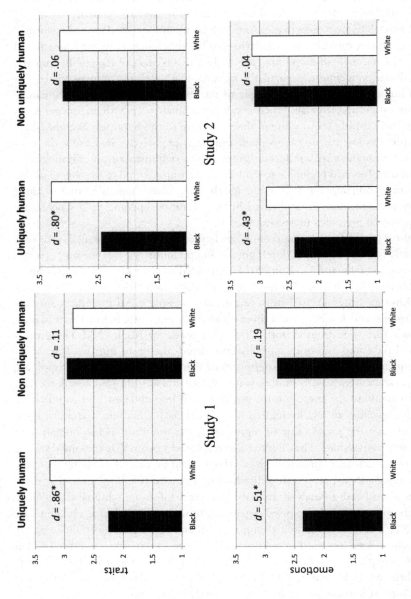

FIGURE 5.1 Dehumanization of Black targets by White children via the differential attribution of traits and emotions generally considered "uniquely human" to White and Black targets. *d*-values represent standard deviation differences between White and Black targets as an indicator of effect size. ★*p* < .05. Based on values from Costello and Hodson (2014a).

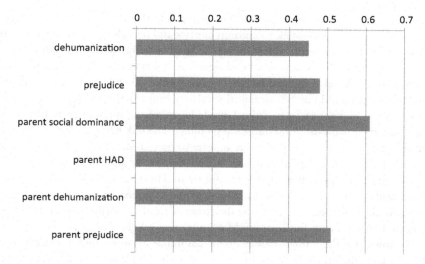

FIGURE 5.2 Correlations between human-animal divide (HAD) perceptions and various human intergroup variables (anti-Black prejudice; dehumanization of Black targets). Scores of 0 would indicate no relation; higher scores reflect stronger positive associations. All *ps* < .05.

Source: Costello & Hodson, 2014a.

dominance orientation, see Sidanius & Pratto, 1999; see also Dhont, Hodson, Leite, & Salmen, this volume). Such findings are consistent with the notion that how people think about animals as "lesser" than humans is associated with thinking that human outgroups are "lesser" than the purportedly more human ingroup, in ways that are systematic and even run in families. Intriguingly, new research shows that 5–10 year old White children also perceive that Black (*vs.* White) children experience less pain (Dore, Hoffman, Lillard, & Trawalter, 2018), in ways that are consistent with people believing that animals are less capable of pain (Singer, 1975). Such findings corroborate the assertion that thinking about animals is associated with thinking about others that are represented in animalistic terms.

Despite the IMP model showing support in university students, children, and adults, it is not intuitive to lay people that how we think about animals impacts how we think about ("animal-like") others. In fact, there is evidence that people actively deny such associations. In a sample of undergraduates at a Canadian university, Costello and Hodson (2014b) asked people the extent to which a host of factors (e.g., closed-mindedness; negative intergroup contact; human-animal divide) were responsible for causing dehumanization and for causing ethnic prejudice. Factors such as closed-mindedness were considered strong causes, and others such as social identity and status were considered less responsible. But only 1 of the 15 potential causes was rated significantly lower than the mid-point on the scale, and thus indicated a denial of this factor as a cause of dehumanization or prejudice: human-animal divide.

Likewise, when asked to rate solutions for solving dehumanization and ethnic prejudice, factors such as increasing intergroup contact and friendships were put forward as strong solutions, whereas others such as religion were rated as less helpful. But 2 of the 10 potential solutions were rated significantly lower than the scale midpoint, representing a denial that these factors could be solutions: highlighting animal-to-human similarity, and human-to-animal similarity. Nonetheless, in this same sample, those scoring higher in human-animal divide nonetheless expressed greater levels of outgroup dehumanization and outgroup prejudice. The link between human-animal relations and human-human relations appears to lie outside of everyday awareness, and worse, is actively denied as a cause of or solution to animalistic dehumanization of other groups and prejudices toward them. This may represent a simple and unmotivated denial (i.e., failure to recognize the relation), or it may represent a rationalization to justify not elevating the status of animals to solve human conflicts, or both. This represents fertile ground for future researchers.

We argue that the human-animal divide has implications for moral concern, not only for animals, but for humans. In a series of studies, Bastian, Costello, Loughnan, and Hodson (2012) pursued these implications using both Australian and Canadian samples. This research showed that those who naturally focus on human-animal similarities (instead of differences) include more animals in their circles of moral concern (i.e., consider a wider range of animals to be worthy of moral consideration). An experiment where participants were induced to think about the similarities of animals to humans or humans to animals further demonstrated that elevating animals "up" (as opposed to lowering humans "down") boosts the number of animal species deemed worthy of moral concern. Consistent with the IMP model, a final study experimentally exposed participants to information that framed animals as similar to humans (or humans as similar to animals). The researchers observed not only a reduction in the willingness to exploit animals but an increase in the degree to which participants would take a moral stand to defend the unfair treatment of marginalized human groups in the host society (e.g., immigrants, Black people). Such findings again demonstrate the powerful links between thinking about animals and thinking about people; the links are so strong that experimentally inducing changes to how we think about animals brings along an accompanying positive change in how we think about groups that are often animalistically dehumanized.

Aversion to being considered animal-like

It is quite evident that people dehumanize other groups in the interest of maintaining or attaining group dominance or other benefits (for reviews see Haslam & Loughnan, 2014; Hodson et al., 2014). By extension, it is also important to look at how people feel and react when they are being (or believe to be) *dehumanized* by others, which has been called *meta-dehumanization*. Kteily, Hodson, and Bruneau (2016) investigated the presumably aversive reaction to meta-dehumanization among a variety of groups, for instance, by testing how Americans feel dehumanized by

Arabs or Muslims, or how Israelis feel dehumanized by Palestinians. The researchers experimentally demonstrated that (falsely) informing people that another group dehumanizes the participant's group resulted in dehumanization of that other group. Put simply, when we learn that others consider us animal-like, we retaliate or reciprocate, thinking of their group as animal-like. Across multiple studies, meta-dehumanization predicted negative outcomes for other groups, such as increased support by Americans for torturing Arabs or Muslims and the use of drone strikes against Muslim countries. Clearly, feeling dehumanized fuels dehumanization in turn that subsequently lowers the bar for negative behaviours toward that group. Ironically, therefore, being thought of as animals unleashes less civilized forms of conduct generally considered "beneath" civilized humans.

Fortunately, Kteily and colleagues (2016) also demonstrated that this process can be reversed and undone. That is, when we learn that others consider our group as particularly human and civilized, we in turn humanize that outgroup. Thus we not only feel negatively about being considered animal-like and retaliate by dehumanizing the other, but other groups are rewarded with the attribution of human qualities when they consider us as particularly good exemplars of being human.

Such interventions hold considerable promise, particularly given that contact with other groups is associated with less tendency to dehumanize the other group (Capozza, Falvo, di Bernardo, Vezzali, & Vistin, 2014), with contact itself being a powerful tool for reducing prejudices and bias (Hodson & Hewstone, 2013; Pettigrew & Tropp, 2011). The more that groups learn about each other, often via contact, on average they should view the outgroup in more humanizing terms, and feel in turn humanized by that group, at least in positive or neutral contact settings.

As we have seen, thinking about animals (e.g., their "distance" from humans) has implications for thinking about human outgroups that are animalized, as detailed in the IMP model. But we have also seen that thinking that others think that we are animal-like has implications for human outgroups, including increased dehumanization of that group and increased willingness to harm or injure that group. Collectively this demonstrates that people generally have a disdain of, and devaluation of, animals, and also dislike those considered to be animal-like (or the accusation of being so natured ourselves). This helps to put into perspective why Frankenstein's monster, comprised of human and animal corpses and thus not fully human, has become one of the most recognized and reviled cultural icons of modern times. But what about the second part of the narrative concerning the disdain not only for animals and those deemed not fully human, but for those who advocate for animals (or at minimum avoid being complicit in animal exploitation)?

Negative reactions toward people who reject animal exploitation

Recently researchers have turned their attention toward reactions toward vegans and vegetarians. Generally, both vegetarians and meat-eaters consider vegetarians to be more "virtuous" albeit less masculine (Ruby & Heine, 2011). Interestingly, there can be negative implications for targets deemed virtuous or moral. For instance, Minson and Monin (2012) examined what they called "do-gooder derogation," how people lash out at morally motivated others. Interestingly, they focused on how meat eaters feel about vegetarians, noting that by refraining from eating meat, vegetarians are seen by meat-eaters to publicly condemn meat consumption. This finding is pertinent to the *meat paradox*, whereby people generally wish no harm to animals but nonetheless harm and exploit animals through consumption (Loughnan, Bastian, & Haslam, 2014; see Piazza this volume). It is psychologically aversive to hold such contradictions in one's mind, and the presence of vegetarians and vegans keeps to the fore of the mind that eating meat is a choice. In Study 1, Minson and Monin asked a sample of undergraduate meat eaters to freely list their thoughts associated with vegetarians. Despite (or presumably due to) seeing vegetarians as more moral than meat eaters, participants readily brought negative terms to mind (e.g., uptight; preachy). More importantly, the more morally superior participants believed vegetarians to consider themselves, the more negative were the descriptions of vegetarians by meat-eaters. In a follow-up experiment, undergraduate meat eaters indicated how they feel vegetarians view them personally, either before or after rating vegetarians along a series of dimensions. As expected, vegetarians were rated more negatively after (*vs.* before) thinking about how vegetarians view the participant (a meat eater). Thus, making salient that "do-gooder" vegetarians would supposedly look down on meat eaters for being less moral caused meat eaters to be more negative in their evaluations of vegetarians. The researchers suggest that thinking about vegetarians poses a threat to one's sense of personal morality, inducing a backlash against vegetarians.

Subsequent research has considered some of the implications of considering non-meat eaters threatening. Intrigued by previous studies showing that right-wing (*vs.* left-wing) adherents are more likely to eat meat, we collected data among two community samples of meat eaters in Belgium (Dhont & Hodson, 2014). Participants completed two measures relevant to right-wing ideology: (a) social dominance orientation (SDO; Sidanius & Pratto, 1999), wherein higher scores reflect greater endorsement of intergroup inequality and hierarchy in group-life generally; and (b) right-wing authoritarianism (RWA; Altemeyer, 1996), wherein higher scores reflect greater conventionalism, respect for traditions, and aggression against norm violators. Political psychologists generally consider such measures to capture much of the left-right divide, with SDO pertaining to the acceptance of inequality on the right, and RWA pertaining to the resistance to change on the right (e.g., Jost, Glaser, Kruglanski, & Sulloway, 2003). The main criteria of interest were participants' meat consumption levels and their endorsement of animal exploitation more generally (e.g., using animals for testing cosmetics).

In both studies (Dhont & Hodson, 2014), those higher in SDO or RWA indeed consumed more meat and were more accepting of practices that exploit animals[6]. But in many ways, the real story concerns *why* this left-right divide exists. In each study participants also reported the extent to which they feel threatened by vegetarianism (e.g., eroding cultural traditions; damaging the economy), and the extent to which they as humans feel superiority (or entitlement) over animals. These two variables largely explained why right-wing adherents, relative to those on the left, eat more meat and endorse animal exploitation. And, in Study 2, these effects were found to be statistically independent from the extent to which participants found hedonic value in eating meat. In other words, right-leaning ideologies about how society should be structured (e.g., RWA, SDO) predict greater animal exploitation and meat consumption because these ideologies encourage beliefs that vegetarians (or vegans) are a cultural threat and that humans are superior and dominant over animals, independent of whether one even enjoys meat.

More recent evidence confirms the association between ideology on the one hand, and vegetarian threat and speciesism on the other. For instance, we collected data from Belgium, the UK, and the USA and found that those scoring higher in RWA, SDO, or political conservatism, hold greater beliefs that vegetarians pose a cultural threat, and endorsed greater speciesism (Dhont, Hodson, & Leite, 2016). These associations were quite strong, often in the $r = .40$ to $.50$ range, very large effects by contemporary standards in psychology (Gignac & Szodorai, 2016). Moreover, recent evidence highlights the importance of vegetarianism threat (i.e., regarding a human group) with regard to animal welfare. Researchers examined, in a 16-month longitudinal study, the extent to which vegetarianism threat and human supremacy predict subsequent moral exclusion of various animal types: food animals (e.g., cows, pigs), appealing wild animals (e.g., chimps, dolphins), unappealing wild animals (e.g., snakes, bats), and companion animals (e.g., cats, dogs) (Leite, Dhont, & Hodson, 2019). Of interest was the degree to which these US participants considered these animals to fall within their moral circles of concern (Laham, 2009; Opotow, 1990); animals that fall outside of one's concern are generally considered fair game for being the brunt of indifference or exploitation. Leite and colleagues found that, even after controlling for initial moral concern regarding animals, beliefs about human supremacy and about vegetarianism threat predicted Time 2 moral exclusion of animal ratings. Specifically, those higher in human supremacy beliefs later demonstrated greater moral exclusion of all types of animals, whereas those higher in vegetarianism threat later demonstrated greater moral exclusion only of food animals (as predicted) and appealing wild animals. Thus beliefs about human supremacy over animals (i.e., human-animal), and about threats posed by vegetarians (i.e., human-human), both have real downstream consequences of the welfare of animals, here determining whether animals deserve moral protections.

Ideologies such as SDO and RWA, and ideologically-motivated beliefs such as human supremacy over animals and vegetarianism threat, are therefore powerful determinants of engaging in behaviors that are harmful to animals. Particularly intriguing is the finding that pushing back against vegetarians as a source of threat is a major factor underpinning practices that are harmful to animals (see Dhont & Hodson, 2014). And related findings confirm this general idea: In a community sample of Americans, variables such as SDO and RWA predicted the denial of climate change and resistance to taking action to curb its effects, in large part through pushback against environmentalists (Hoffarth & Hodson, 2016). Here we again see that viewing human outgroups (e.g., vegetarians; environmentalists) as threatening plays a role in promoting behaviors wherein human interests are put before those of animals and nature more generally. Thinking about animals and humans appears again intertwined.

But what about predicting prejudice toward vegetarians or vegans (henceforth veg*ns), in addition to feeling threat from such groups? Anecdotally, most veg*ns can recount many experiences of negativity and even discrimination. This encouraged Chin, Fisak, and Sims (2002) to develop the Attitudes Toward Vegetarians scale, tapping attitudes toward veg*n behaviours, beliefs, health/mental effects, and how to treat the group. A sample item reads: "Vegetarians preach too much about their beliefs and eating habits." The authors found that those higher in authoritarianism (but not conservatism) expressed significantly more negative attitudes toward vegetarians. Contrary to their expectations however, scores overall reflected relatively positive attitudes. As the authors note, this might be attributable to the fact that the sample was a university sample (i.e., relatively liberal and young) and 81% female. Moreover, 84% knew a vegetarian personally, and indeed, many veg*ns may have been included in the analysis because the authors did not report dietary habits of respondents.

More recently MacInnis and Hodson (2017) examined the nature of anti-veg*n attitudes, and their consequences, more closely. Their first two samples were adult Americans (MTurk workers), with roughly equal numbers of men and women, and were all meat eaters. In these studies, attitudes were assessed with ratings thermometers (i.e., one's feelings toward the group in question). In Study 1, both vegetarians and vegans were rated significantly more negatively than Black people, and equivalently to immigrants, asexuals, and atheists. Of the groups examined, only drug addicts were rated more negatively than vegetarians and vegans. There was little indication, however, of overall support for discriminatory policies in terms of hiring or renting an apartment to veg*ns. As expected, those higher in RWA, SDO, or conservatism were significantly more negative (and discriminatory) toward vegetarians or vegans. Moreover, and related to our earlier discussions of vegetarian threat, those on the right (vs. left) were more prejudiced and discriminatory *because* they considered vegetarians and vegans threatening.

Study 2 of MacInnis and Hodson (2017) found that meat eaters were biased against vegetarians and vegans depending on the *reasons* given for their dietary choice. Specifically, more negative attitudes were expressed toward those veg*n for *animal rights* reasons than for health or environmental reasons. Clearly veg*ns are more disliked when their food choices are based in morality, consistent with the view that vegetarians are "do-gooders" who highlight the moral imperfections in meat eaters (see Minson & Monin, 2012). In Study 1 of MacInnis and Hodson, male veg*ns were also significantly more disliked than their female counterparts, presumably due to their violation of masculinity norms, again highlighting that those avoiding meat are disliked for the specific nature of what they represent (and not merely as an "other" type of person).

And there is evidence that veg*ns feel the sting of stigma and discrimination. In Study 3 of MacInnis and Hodson (2017) the researchers actively recruited adult veg*ns through classified ads and social media groups dedicated to these groups. One quarter of vegans reported distancing from friends after disclosing being vegan, with 10% even observing such distancing by family members. A third of both vegans and vegetarians reported anxiety about revealing being veg*n to others, and understandably so: 46% of vegetarians, and 67% of vegans, reported some level of every-day discrimination in their lives. Not surprisingly, more than half of each group also reported engaging in coping strategies to cope with the discrimination and alienation. And consistent with the notion that threat reactions underlie some of the bias, vegans reported significantly more negative experiences than did vegetarians. Thus, the more a person rejects animal exploitation, the more they are rejected by meat-eaters; it is not simply the case that meat-eaters dislike veg*ns, but they dislike veg*ns the more they restrict or eliminate their animal consumption and use.

The studies of MacInnis and Hodson (2017) clearly demonstrate that veg*ns are disliked, as much if not more than, many of the other marginalized social groups (e.g., Blacks; immigrants). And this dislike is targeted and nuanced: vegans are disliked and rejected more than vegetarians; veg*n men are disliked more than their female counterparts; those who are veg*n for animal justice reasons are more disliked than those claiming environmental reasons or health reasons for their non-meat diets. Moreover, this negativity is psychologically experienced by veg*ns, as they report anxiety, discrimination, social distancing, and the need for coping mechanisms. Here again we witness pushback against those not eating meat, and the "moral threats" posed by veg*ns as they demonstrate that animal exploitation is not necessary or needed.

Recent research by Judge and Wilson (2019) supports many of the findings of MacInnis and Hodson (2017) but within a New Zealand context. In a large sample of more than 1300 citizens, they found that overall attitudes toward veg*ns were somewhat positive, that vegans were more disliked than vegetarians, and men (*vs.* women) show stronger biases against those not eating meat. They also confirmed that those scoring higher in RWA or SDO expressed more negative attitudes toward both vegans and vegetarians, as did those who consider the world dangerous or competitive. These findings nicely fit the central theme of this section of the chapter: those feeling rattled about the world in terms of instability and/or competition, and those more strongly endorsing right-leaning ideologies such as RWA and SDO, express consistently more negative attitudes toward vegans and vegetarians.

Another approach has been to consider the relation between one's personal meat consumption and anti-veg*n bias. Would the amount of meat (e.g., beef) one consumes be related to the extent to which one distances from veg*ns? An ambitious study by Ruby and colleagues (2016) asked questions of relatively large numbers of people in four heavy beef-consuming nations: Argentina, Brazil, France, and the USA. Overall, men craved and consumed more beef than did women. Describing overall trends, the authors concluded that people generally show pro-beef attitudes across nations, with attitudes toward vegetarians being fairly neutral in valence. Using these rich data, Earle and Hodson (2017) conducted additional analyses to further determine whether a general pro-beef attitude, indicated by a stronger desire for and higher consumption and liking of beef, predicts general anti-vegetarian prejudice, indicated by feeling more bothered by, lesser admiration of, and lesser willingness to date vegetarians. This analysis revealed a very strong pattern: the more a person is pro-beef, the more negative their anti-vegetarian prejudice. Although statistically significant in all countries, the percentage of variance in anti-vegetarian prejudice explained by pro-beef orientations differed by country. Particularly remarkable is the finding that 43 percent of the variance in American anti-vegetarian attitudes was explained by personal pro-beef attitudes. Thus, meat-eaters who enjoy beef do not simply dislike vegetarians as a group, but the strength of their dislike is systematically and strongly linked to the degree that they personally enjoy beef. Such patterns are very consistent with the notion that meat-eaters pushback against non-meat eaters in light of the threat that such individuals pose to the meat-eater personally (and presumably morally).

If meat-eaters are more negative toward veg*ns as a function of how much meat they personally consume, this suggests sensitivity to personal choices and behaviors (e.g., guilt induction) that then becomes directed outward toward those disavowing meat consumption. This begs a related but distinct question: Might reminders that meat originates from animals impact attitudes toward veg*ns? A recent set of studies by Earle, Hodson, Dhont, and MacInnis (2019) addressed this question in an experimental context, exposing some participants to visual advertisements where animals were paired with meat (e.g., a lamb paired with a lamb chop) and others only to the meat image (e.g., a lamb chop). These images can be found at https://osf.io/25jfr. This manipulation exerted several interesting effects. Across studies, meat-animal reminders (*vs.* control) boosted empathy for the animal in question, and elevated distress about eating meat; in Study 2 it also boosted disgust at the thought of eating meat. Such reactions are widely known by the meat industry, who are generally careful to limit pairings of meat with animals to avoid putting off potential meat customers. But these psychological reactions, such as elevated empathy for animals paired with images of their meat "product," exerted interesting knock-on effects, not only reducing participants' willingness to eat meat, but also reducing biases against veg*ns. For instance, in Study 2, exposure to meat-animal reminders: (a) induced greater empathy for the meat-animal, which in turn lowered prejudice toward veg*ns; and (b) generated greater meat distress, which in turn lowered perceptions of veg*ns as culturally threatening. Thus, reminding meat-eaters that meat originates from living, sentient animals can reduce willingness to eat meat but also lower biases against those who refrain from eating meat.

In the title of their paper, MacInnis and Hodson (2017) expressed that "*It ain't easy eating greens*." This was not in reference to the difficulty in giving up meat and eating a plant-based diet, but rather reflect the marginalization by the meat-eating majority. We argue that better understanding anti-veg*n attitudes, and learning how to develop interventions, are key goals moving forward. Notably, veg*ns are disliked not for what they do, but for what they fail to do (and what those inactions represent). By failing to endorse mainstream ideologies and behaviors, these groups are targeted for failing to uphold the status quo and its legitimizing rationalizations (MacInnis & Hodson, 2017). In this way, veg★ns are not targeted because they eat the wrong forms of meat but because they disavow meat altogether, drawing attention to the meat paradox (i.e., that we care about animals but exploit them for food and other uses). And as our review has illustrated, vegetarians and vegans are considered a threat to society, against which the dominant meat culture pushes back strongly. We have also shown how such threats predict prejudices toward veg★ns, particularly among those with right-leaning ideologies and those who consume more meat. These findings support the assertion that people are threatened by their own pro-meat behaviors and beliefs that conflict with their positive self-image, who then lash out at others who highlight the moral shortcomings associated with eating meat.

Thinking about our relations with non-human animals

Anthony Bourdain, host of CNN's *Parts Unknown*, was arguably one of America's best-known food critics. He was not shy about his love of meat and his related disdain for veg★ns. We can let his words speak for themselves ("Vegans vs Anthony Bourdain?", n.d.):

> Vegetarians, and the Hezbollah-like splinter faction, the vegans, are a persistent irritant to any chef worth a damn... Vegetarians are the enemy of everything good and decent in the human spirit, an affront to all I stand for, the pure enjoyment of food.

The threat supposedly posed by veg★ns and the implications of a meat-free world is palpable, and mirror much of the revulsion expressed toward Frankenstein's "monster," a despised creature who seeks a vegan existence. As reflected in Bourdain's words, contemporary interpretations of veg★ns as supposed "terrorists" and evil-doers serve a similar purpose: to vilify and alienate those who do not buy into and endorse a culture that considers the exploitation of animals as normal, natural, and necessary. Veg★ns may therefore be disliked, in part, because they "elevate" animals to human or near-human status, meaning that animals are given greater moral consideration. This can foster a backlash against veg★nism because the veg★n perspective not only makes people feel negatively for exploiting animals (who now would have moral consideration), but society loses a justification for violence toward and bias against groups deemed animal-like (also making society feel negatively).

Much of this would not have been lost on Mary Shelley who, along with her famous husband Percy, were part of the Romantic vegetarian movement (Adams, 2015) and keen to make a profound statement about human nature and social exclusion. Thus her deliberate depiction of the creature as the *ultimate other* by virtue of his animal origins and vegan ideals offered a biting commentary on society's treatment of animals (and people). That we then feared the creature, and continue to do so 200 years later, is presumably a reflection of the undying human ambivalence about loving but exploiting animals. As noted by Cambridge professor Ottoline Leyser, such popular fiction often reflects society's concerns of the day (Sample, 2017). For example, the Peter Parker character who subsequently became Spider-Man was originally bitten by a *radioactive* spider, with radioactivity a top concern at the time. But in later incarnations Parker was bitten by a *genetically modified* spider, at a time when concerns were being raised about genetically modified organisms. That the Frankenstein creature has remained the quintessential villain of popular culture for centuries presumably reflects society's largely unchanged revulsion at both animal nature and the moral dilemmas faced by a species that supposedly loves yet exploits animals.

Implications

When we examine how we think about animals, animal-like humans, and humans who protect (or at minimum avoid exploiting) animals we recognize the complexity involved. We also recognize that much of this thinking is mired in rationalizations and ambivalent feelings. There are several implications of these findings. In terms of our scientific theories, the field needs to better recognize the links between human-animal relations and human-human relations. A related implication is that the field might need to move away from focusing so heavily on *attitudes* or evaluations of animals. Most people "like" animals but this seems disassociated from whether or not they will exploit or protect animals – the point at which the proverbial rubber hits the road. In the same way, most men "like" women but do little to fight for equal rights and pay for women. What matters most in predicting outcomes is not likely whether we like the other in question (e.g., animals; women; racial minorities), but rather our thoughts about their perceived value, and in particular, their *exploitation value* (see Hodson, 2017). That is, people endure the aversion induced by the meat paradox because the personal and cultural value they see in exploiting animals outweighs whether or not the target is likeable or whether such exploitation reflects poorly on one's own moral character. Here again we see how rationalizations and dominance motives shape thinking about both humans and about animals. Consider how people will dehumanize other human groups, likening them to animals in ways that remove protections. The success of this process itself gains further social value and traction. Humans also engage in the "*dehumanization*" *of animals* (Hodson, 2017), as when Indians recently made moves to reclassify peacocks as vermin to justify their easier elimination as a nuisance. The more that another group or species is distanced from humans, either from humans to animal, or from animals to vermin, the less people show concern for that species and a willingness to engage protections.

There are also implications in terms of the development of bias interventions. In order for us to fully understand human-human prejudices, we must consider the overlap of these prejudices with speciesism and animal attitudes (Dhont et al., 2014, 2016), and that some human-human prejudices, especially those linked with animalistic dehumanization, find their roots in human–animal relations (Hodson et al., 2014; see also Dhont et al., this volume). This opens up tremendous potential for theorists and practitioners alike. Consider how reducing the human–animal divide lowers prejudice toward immigrants, even among those higher in SDO and generally anti-immigrant (Costello & Hodson, 2010). Interventions that target human–animal relations on the surface, but human–human relations underneath, might offer backdoor pathways to reducing conflict among human groups.

In keeping with the theme of the present book, there needs to be greater contact and connection between the researcher and theorist on the one hand, and the advocate on the other. Many researchers, particularly in psychology, have been exceptionally slow to recognize these human–animal connections and have much catching up to do. Much of what we have discussed in terms of research findings here has undoubtedly been long "known" but not empirically demonstrated by advocates: that animals are devalued relative to humans; that systems of inter-species oppression are linked under dominance motives. But changes are afoot in the science of psychology, as theorists and researchers have pushed the psychology of thinking about animals to the fore, with several recent publications on the psychology of speciesism and human–animal relations in top psychology journals (e.g., Amiot & Bastian, 2015; Caviola, Everett, & Faber, 2019). Psychologists entering the discussion bring with them the tools (empiricism) and the outlets (journals, conferences, classrooms) to shape thinking about animals in meaningful ways.

Despite these advances in our understanding about human–animal relations, there are limitations to the research conducted thus far. Although much of the literature we have discussed is experimental (e.g., Bastian et al., 2012; Costello & Hodson, 2010, 2011; Kteily et al., 2016), the evidence is often self-report in nature. Large-scale studies are needed with both behavioral and observational components. Longitudinal approaches (e.g., Leite et al., 2019) are also helpful in determining the flow of process over time. We are quick to defend self-report research as being extremely valuable, particularly when exploring the structural relations among constructs such as that between ethnic prejudice and speciesism (see Dhont et al., 2014, 2016). Complex questions about animals and human–animal relations will require a diverse range of methodologies.

Like most psychological research, the vast majority of the participants in these studies are considered WEIRD (Western, Educated, Industrialized, Rich, Democratic; see Henrich, Heine, & Norenzayan, 2010). Although such people make up the vast majority of psychological research, they are atypical compared to the rest of the world. This point seems particularly poignant when it comes to human–animal relations, which are deeply rooted in local culture and tradition. A better understanding of how people think about animals and animal-like or animal-protecting others will only be possible when we expand the range of human groups considered and the range of animals of focus. Cross-cultural research is sorely needed in this discipline, and recent efforts have already proven fruitful (e.g., Ruby et al., 2016).

Take home message

Our closeness to animals is a source of considerable psychological angst, exacerbated when reminded how we routinely exploit and harm animals. Consider the ambivalence experienced when Queen Victoria observed her first orangutan in a public zoo. Despite being captivated by the creature she was nonetheless deeply disturbed, referring to the creature as "frightful and painfully and disagreeably human" (Lemonick & Dorfman, 2006). Relatedly, Desmond Morris relayed to Frans de Waal (2016) his experience working at the London Zoo, where apes had been trained to play in tea parties in front of visitors. The apes, however, were so adept at handling the basic tools (spoons, cups, teapots) that this disturbed the English public, who considered the act of taking tea as one that marks civilization. Incredibly, the apes were then retrained to throw food and mishandle the tools, which greatly relieved the public's anxiety.

Such are the frailties of the human psyche. Humans go to considerable lengths to distinguish themselves from animals (e.g., de Waal, 2016; Hodson et al., 2014). In our quest for supremacy and self-overvaluing, we even seek to distance ourselves from other humans, and not simply those from other races. Lay people tend, for instance, to think of *homo sapiens* as virtually synonymous with "human," despite there in fact being multiple subspecies of human: *homo sapiens*, Neanderthals, and Denisovans. Interestingly, our supposed superiority within this highly select group is increasingly drawn into question. Recent discoveries of Palaeolithic art in Spain clearly demonstrate that Neanderthals were also capable of complex symbolic thought previously considered unique to *homo sapiens* (The Guardian, 2018). This comes after a barrage of evidence reveals that non-human animals are very much capable of intelligent thinking, moral behavior, and supposedly "human" emotions such as empathy (for reviews see de Waal, 2009, 2016). As such discoveries creep into the public consciousness, we anticipate a pushback that emphasizes the uniqueness of *homo sapiens* as distinct from other humans, as well as humans as distinct from other animals. After all, we are adept at moving the goal-posts to keep racial groups at a disadvantage, and we similarly move the goal-posts to distance ourselves from animals to give humans the advantage (de Waal, 2016). These psychological exercises that move the parameters in our favor all comes at a cost, it is worth remembering, to beings characterized as "the other."

The take home message is that human psychology functions in ways that make life difficult for animals, people seen as more animal-like, and even for those not eating or otherwise exploiting animals. Relative to humans, animals are devalued, hold few rights, and are easy targets for exploitation as a result. Dehumanized humans suffer many of the same indignities. Moreover, veg*ns are a social minority who suffer prejudice, discrimination, and alienation, despite actively avoiding causing others harm objectively. Indeed, a perceived lack of social support is a strong predictor in explaining why people often lapse back to eating meat (Hodson & Earle, 2018). Addressing such issues will require societal change, and dare we say, enlightenment. Humans will need to scale back their exclusive dominance on the planet, even if only to save themselves. Animals will increasingly require strong legally-enforceable rights to protect them as they continue to be exploited and natural resources become more finite and land/sea becomes more compromised. An ideal starting point is to reduce or end the consumption of animals and shift to plant-based diets; a side benefit would be a dramatic reduction in greenhouse gas emissions that fuel climate change (Springman, Godfray, Rayner, & Scarborough, 2016). The meat paradox, a phenomenon rife with ambivalence, can become a useful tool in this endeavor instead of an obstacle. People already *like* animals, so harnessing this pre-existing positivity will be critical. Such actions will arguably become easier as veg*n populations continue to grow and new norms develop.

At the beginning of this chapter we contemplated whether Frankenstein's monster is deemed too inhuman or too human. The answer, based on the research reviewed here, may lead us to the conclusion that he is both. Moreover, perhaps the real horror is that society recognizes itself in the monster – both too inhuman, as reflected in modern factory-farming practices, and too human, as reflected in our powerful ability to rationalize and justify our actions from our position of dominance and privilege.

Notes

1 We recognize that humans are animals; we use the term *animals* as a shorthand for *non-human animals*.
2 Keen observers will note that Frankenstein's monster craved and consumed milk and cheese (p. 92) and pushed a team of sled dogs at great speed (p. 185) when hunted angrily by his creator. But the monster's philosophy and preferences were arguably vegan in nature. Indeed, prior to discovering and consuming these dairy products he nearly starved searching for acorns to eat. His vision for the world was one of social inclusion and existence without reliance on animal exploitation.
3 For an alternative take that crosses this vertical Value dimension with a horizontal Threat Potential dimension (i.e., the ability to inflict harm), we refer the reader to Hodson and colleagues (2014). For a related discussion on how people rate animals in terms of their warmth versus competence, see Sevillano and Fiske (this volume).
4 Fiske and colleagues also argue that people considered both low in warmth and competence as dehumanized, in light of their neuroscientific evidence such people are mentally processed as objects not people (see Sevillano & Fiske, this volume, for details).
5 We also recognize that experimental manipulations of dehumanization might drive or influence human-animal divide perceptions. It is also possible that concepts such as dehumanization and divide perceptions are "downstream" consequences of hierarchical thinking (see Costello & Hodson, 2010, Study 1; for related points on dominance and hierarchy playing a more causal role, see Dhont, Hodson, Costello, & MacInnis, 2014; Dhont, Hodson, & Leite, 2016; Dhont, Hodson, Leite, & Salmen, this volume).
6 On average the correlation between these ideology variables and meat consumption was approximately $r = .33$, a rather sizeable correlation, considered "large" in the individual differences field (Gignac & Szodorai, 2016). There are several ways of thinking about the meaning of this correlation. By one account, 11% (i.e., .33 x .33) of the variability in meat consumption was explained by SDO or RWA. Another way is to express it as a binomial effect size display: Among those above the median in SDO or RWA, approximately two-thirds were above the median in meat consumption.

References

Adams, C.J. (2015). *The Sexual Politics of Meat: A Feminist-Vegetarian Critical Theory* (25th anniversary edition). New York: Bloomsbury Academic.

Altemeyer, B. (1996). *The Authoritarian Specter*. Cambridge, MA: Harvard University Press.

Amiot, C., & Bastian, B. (2015). Toward a psychology of human-animal relations, *Psychological Bulletin, 141*, 6–47. doi: 10.1037/a0038147

Bastian, B., Costello, K., Loughnan, S., & Hodson, G. (2012). When closing the human-animal divide expands moral concern: The importance of framing. *Social Psychological and Personality Science, 3*, 421–429. doi: 10.1177/1948550611425106

Brandt, M. J., & Reyna, C. (2011). The chain of being: A hierarchy of morality. *Perspectives on Psychological Science, 6*, 428–446. doi:10.1177/1745691611414587

Capozza, D., Falvo, R., Di Bernardo, G. A., Vezzali, L., & Visintin, E. P. (2014). Intergroup contact as a strategy to improve humanness attributions: A review of studies. *TPM: Testing, Psychometrics, Methodology in Applied Psychology, 21*, 349–362. doi:10.4473/TPM21.3.9

Caviola, L., Everett, J. A. C., & Faber, N. S. (2019). The moral standing of animals: Towards a psychology of speciesism. *Journal of Personality and Social Psychology, 116*(6), 1011–1029. doi: 10.1037/pspp0000182

Ceballos, G., Ehrlich, P. R., & Dirzo, R. (2017). Biological annihilation via the ongoing sixth mass extinction signaled by vertebrate population losses and declines. *Proceedings of the National Academy of Sciences, 114*, E6089–E6096. www.pnas.org/cgi/doi/10.1073/pnas.1704949114

Chin, M. G., Fisak Jr., B., & Sims, V. K. (2002). Development of the Attitudes Toward Vegetarians Scale. *Anthrozoos, 15*, 332–342. doi: 10.2752/089279302786992441

Connelly, K. (2018). VW condemned for testing diesel fumes on humans and monkeys. *The Guardian.* Retrieved from www.theguardian.com/business/2018/jan/29/vw-condemned-for-testing-diesel-fumes-on-humans-and-monkeys

Costello, K., & Hodson, G. (2010). Exploring the roots of dehumanization: The role of animal-human similarity in promoting immigrant humanization. *Group Processes and Intergroup Relations, 13*, 3–22. doi: 10.1177/1368430209347725

Costello, K., & Hodson, G. (2011). Social dominance-based threat reactions to immigrants in need of assistance. *European Journal of Social Psychology, 41*, 220–231. doi: 10.1002/ejsp.769

Costello, K., & Hodson, G. (2014a). Explaining dehumanization among children: The interspecies model of prejudice. *British Journal of Social Psychology, 53*, 175–197. doi:10.1111/bjso.12016

Costello, K., & Hodson, G. (2014b). Lay beliefs about the causes of and solutions to dehumanization and prejudice: Do non-experts recognize the role of human-animal relations? *Journal of Applied Social Psychology, 44*, 278–288. doi: 10.1111/jasp.12221

de Waal, F. (2009). *The Age of Empathy: Nature's Lessons for a Kinder Society*. New York: Harmony Books.

de Waal, F. (2016). *Are We Smart Enough to Know How Smart Animals Are?* New York: Norton.

Dhont, K., & Hodson, G. (2014). Why do right-wing adherents engage in more animal exploitation and meat consumption? *Personality and Individual Differences, 64*, 12–17. doi: 10.1016/j.paid.2014.02.002

Dhont, K., Hodson, G., Costello, K., & MacInnis, C. C. (2014). Social dominance orientation connects prejudicial human-human and human-animal relations. *Personality and Individual Differences, 61–62*, 105–108. doi: 10.1016/j.paid.2013.12.020

Dhont, K., Hodson, G., & Leite, A. C. (2016). Common ideological roots of speciesism and generalized ethnic prejudice: The social dominance human-animal relations model (SD-HARM). *European Journal of Personality, 30*, 507–522. doi: 10.1002/per.2069

Dhont, K., Hodson, G., Leite, A., & Salmen, A. (this volume). The psychology of speciesism. In K. Dhont & G. Hodson (Eds.), *Why We Love and Exploit Animals: Bridging Insights from Academia and Advocacy.* Abingdon: Routledge.

Dore, R. A., Hoffman, K. M., Lillard, A. S., & Trawalter, S. (2018). Developing cognitions about race: White 5- to 10-year-olds' perceptions of hardship and pain. *European Journal of Social Psychology, 48,* 121–132. doi:10.1002/ejsp.2323

Earle, M., & Hodson, G. (2017). What's your beef with vegetarians? Predicting anti-vegetarian prejudice from pro-beef attitudes across cultures. *Personality and Individual Differences, 119,* 52–55. doi: 10.1016/j.paid.2017.06.034

Earle, M., Hodson, G., Dhont, K., & MacInnis, C. C. (2019). Eating with our eyes (closed): Effects of visually associating animals with meat on anti-vegan/vegetarian attitudes and meat consumption willingness. *Group Processes and Intergroup Relations, 22,* 818–835. DOI: 10.1177/1368430219861848.

Gignac, G.E., & Szodorai, E.T. (2016). Effect size guidelines for individual differences researchers. *Personality and Individual Differences, 102,* 74–78. doi: 10.1016/j.paid.2016.06.069

The Guardian. (2018). The Guardian view on Neanderthals. *The Guardian.* Retrieved from www. theguardian.com/commentisfree/2018/feb/25/the-guardian-view-on-neanderthals- we-were-not-alone

Haslam, N., & Loughnan, S. (2014). Dehumanization and infrahumanization. *Annual Review of Psychology, 65,* 399–423. doi: 10.1146/annurev-psych-010213-115045

Henrich, J., Heine, S. J., & Norenzayan, A. (2010). The weirdest people in the world? (Target Article). *Behavioral and Brain Sciences, 33,* 61–83. doi:10.1017/S0140525X0999152X

Hodson, G. (2017). What is the pressing "animal question" about? Thinking/feeling capacity or exploitability? (Invited Commentary on Marino, 2017). *Animal Sentience, 2*(17), #12, pp.1–4. Retrieved from http://animalstudiesrepository.org/animsent/vol2/ iss17/12/

Hodson, G., & Earle, M. (2018). Conservatism predicts lapses from vegetarian/vegan diets to meat consumption (through lower social justice concerns and social support). *Appetite, 120,* 75–81. doi: https://doi.org/10.1016/j.appet.2017.08.027

Hodson, G., & Hewstone, M. (Eds.) (2013). *Advances in Intergroup Contact.* London: Psychology Press.

Hodson, G., MacInnis, C.C., & Costello, K. (2014). (Over)Valuing "humanness" as an aggravator of intergroup prejudices and discrimination. In P.G. Bain, J. Vaes, & J.-Ph. Leyens (Eds.), *Humanness and Dehumanization* (pp. 86–110). London: Psychology Press.

Hoffarth, M. R., & Hodson, G. (2016). Green on the outside, red on the inside: Perceived environmentalist threat as a factor explaining political polarization of climate change. *Journal of Environmental Psychology, 45,* 40–49. doi: 10.1016/j.jenvp.2015.11.002

Joffe, A. R., Bara, M., Anton, N., & Nobis, N. (2016). The ethics of animal research: A survey of the public and scientists in North America. *BMC Medical Ethics, 17,* 1–12. doi:10.1186/s12910-016-0100-x

Jost, J. T., Glaser, J., Kruglanski, A. W., & Sulloway, F. J. (2003). Political conservatism as motivated social cognition. *Psychological Bulletin, 129,* 339–375. doi: 10.1037/ 0033-2909.129.3.339

Judge, M., & Wilson, M. S. (2019). A dual-process model of attitudes toward vegetarians and vegans. *European Journal of Social Psychology, 49,* 169–178. doi: 10.1002/ejsp.2386

Kramer, L.A., & Greek, R. (2018). Human stakeholders and the use of animals in drug development. *Business and Society Review, 123,* 3–58. doi: 10.1111/basr.12134Laham, 2009

Kteily, N., Hodson, G., & Bruneau, E. (2016). They see us as less than human: Meta-dehumanization predicts intergroup conflict via reciprocal dehumanization. *Journal of Personality and Social Psychology, 110,* 343–370.

Laham, S. M. (2009). Expanding the moral circle: Inclusion and exclusion mindsets and circle of moral regard. *Journal of Experimental Social Psychology, 45,* 250–253. doi: 10.1016/j.jesp.2008.08.012

Leite, A. C., Dhont, K., & Hodson, G. (2019). Longitudinal effects of human supremacy beliefs and vegetarian threat on moral exclusion (*vs.* inclusion) of animals. *European Journal of Social Psychology, 49,* 179–189. https://doi.org/10.1002/ejsp.2497

Lemonick, M., & Dorfman, A. (2006). What makes us different? *Time, 168,* 44–53.

Loughnan, S., Bastian, B., & Haslam, N. (2014). The psychology of eating animals. *Current Directions in Psychological Science, 32,* 104–108. doi: 10.1177/0963721414525781

Loughnan, S., Haslam, N., & Kashima, Y. (2009). Understanding the relationship between attribute-based and metaphor-based dehumanization. *Group Processes & Intergroup Relations, 12,* 747–762. doi: 10.1177/1368430209347726

MacInnis, C.C., & Hodson, G. (2017). It ain't easy eating greens: Evidence of bias toward vegetarians and vegans from both source and target. *Group Processes and Intergroup Relations, 20,* 721–744. doi: 10.1177/1368430215618253

Minson, J. A., & Monin, B. (2012). Do-gooder derogation: Disparaging morally motivated minorities to defuse anticipated reproach. *Social and Personality Psychological Science, 3,* 200–207. doi: 10.1177/1948550611415695

Opotow, S. (1990). Moral exclusion and injustice. An introduction. *Journal of Social Issues, 46,* 1–20. doi: 10.1111/j.1540–4560.1990.tb00268.x

Pettigrew, T. F., & Tropp, L. R. (2011). *When Groups Meet: The Dynamics of Intergroup Contact.* New York: Psychology Press.

Piazza, J. (this volume). Why people love animals yet continue to eat them. In K. Dhont & G. Hodson (Eds.), *Why We Love and Exploit Animals: Bridging Insights from Academia and Advocacy.* Abingdon: Routledge.

Rosenblatt, K. (2017). Do apes deserve "personhood" rights? Lawyer head to N.Y. supreme court to make case. *NBC News.* Retrieved from www.nbcnews.com/news/us-news/do-apes-deserve-personhood-rights-lawyer-heads-n-y-supreme-n731431

Ruby, M. B., Alvarenga, M. S., Rozin, P., Kirby, T. A., Richer, E., & Rutsztein, G. (2016). Attitudes toward beef and vegetarians in Argentina, Brazil, France, and the USA. *Appetite, 96,* 546–554. doi: 10.1016/j.appet.2015.10.018

Ruby, M. B., & Heine, S. J. (2011). Meat, morals, and masculinity. *Appetite, 56,* 447–450. doi:10.1016/j.appet.2011.01.018

Sample, I. (2017, Dec 27). Frankenpod 200: Celebrating Mary Shelley's masterpiece – *Science Weekly* podcast. www.theguardian.com/science/audio/2017/dec/27/frankenstein-frankenpod-200-celebrating-mary-shelleys-masterpiece-science-weekly-podcast [see 30:40–31:45]

Sevillano, V., & Fiske, S. T. (this volume). Animals as social groups: An intergroup relations analysis of human-animal conflicts. In K. Dhont & G. Hodson (Eds.), *Why We Love and Exploit Animals: Bridging Insights from Academia and Advocacy.* Abingdon: Routledge.

Shelley, M. (1818/2016). *Frankenstein.* London: Collins Classics.

Sidanius, J., & Pratto, F. (1999). *Social Dominance: An Intergroup Theory of Social Hierarchy and Oppression.* Cambridge: Cambridge University Press.

Singer, P. (1975). *Animal Liberation: A New Ethics for Our Treatment of Animals.* New York: HarperCollins.

Springmann, M., Godfray, H. J., Rayner, M., & Scarborough, P. (2016). Analysis and valuation of the health and climate change co-benefits of dietary change. *Proceedings of The National Academy Of Sciences Of The United States, 113,* 4146–4151. doi:10.1073/pnas.1523119113

Totenberg, N. (2014). When did companies become people? Excavating the legal evolution. *NPR*. Retrieved from www.npr.org/2014/07/28/335288388/when-did-companies-become-people-excavating-the-legal-evolution

Vegans vs. Anthony Bourdain? (n.d.) *PETA*. Retrieved from www.peta.org/living/food/vegans-vs-anthony-bourdain/

Weiss, A. (2017). Personality traits: A view from the animal kingdom. *Journal of Personality, 86*, 12–22. doi: 10.1111/jopy.12310

6

KITTENS, PIGS, RATS, AND APES

The psychology of animal metaphors

Nick Haslam, Elise Holland, and Michelle Stratemeyer

Abstract

People employ a large menagerie of animal names to refer to human attributes and identities. Animals present a rich metaphorical domain that we can use to praise or to vilify, to express love or hatred, and to humanize and dehumanize. Although animal metaphors carry diverse meanings and serve varied ends, the more general concept of animality tends to have a negative connotation, representing immorality, stupidity, savagery, and primitiveness. This idea of animality as a devalued contrast to humanness increasingly appears to be a major dimension of social perception. Consciously or nonconsciously, blatantly or subtly, some ethnic, racial, and gender groups are often judged to be less human and more animal-like than others. This chapter explores the workings of animal metaphors as they appear in recent social psychological work, with a special focus on research into dehumanization. We argue that although the use of animal metaphors is not invariably dehumanizing, the belief that some humans are particularly animal-like is both dangerous and troublingly prevalent.

Half a century ago, the anthropologist Stanley Tambiah (1969) wrote that "Animals are good to think and good to prohibit." Referring to his fieldwork in a Thai village, he revealed a complex web of symbolic connections between animals and cultural rules governing diet, sex, and social organization. Villagers used canine expressions when talking about incest, refrained from eating ducks at marriage feasts because they are poor examples of maternal behavior for new wives, and likened heroes and bandits to elephants and tigers, respectively. "[A]nimals," Tambiah concludes, "are effective vehicles for embodying highly emotionally charged ideas" and are particularly "appropriate ... to objectify human sentiments and ideas" (p. 457).

Tambiah's claims about the symbolic affinity between animals and human affairs are hard to dispute. The vast diversity of nonhuman animals and their almost-humanness make them a fertile source of metaphors for human qualities and groups. Animal terms can be used to describe valued and devalued attributes, such as having aquiline vision or feline stealth, or serve as a zoological shorthand for personality traits, as with timid mice and conformist sheep. They can be the totems of our sporting teams, so that a fan can proudly calls herself a Dolphin or an Eagle. Animal nouns can function as insults, such as pig or cow, or as superficially flattering but possibly questionable expressions, as when a young woman is called a fox. In allegory, animals can stand in for specific humans, as Napoleon the pig did for Joseph Stalin in Orwell's *Animal Farm*. And expressions used to describe animal behavior can be extended figuratively to humans, as when a xenophobic politician refers to a swarm of refugees approaching their shores. As Tambiah argued, animals are indeed good to think (with), and also good to speak, praise, and insult.

In this chapter we will review the psychological literature on animal metaphors. This area of research is not very extensive if we confine ourselves to explicit metaphors, where people directly liken humans to nonhuman animals in language. However, the literature becomes substantially larger if we extend it to cover implicit metaphors, where humans are perceived in ways that are not overtly expressed through bestial metaphors but that nevertheless picture people as animal-like. A rapidly growing body of research on dehumanization shows how subtly animalistic perceptions of people are widespread and have an assortment of troubling implications. Our chapter begins with an examination of explicit animal metaphors before moving on to explore dehumanization.

Explicit animal metaphors

The psychological research literature on animal metaphors is somewhat limited in scope. However, it points to the diverse meanings carried by those metaphors, as well as to a few general trends in those meanings. A majority of the work has been conducted in English, although some studies in other languages have been reported. For instance, in Persian, owls are not wise (Talebinejad & Dastjerdi, 2005). The research conducted to date also suggests that the meaning of animal metaphors is not entirely intrinsic to particular animal comparisons, but depends in important ways on the context and manner in which a specific metaphor is expressed. In this section, we review several studies that explore the connotations of animal metaphors, their role in insults, and the factors – both intrinsic to their content and contained in the contexts in which they are used – that contribute to their offensiveness.

One of the most thorough examinations of English-language animal metaphors was conducted by Sommer and Sommer (2011), who referred to these metaphors as "zoomorphs." These researchers focused on the personality characteristics communicated by specific zoomorphs, and whether these expressions primarily display complimentary or uncomplimentary meanings. They also examined whether the metaphors are applied selectively to women or men, or to people of particular ages. In a series of studies, Sommer and Sommer asked participants to make judgments about the favorability, and gender and age associations, of sets of mammals (Studies 1 & 2) or fish, insects, and birds (Study 3). The studies generated several consistent findings. First, zoomorphs were predominantly seen as uncomplimentary, with only 14 of 36 of the mammalian metaphors being judged more favorable than unfavorable. The complimentary minority included creatures associated with virtues such as bravery (lion), intelligence (fox), and industriousness (horse). Second, many zoomorphs were strongly gendered, with a preponderance judged to be more applicable to men (e.g., lion, bear, gorilla) than to women (e.g., cat, deer, mouse). Interestingly, gender and favorability were linked: all of the ten most uncomplimentary zoomorphs were seen as predominantly male in application. Third, most zoomorphs were seen as relevant chiefly to adults, although a few diminutive or otherwise "cute" animals (e.g., mouse, rabbit, squirrel, monkey) were adjudged especially applicable to children or teens, and few larger animals (e.g., bear, camel, elephant, rhinoceros) were ascribed especially to older adults. Finally, Sommer and Sommer found that non-mammalian zoomorphs were rarer than mammalian ones, and that fish were especially scarce. By implication, animal species tend to be more metaphorically applicable to humans as they resemble us taxonomically.

Sommer and Sommer (2011) approached the study of animal metaphors in a relatively structured way, asking participants to reflect about the connotations of lists of zoomorphs, whether positive or negative. Another key study adopted a different approach, exploring the uncomplimentary animal expressions that people spontaneously produce in response to a provocation. In a cross-cultural study of insults, Van Oudenhoven and colleagues (2008) asked people from 11 countries to state what they would say if a man or woman rudely bumped into them without apologizing. The 12,000 expressions generated in response to this prompt were very diverse, including an assortment of evocative terms referring to body parts, sexual and excretory acts, blasphemies, and deficiencies of intellect, culture, hygiene, or breeding. However, animal metaphors also made a strong showing, including many expressions involving goats (for men), cows (for women), and the generic "animal" (both). Although Sommer and Sommer (2011) found that more zoomorphs apply to men than to women, Van Oudenhoven and colleagues showed that women were more likely to receive an animal insult than men in nine of the eleven cultures. That disproportion would have been greater if they had counted "bitch" (female dog) as an animal metaphor (Rodriguez, 2009), which they unaccountably failed to do. Without including it, animal terms represented the seventh most common class of insults, but with it they would have come second only to genital expressions.

Numerous feminist researchers have noted the special relevance of animal metaphors to perceptions of women. Some, like "cow," "pig," and "bitch," are straightforwardly derogatory and sexist (Dunayer, 1995; Roach, 2003), often conveying an impression of physical ugliness, dirtiness, or being overweight. Others, like "pet" or "kitten," are at least superficially benevolent but patronizing, and are used as endearments to refer primarily to younger women who are cute and perceived as tame, implying "pleasurable domestication" (Marley, 2007, pp. 61–62). "Kitten" implies a lack of aggression and active sexuality, whereas women who do possess these dispositions are more often designated as feral felines or foxes: "wildcats," "tigers," "cougars," and "vixens." Tipler and Ruscher (2014) discuss this flipping between prey and predator metaphors, and how predatory animal metaphors addressing women implicate not only sexuality but also competition with other women, as in the "catfight" where the "claws come out." In follow-up work (Tipler & Ruscher, 2019), they further show that people who read a news media story about voting behavior that was laden with predatory animal metaphors about women subsequently expressed more hostile sexist attitudes towards them, showing that these metaphors can have troublesome consequences.

Van Oudenhoven et al.'s (2008) work on insults showed that animal metaphors are potent and cross-culturally widespread ways of demeaning others who violate social norms, and perhaps especially used to denigrate women. Together with the work of Sommer and Sommer (2011) it demonstrates that animal metaphors often convey derogatory meanings and do so in ways that are gendered. However, neither study explores the potentially varied meanings of these metaphors or the factors that contribute to their offensiveness. It is possible, for example, that animal metaphors communicate negative evaluations in several qualitatively distinct ways, and that their degree of negativity depends on the specific features of those metaphors or the contexts in which they are employed, such as their target and the manner of their expression. These possibilities were examined in a pair of studies by Haslam, Loughnan, and Sun (2011), who attempted to map the variations of a large set of animal metaphors and explore the influences on their offensiveness.

In their first study, Haslam et al. (2011) presented participants with 40 common zoomorphs and asked them to rate each metaphor's offensiveness, and the personality and evaluative traits it implied about its target. Participants also rated several other abstract characteristics of the metaphor that might predict its offensiveness, such as whether the traits conveyed by each metaphor accurately represented those of the animal itself, how negatively the animal was evaluated, how genetically dissimilar the animal was from humans, and how much the user of each metaphor would perceive its target as less than human. The findings of the study revealed that animal metaphors carry a diversity of meanings that are not well captured by a single dimension of favorability. Specific metaphors captured both extremes of all but one of the five primary factors of human personality: high and low Agreeableness (puppy versus snake), Extraversion (fox versus mouse), Neuroticism (chicken versus owl), and Openness (monkey versus sheep). The remaining factor, Conscientiousness, represents self-control, diligence, and orderliness, indicating that none of the animal metaphors examined symbolized the exemplary presence or absence of this trait, perhaps because it is understood as relatively unique to humans (Hodson & Costello, 2007). In addition to representing extremes on the four personality factors, specific animal metaphors also stood for three of the four primary dimensions of negative evaluation (i.e., depravity, stupidity and worthlessness). Although specific animal metaphors therefore represented a wide assortment of meanings, both positive and negative, on average they were derogatory. The average animal metaphor pictured the person described by it as depraved, disagreeable, and unintelligent.

The study also clarified some of the aspects of animal metaphors that contribute to their greater or lesser offensiveness. More offensive metaphors tended, perhaps unsurprisingly, to involve negatively evaluated animals, comparisons between animals and humans that were seen as dehumanizing, and expressions that implied that the metaphor's target was literally rather than merely figuratively bestial. Intriguingly, analysis pointed to two distinct pathways to offensiveness for animal metaphors. Some of the most offensive animal metaphors (e.g., rats, pigs, leeches) likened people to taboo or disgusting animals, and appear to involve a symbolic transfer of reviled properties from the animal to the person. Other highly offensive metaphors (e.g., apes, dogs) do not involve reviled or disgusting animals, but make a comparison that seems intended to diminish the person's humanity. The former class of metaphors appears to be grounded in disgust whereas the latter is grounded in degradation. It is noteworthy that some of the best-known dehumanizing metaphors employed in the context of genocide and colonialism illustrate these two types. Nazi images of Jews as vermin, and Hutu images of Tutsis as cockroaches exemplify the disgust type, as do common metaphors of immigrants as disease-bearing rats or bacteria (O'Brien, 2003). In contrast, colonial era images of Africans and other "primitives" exemplify the degradation type. Research by Volpato, Dirante, Gabbiadini, Andrighetto, and Mari (2010) on dehumanizing imagery in an Italian Fascist magazine between 1938 and 1943 confirmed that Africans were metaphorically linked to apes, and Jews to spiders, vipers, parasites, and microbes.

Haslam et al.'s (2011) first study clarifies some of the complexities inherent in the meaning of animal metaphors, but their second study demonstrates how those meanings depend in part on the context of their use rather than inhering in the metaphors themselves. Participants read a scenario where one person refers to another using an animal metaphor and the speaker's tone (hostile versus jocular), the target's gender (male versus female), and group membership (ingroup versus outgroup) are varied. The same zoomorph was judged to be more offensive when the tone was described as hostile, the context inter-group, and the target female. Variations in the perceived offensiveness of the metaphor use across the 24 versions of the scenario were strongly associated with the extent to which the speaker was judged to be denying distinctively human attributes to the target. Thus, the offensiveness of animal metaphors appears to rest on how much their use, in its interpersonal context, dehumanizes the person to whom they refer. Animal metaphors are diverse in their connotations, but there are some common factors that contribute to those connotations becoming objectionable.

Implicit animal metaphors

The research introduced up to this point has investigated the overt expression of animal metaphors in language, where people are explicitly named as nonhuman animals. However, there are several ways in which animal metaphors operate at a covert or implicit level, organizing social perception without animal terms being expressed. Psychological studies have shown that the human–animal distinction is implicated in how people judge one another, that images of specific animals guide social perception in nonconscious ways, that people associate members of some groups more than others with animals, and that language or visual imagery that is subtly rather than overtly animalistic may color the impressions we form of others. Much of this work has been carried out in the past two decades as part of a renaissance in the study of dehumanization.

The new psychology of dehumanization

Prior to around the year 2000 psychological theory and research on dehumanization focused on its most blatant expressions. Dehumanization was understood to be something abnormal and overt, exemplified by explicit portrayals of groups as subhuman beasts in genocidal propaganda or by the most extreme forms of cruelty in wartime, and was linked primarily to the perpetration of violence. Dehumanizing others was theorized as a way to exclude them categorically from the human community and consequently remove the usual restraints on aggression and killing. Early studies of dehumanization (e.g., Kelman, 1976; Staub, 1989) therefore emphasized historical events in which groups of people were depicted using derogatory animal metaphors in the context of intense conflict.

Beginning in the new millennium, a more expansive understanding of dehumanization was developed by a team of Belgian psychologists. Jacque-Philippe Leyens and colleagues (2003) coined the term "infrahumanization" to describe subtle and everyday phenomenon in social perception by which people reserved uniquely human attributes for members of their ingroup, such as the fellow citizens of their country. Empirical support for this effect came from studies in which participants judged the emotional repertoires of ingroup and outgroup members. The work hinged on a distinction between primary and secondary emotions that is recognized in Romance languages (e.g., *emotions* versus *sentiments* in French) as well as in the psychology of emotions: primary emotions such as anger and fear are conceptualized as being shared with many nonhuman animals, whereas secondary emotions, such as nostalgia and guilt, which typically involve complex cognition and social learning, are conceptualized as unique to *Homo sapiens*. Infrahumanization researchers repeatedly found that people tended to ascribe primary emotions equally to members of groups they belong to and identify with (the "ingroup") and to those of other groups ("outgroups") but reserved secondary emotions for their ingroup. In effect, they saw their group as subtly more human than outgroups, where "human" is defined as the attributes that distinguish people from nonhuman animals. The infrahumanization phenomenon differs from the earlier understanding of dehumanization in that it is not inevitably accompanied by intense derogation of the outgroup, does not occur only under conditions of conflict, and takes place out of conscious awareness. Outgroup members are not explicitly recognized as subhuman by the use of demeaning animal epithets, but simply seen as lower on a human-animal continuum than one's own tribe.

Infrahumanization was subsequently shown to be a robust phenomenon that could be demonstrated in mutual perceptions across national and ethnic boundaries. Infrahumanization has a range of behavioral consequences, such as reduced willingness to help outgroup members (e.g., Vaes, Paladino, Castelli, Leyens, & Giovanazzi, 2003), and can be shown with an assortment of research methodologies. People tend to rate their ingroup as having more uniquely human traits as well as emotions, and automatically associate their ingroup with such characteristics as well. The infrahumanization effect shows that the human-animal distinction functions as a fundamental dimension on group perception, evident in everyday life. It also makes plain that the category "animal" serves as a sort of hidden metaphor against which humanness is defined contrastively, in ways that appear to have some degree of cross-cultural validity (Haslam, Kashima, Loughnan, Shi, & Suitner, 2008). It is arguably one of two such dominant metaphors, the other being the machine or inanimate object (Haslam, 2006; see Haslam & Loughnan, 2014, for a review).

Animal metaphors and ethnicity or race

Many demonstrations of infrahumanization have involved comparisons between national or ethnic groups. Findings have consistently shown that ethnic outgroup members are judged to have fewer uniquely human emotions or traits than ingroup members. This work supports the view that ethnocentrism is rooted not only in a preference for one's own kind, but also in the subtle, generally unexpressed belief that only one's kind is fully human, possessing the characteristics that distinguish humans from the rest of creation. However, although the infrahumanization of outgroups, based on a broad distinction between humans and other animals, appears to be a general phenomenon, there is also evidence that the more specific simian metaphor of humans as apes or monkeys underpins perceptions of certain groups.

Powerful work by Philip Atiba Goff and colleagues (e.g., Goff, Eberhardt, Williams, & Jackson, 2008) points to an enduring metaphor of people of African descent as ape-like, a metaphor that dates to the bogus "racial science" of the colonial era that portrayed sub-Saharan Africans as incompletely evolved. The persistence of this metaphor is by no means uniquely American, documented in explicit form in the recent past in Arab militia communications in Darfur (Hagan & Rymond-Richmond, 2008) and in occasional shameful displays by fans at European soccer matches when Black players appear, such as throwing bananas onto the field. However, Goff and colleagues argued that it might be present among White Americans even if they were unaware of holding a "Black-Ape" association.

In a series of studies involving White participants, Goff and colleagues found clear evidence for this association: viewing Black faces facilitated the identification of ape images, and activating the concept of "ape" shifted visual attention towards Black faces. The association was specifically simian rather than broadly animalistic, as there was no association between Blacks and big cats. Crucially, this association had consequences. Study participants who were first primed to think about apes, and then viewed a short video in which a racially ambiguous suspect is being beaten by police officers, believed the violent treatment was more justified when told the suspect was Black than when told he was White. When the Black–Ape association was made salient and participants believed the suspect to be Black, participants tended to see the suspect as wild and dangerous, and thus in need of forceful subduing.

More strikingly still, Goff et al. showed that bestial words consistent with the ape metaphor often accompanied news media communications about murders committed by African Americans in Philadelphia. The presence of such language in media reports was associated with African American suspects receiving death penalty verdicts, holding other aspects of the crime constant. Goff et al.'s (2008) work reveals how a particularly noxious animal metaphor with deep historical roots remains present and influential in contemporary American minds and in the workings of the US justice system. Versions of the same metaphor may also be present elsewhere, with one Italian study finding evidence of an unconscious association between apes and southern Italians held by northern Italian participants (Boccato, Capozza, Falvo, & Durante, 2008).

One reading of research on the Black–Ape metaphor, which accords with its possible extension to regional perceptions in southern Europe, is that it is an example of a broader class of metaphors in which some people are perceived as less evolved or more primitive than others, whether or not that primitivity is expressed with an animal metaphor. This reading is consistent with the work of Saminaden, Loughnan, and Haslam (2010), who used a computer-based task to examine nonconscious associations between people from traditional societies, such as indigenous people, and animals. Study participants viewed faces of a multi-ethnic sample of people from traditional or industrialized societies and words representing humans or animals, adults or children, and uniquely human or non-uniquely human personality characteristics. The extent to which they automatically associated one set of faces with particular sets of words was assessed. Consistent with the view that people from traditional societies are viewed as less evolved and developed, their faces were associated with animals, children, and non-uniquely human traits more than were the faces of people from industrialized societies.

Several aspects of this work are noteworthy. First, unlike the Black–Ape association documented by Goff and colleagues, the associations it found were not intrinsically negative. Traditional people were not viewed as more dangerously bestial but simply as less developed, as they are pictured in romanticized notions of the 'noble savage': uncivilized but also uncontaminated by modernity (Jahoda, 1999). Second, the work is not directly about race, as the two sets of faces were both equally multi-racial in composition. Finally, the work indicates that animal and child metaphors may be linked, such that groups seen as more animal-like may be seen as developmentally as well as evolutionarily backward.

This idea of animality as backwardness is captured well by a new program of research (e.g., Kteily, Bruneau, Waytz, & Cotterill, 2015) that assesses dehumanization using a variant of the familiar "Ascent of man [sic]" image, where a series of figures processes from left to right. The leftmost figure is a knuckle-dragging ape, the rightmost is a modern human, and in between are loping intermediates who include a club-wielding Neanderthal. Kteily and colleagues have asked participants to rate a variety of human groups on where they fall on this continuum, on a sliding scale that ranges from 0 (ape) to 100 (modern human). Their work using this vivid visualization of the animal-human distinction has yielded several confronting findings. First, people are remarkably willing to judge an assortment of groups as less than fully human, giving mean ratings often in the 50s to 70s for pariah groups. These ratings represent a blatant form of dehumanization, whether or not the raters sincerely believe that the group in question is literally subhuman. Second, ratings on the Ascent scale are strongly related to a wide variety of important judgments about social groups. For example, dehumanizing judgments of Muslims by American participants predicted lack of support for Arab immigration, indifference to anti-Arab discrimination, acceptance of anti-Arab media portrayals, and support for aggressive counter-terrorism tactics, such as the use of torture and pre-emptive bombings.

Kteily and colleagues have obtained similar findings in several countries. For example, ratings of Roma by non-Roma Hungarians predicting support for an assortment of hostile anti-Roma policies. Third, rating a group as more backward or simian on the Ascent scale is not equivalent to being prejudiced towards it in the traditional sense of negative evaluation or animosity. Although dehumanization and dislike are associated, dehumanization is uniquely linked to aggressive feelings and responses to the group concerned, independent of simple animosity. For example, in a study of Israelis and Palestinians during the 2014 Gaza war, Bruneau and Kteily (2017) found that one group's dehumanization of the other was associated with several forms of hostility independently of the participants' general hawkishness. This finding demonstrates the sobering additional point that dehumanization is often reciprocal, producing a vicious circle of escalating animalistic perceptions and brutal realities.

Animal metaphors and gender

Although a considerable amount of research has explored how perceptions of racial and ethnic groups are influenced by implicit animal metaphors, a smaller body of work has addressed perceptions of gender. Just as studies of explicit animal metaphors showed that these are commonly applied to women, and often in a demeaning way, studies of implicit metaphors indicate that these may also have troublesome implications. For example, studies of nonconscious associations between women and animals suggest that these may be associated with sexual objectification and violence. Vaes, Paladino, and Puvia (2011) showed participants photos of men and women who were either scantily clad or fully clothed, and using a similar procedure to Saminaden et al. (2010) assessed which set of photos were most associated with words representing animals (e.g., paw, snout, nature) or humans (e.g., culture, values, tradition). Their findings, obtained both with male and female participants, indicated that sexualized women were more strongly associated with the animal words than were non-sexualized women or either group of men. Recently the same conclusion has been drawn by Morris, Goldenberg, and Boyd (2018), who found that when women were objectified through a focus on their sexuality they were perceived to lack uniquely human characteristics.

The tendency to associate sexualized women with animals is conceivably linked to a propensity to exploit or harm them among some men, given that animals are often believed to merit less moral concern than humans. That proposition was supported in studies conducted by Rudman and Mescher (2012), who used a similar method to Vaes and colleagues to assess the degree to which young male participants implicitly associated women with animals. They found that men who held this association to a greater degree showed a greater willingness to sexually harass women and a greater likelihood of perpetrating sexual assault. Feminist theorists would be unsurprised by findings such as these. The psychological linkage of femaleness and animality could be viewed as a case of the more fundamental alignment of women with nature (and men with culture).

This relationship was recognized by Sherry Ortner (1972) as a function of women's close, embodied connection with the processes of reproduction, and by Simone de Beauvoir (1974) as the root of women's subordination. It is also arguably encouraged by the advertising and pornography industries. A study of magazine advertising by Plous and Neptune (1997), relatively old but by no means completely obsolete, shows that women, and Black women in particular, are sometimes pictured as animals, occasionally caged or shackled to signify their need to be tamed or placed in animal-like postures. Bongiorno, Bain, and Haslam (2013) explored how this linking of sexualized woman to animals operates in advertising for animal rights organizations such as PETA, finding that it typically backfired, reducing rather than building support for the ethical cause. Thus animal metaphors for women, both hidden from conscious awareness in our automatic associations and also visually present in our consumer culture, appear to have an ongoing relevance to perceptions and treatment of women.

Animal metaphors in other domains

The bulk of psychological research on the implicit animal metaphors has addressed ethnicity, race, and gender, but a smaller quantity of work has explored other domains. Some domains in which these metaphors might operate were suggested by the work of Gustav Jahoda (1999), whose work on "images of savages," although focused on animalistic perceptions of indigenous people and others living in traditional societies, also speculated on other contemporary groups that might be perceived in similar ways.

> During the second half of the 19th century ... the 'otherness' of the savage came to be extended to a whole series of European 'Others', as viewed from the pedestal of the educated male middle class ... Savages, in this context,

form part of a cluster that includes not only children but also the rural and urban poor, criminals, the mentally ill, and even women.

p. 237

Jahoda's speculation about the lasting mental association between women and animals has been amply supported by research on explicit and implicit animal metaphors. More recently, Loughnan, Haslam, Sutton, and Spencer (2014) addressed his suspicion that the poor, or at least people of low socioeconomic status, might be viewed through the lens of animality. Loughnan and colleagues explored stereotypes of lower-class people in the USA ("white trash"), the United Kingdom ("chavs"), and Australia ("bogans"), as judged by citizens of the respective countries. They found that the attributes that composed each class stereotype overlapped substantially with those that composed the ape stereotype. The traits seen as differentiating lower from higher class people were very similar to those seen as differentiating apes from humans. By implication, social class stereotypes may be underpinned by an unacknowledged animal metaphor of the same kind that supports demeaning perceptions of subordinate racial and ethnic groups, as Jahoda (1999) suggested.

Another domain in which researchers have examined relatively subtle expressions of animal metaphor is representations of terrorists. Although there is an ethnic or racial underlay to these representations, at least some of their elements relate to the violent actions of terrorists and attempts to pursue them. Research by Steuter and Wills (2008) demonstrates that media reports on this pursuit frequently rely on hunter metaphors that picture the pursued as animal quarry. This metaphor is expressed indirectly through language that does not explicitly label terrorists as animals. The hunters engage in "snaring," "caging," "trapping," "netting," "smoking out," "bagging," "flushing out," and "picking up the scent," while the hunted "slither," "scurry," and secrete themselves in "nests" or "lairs." This is not to say that explicit animal metaphors are absent. As Spencer (2012) observed, newspaper articles often directly refer to terrorist fighters as subhuman or inhuman, and Steuter and Wills (2009) report comparisons between terrorists and assorted vermin, including rats, snakes, spiders, and cockroaches.

This section shows that animal metaphors operate in domains outside the classic fields of ethnicity, race, and gender. A small quantity of work on perceptions of social class and terrorism has been revealing, but much more remains to be done by creative researchers. The extent to which the economically disadvantaged and offender groups are implicitly judged to be uncivilized and feral has barely been examined and might have implications for people's support for income redistribution, other forms of social spending, and punitive rather than rehabilitative responses to crime. Animal metaphors are also clearly implicated in perceptions of children, and although these metaphors are often humorous or loving diminutives, they might also have an undesirable aspect that warrants investigation. Regardless of whether and how such research is conducted, the key point to remember is that animal metaphors may be influencing and organizing how we think, feel, and behave towards others even if animal labels are never overtly attached to them.

Recognizing this point is the key to overcoming or at least reducing the negative consequences of animal metaphors. Drawing public attention to the problematic and often dehumanizing nature of animal comparisons, and promoting social norms against their use, as in the general taboo against the use of the word "bitch" in many Anglophone contexts, is probably the best way to minimize their adverse effects.

Implications and conclusions

In this chapter we have attempted to show that animal metaphors color how we perceive and behave towards other people in myriad ways. These metaphors are diverse in their connotations and highly compelling for that reason, as Tambiah (1969) recognized when he observed how animals are "good to think." At the same time, there are some common threads in the use of animal metaphors that make them frequently troublesome in their social implications. Here we lay out some of the key messages that the psychology of animal metaphors teaches us.

First, explicit, linguistically expressed animal metaphors can refer to almost any human attribute or group, whether positive or negative. Nonhuman animals are believed to possess, or at least to represent, an assortment of human virtues, so to apply an animal label to a person is not invariably degrading, demeaning, or dehumanizing. Nevertheless, the typical or average content of explicit animal metaphors, and the content of the overarching "animal" metaphor itself, tends to be negative, often imputing the vices of stupidity and immorality in particular or, especially in the context of gender, ugliness, childish innocence, or sexual disapproval. For this reason, animal metaphors are popular insults in many languages, and they are offensive to the degree that they are seen as debasing their targets and stripping them of their humanity. As much as humans have deep, often affectionate ties to nonhuman animals, we should therefore be vigilant for adverse implications when the two are linked by metaphor. These links might appear to bridge the boundary between human and nonhuman, but they are just as likely to reinforce it by placing some humans on the other side.

Second, although explicit animal metaphors feature in everyday affectionate names for loved ones, in insults for disliked ones, and in genocidal propaganda during extreme conflicts, implicit metaphors may be even more psychological important. There is now ample evidence that the human/animal distinction plays a role in the perception of social groups, and that in the absence of conflict many people subtly view members of other groups as closer to animals than they view themselves. Similarly, it is increasingly clear that people unconsciously or automatically equate some groups with animals, and that simian metaphors, in particular, distort perceptions of human groups. These implicit animal metaphors are insidious and difficult to challenge or uproot because they operate outside of awareness. Despite working undercover in this way, research demonstrates that such metaphors are associated with a range of deeply problematic behavior, from sexual harassment to justifying police violence to punitive sentencing of criminal suspects to class prejudice. Implicit animal metaphors show how the fallacious belief that some humans are more evolved and cultured than others – a fallacy long overturned by biology and anthropology – persists to this day, with dire consequences. The psychology of animal metaphors helps us to understand how and why this dangerous idea endures.

References

Boccato, G., Capozza, D., Falvo, R., & Durante, F. (2008). The missing link: Ingroup, outgroup and the human species. *Social Cognition, 26*, 224–234.

Bongiorno, R., Bain, P., & Haslam, N. (2013). When sex doesn't sell: Using sexualized images of women reduces support for ethical campaigns. *PLoS ONE, 8*, e83311. doi: 10.1371/journal.pone.0083311

Bruneau, E., & Kteily, N. (2017). The enemy as animal: Symmetric dehumanization during asymmetric warfare. *PLoS ONE, 12*(7), e0181422.

de Beauvoir, S. (1974). *The Second Sex*. London: Penguin.

Dunayer, J. (1995). Sexist words, speciesist roots. In C. Adams & J. Donovan (Eds.), *Animals and Women: Feminist Theoretical Explorations* (pp. 11–32). Durham, NC: Duke University Press.

Goff, P. A., Eberhardt, J. L., Williams, M. J., & Jackson, M. C. (2008). Not yet human: Implicit knowledge, historical dehumanization, and contemporary consequences. *Journal of Personality and Social Psychology, 94*, 292–306.

Hagan, J., & Rymond-Richmond, W. (2008). The collective dynamics of racial dehumanization and genocidal victimization in Darfur. *American Sociological Review, 73*, 875–902.

Haslam, N. (2006). Dehumanization: An integrative review. *Personality and Social Psychology Review, 10*, 252–264.

Haslam, N., Kashima, Y., Loughnan, S., Shi, J., & Suitner, C. (2008). Subhuman, inhuman, and superhuman: Contrasting humans and nonhumans in three cultures. *Social Cognition, 26*, 248–258.

Haslam, N., & Loughnan, S. (2014). Dehumanization and infrahumanization. *Annual Review of Psychology, 65*, 399–423.

Haslam, N., Loughnan, S., & Sun, P. (2011). Beastly: What makes animal metaphors offensive? *Journal of Language and Social Psychology, 30*, 311–325.

Hodson, G., & Costello, K. (2007). Interpersonal disgust, ideological orientations, and dehumanization as predictors of intergroup attitudes. *Psychological Science, 18*, 691–698.

Jahoda, G. (1999). *Images of Savages: Ancient Roots of Modern Prejudice in Western Culture*. London: Routledge.

Kelman, H. (1976). Violence without restraint: reflections on the dehumanization of victims and victimizers. In G. Kren & L. Rappoport (Eds.), *Varieties of Psychohistory* (pp. 282–314). New York: Springer.

Kteily, N., Bruneau, E., Waytz, A., & Cotterill, S. (2015). The ascent of man: Theoretical and empirical evidence for blatant dehumanization. *Journal of Personality and Social Psychology, 109*, 901–931.

Leyens, J.-Ph., Cortes, B., Demoulin, S., Dovidio, J., Fiske, S., Gaunt, R., et al. (2003). Emotional prejudice, essentialism, and nationalism: The 2002 Tajfel lecture. *European Journal of Social Psychology, 33*, 703–717.

Loughnan, S., Haslam, N., Sutton, R., & Spencer, B. (2014). Dehumanization and social class: Animality in stereotypes of 'white trash', 'chavs', and 'bogans'. *Social Psychology, 45*, 54–61.

Marley, C. (2007). Metaphors of identity in dating ads and newspaper articles. *Text and Talk, 27*, 55–78.

Morris, K. L., Goldenberg, J., & Boyd, P. (2018). Women as animals, women as objects: Evidence for two forms of objectification. *Personality and Social Psychology Bulletin, 44,* 1302–1314.

O'Brien, G.V. (2003). Indigestible food, conquering hordes, and waste materials: Metaphors of immigrants and the early immigration restriction debate in the United States. *Metaphor and Symbol, 18,* 33–47.

Ortner, S. (1972). Is female to male as nature is to culture? *Feminist Studies, 1,* 5–31.

Plous, S., & Neptune, D. (1997). Racial and gender biases in magazine advertising: A content-analytic study. *Psychology of Women Quarterly, 21,* 627–644.

Roach, M. (2003). *Mother/Nature.* Bloomington, IN: Indiana University Press.

Rodriguez, I. L. (2009). Of women bitches, chickens and vixens: Animal metaphors for women in English and Spanish. *Culture, Language and Representation, 7,* 77–100.

Rudman, L. A., & Mescher, K. (2012). Of animals and objects: Men's implicit dehumanization of women and likelihood of sexual aggression. *Personality and Social Psychology Bulletin, 38,* 734–746.

Saminaden, A., Loughnan, S., & Haslam, N. (2010). Afterimages of savages: Implicit associations between 'primitives', animals and children. *British Journal of Social Psychology, 49,* 91–105.

Sommer, R., & Sommer, B. A. (2011). Zoomorphy: Animal metaphors for human personality. *Anthrozoos, 24,* 237–248.

Spencer, A. (2012). The social construction of terrorism: Media, metaphors and policy implications. *Journal of International Relations and Development, 15,* 393–419.

Staub E. (1989). *The Roots of Evil: The Origins of Genocide and Other Group Violence.* New York: Cambridge University Press.

Steuter, E., & Wills, D. (2008). *At War with Metaphor: Media, Propaganda, and Racism in the War on Terror.* Plymouth: Lexington Books.

Steuter, E., & Wills, D. (2009). Discourses of dehumanization: Enemy construction and Canadian media complicity in the framing of the war on terror. *Global Media Journal: Canadian Edition, 2*(2), 7–24.

Talebinejad, M. R., & Dastjerdi, H.V. (2005). A cross-cultural study of animal metaphors: When owls are not wise! *Metaphor and Symbol, 20,* 133–150.

Tambiah, S. J. (1969). Animals are good to think and good to prohibit. *Ethnology, 8,* 423–459.

Tipler, C., & Ruscher, J. B. (2019). Dehumanizing representations of women: The shaping of hostile sexist attitudes through animalistic metaphors. *Journal of Gender Studies, 28,* 109–118.

Tipler, C., & Ruscher, J. B. (2014). Agency's role in dehumanization: Non-human metaphors of outgroups. *Social and Personality Psychology Compass, 8,* 214–228.

Vaes, J., Paladino, M. P., Castelli, L., Leyens, J.-Ph., & Giovanazzi, A. (2003). On the behavioral consequences of infrahumanization: The implicit role of uniquely human emotions. *Journal of Personality and Social Psychology, 85,* 1016–1034.

Vaes, J., Paladino, M. P., & Puvia, E. (2011). Are sexualized females complete human beings? Why males and females dehumanize sexually objectified women. *European Journal of Social Psychology, 41*(6), 774–785.

Van Oudenhoven, J. P., de Raad, B., Askevis-Leherpeux, F., Boski, P., Brunborg, G. S., Carmona,...Woods, S. (2008). Terms of abuse as expression and reinforcement of cultures. *International Journal of Intercultural Relations, 32,* 174–185.

Volpato, C., Durante, F., Gabbiadini, L., Andrighetto, L., & Mari, S. (2010). Picturing the other: Targets of delegitimization across time. *International Journal of Conflict and Violence, 4,* 269–287.

7

UNCANNY VALLEY OF THE APES

Vanessa Woods and Brian Hare

Abstract

Humans have a complex emotional relationship with the other members of our great apes family. Great apes are appealing because of the close resemblances we share, but these resemblances can cause feelings of aversion and disgust. We propose that these feelings may stem from the *uncanny valley* – the point at which someone or something becomes almost, but not quite human. We argue that the uncanny valley has made great apes an effective tool for a particular type of dehumanization, known as simianization. Simianization has been especially prolonged and pervasive towards black people, dating from the transatlantic slave trade and continuing today. In the reverse direction, the humanness of great apes has made them a source of fascination, but has not saved them from extensive population decline and may have facilitated their exploitation. Preliminary research into decreasing the human-animal divide shows promise for mitigating the negative impact of simianization of people and for improving attitudes towards, and the welfare of, great apes.

The first bonobo that Claudine André saved had an unusual story. He arrived in 1998, at the war-ravaged Kinshasa Zoo in the Democratic Republic of Congo, where Claudine and her friend, Denise, volunteered, trying to keep the animals from starvation. The bonobo was more dead than alive, but Denise was an Emergency Room (ER) nurse and Claudine asked her to help. Denise took the little bonobo home to live with her and her two-and-a-half-year-old grandson. Denise's daughter was murdered by rebels in Angola, and her grandson was found alive, beside his mother's body. Distraught, Denise brought the boy to live with her in Kinshasa.

Against all expectations, Denise nursed the baby bonobo back to health. The bonobo and Denise's grandson were about the same age. The bonobo had also seen his mother killed, and was a long way from home. Soon the bonobo and the boy were inseparable. They played together in Denise's garden, providing each other with companionship and comfort. But not long after, the boy began to cry like a bonobo and spent all day in the trees. The bonobo dressed in clothes, and followed Denise around like a toddler. Worried for the mental health of the boy, Claudine begged Denise to allow the bonobo to join the other orphan bonobos, who were slowly arriving from the interior of the Congo Basin. Denise reluctantly agreed.

The bonobo settled in with his new bonobo family, living and playing in a forest that was as close to his original home as possible. Eventually, Claudine went on to found the world's only bonobo sanctuary, a 100-acre forest called Lola ya Bonobo, where 60 orphan bonobos lived in large social groups, and some of whom were eventually released back into the wild. Denise took her grandson back to Belgium, where she hoped that time and distance would heal their wounds (André, 2006). The two playmates each returned to live among their own species, but the time they spent together highlights the fluidity between humans and great apes.

This fluidity exists in many cultures, especially in the range countries of great apes. The Malay word for orangutan is "man of the forest." Ugandans do not eat great apes, because they believe it is too similar to cannibalism. The Bantu of Congo tell of bonobos coming to live in villages and taking human "wives." This fluidity is what makes our relationship to great apes unique. Anyone who has spent time with great apes has been struck by the way they are, as Robert Yerkes wrote, "incomparably similar" to us in every detail (Yerkes, 1925); from their facial expressions to their fingernails to their cognition.

And yet, as much as great apes are compared to people, or anthropomorphized, the reverse direction, comparing people to primates, particularly great apes, or *simianization* (from the Latin *simia* "ape"), has caused significant controversy in recent years. In 2018, actress Roseanne Barr's show was cancelled after she tweeted if "muslim brotherhood & planet of the apes had a baby=vj," referring to Valerie Jarrett, the former aide to Obama (Koblin, 2018). In 2008, Le Bron James was photographed for Vogue, screaming and clutching the slender waist of white supermodel Gisele Budchen, which many took as a reference to the fictional gorilla, King Kong.

Researchers have commented on the long history of simianization of black people as a form of dehumanization (Hund, Mills, & Sebastiani, 2015). But few people have addressed why simianization is an effective tool for dehumanization in the first place. Usually, the animals used to dehumanize people have a reputation for being unclean, like rats or pigs. In contrast, great apes[1] are celebrated, not just in their range states, but in popular Western culture. Starting in the 1960s and through the 1980s, the general public was enraptured by "language"-trained apes who could use dozens of symbols, and Kanzi the bonobo could understand novel sentences in English that communicated an imperative (Savage Rumbaugh et al., 1993). Many of these apes were raised in human homes, like children, dressing in clothes, eating at the table with a knife and fork. The media enthusiastically captured the child-like quality of these great apes "talking," hugging, and playing with their "parents."

At the same time, three women were conducting ground-breaking research on great apes in the wild: Jane Goodall on chimpanzees, Diane Fossey on gorillas, and

Birute Galdikas on orangutans. They made the first observations of tool making, complex friendships, political fights, and cooperative hunting. National Geographic made not only these women famous, but also the great apes they studied, emphasizing their human-like intelligence and social lives as complicated as our own. Great apes blurred the line between humans and animals, in a manner that was sympathetic to the growing appreciation of nature during the birth of the environmental movement in the 1960s and 1970s.

Even in the last few decades, the perceived gap between humans and great apes has been closing. Genetic evidence placed chimpanzees and bonobos as more closely related to us than to their other great ape relatives, gorillas and orangutans (Hobolth, Christensen, Mailund, & Schierup, 2007). There is evidence for dozens of cognitive abilities in other great apes that were once thought to be unique to humans, for example forms of mental time travel, theory of mind, cooperation, negotiation, and prosocial sharing (Carpenter, Nagell, & Tomasello, 1998; Hare, 2011; Hare, Call, Agnetta, & Tomasello, 2000; Hare & Kwetuenda, 2010; Kano & Call, 2014; Melis, Hare, & Tomasello, 2006, 2009; Tomasello & Call, 1997).

Perhaps the bewildered reaction of Rosanne Barr, Vogue, and everyone else accused of simianization in the twenty-first century is because they were so familiar with the humanized version of great apes, that it never occurred to them that great apes would be used as a tool to dehumanize (see also the Interspecies Model of Prejudice, Hodson, MacInnis, & Costello, 2014).

On the contrary, our hypothesis is that great apes are, and always will be, a perfect tool for dehumanization because they fall into the uncanny valley – the point at which something or someone becomes almost, but not quite human. In 1970, roboticist Masahiro Mori hypothesized that robots would become more appealing as they became more human like, but there is a point at which robots appear almost too similar to humans, abruptly eliciting revulsion and disgust (Mori, 1970). Mori called this range the "uncanny valley." We argue that like robots, the uncanny valley can equally be applied to great apes, and that as a result, simianization is a particularly powerful form of dehumanization. Because unlike rats and dogs, who are reviled in different cultures for being unclean, great apes are disliked because their almost perfect reflection of us falls right into the uncanny valley.

There is a general awareness that simianization of black people dates back to the slave trade. There is also a general awareness that simianization dehumanizes because it classes someone as subhuman, or bestial. But few people distinguish between simianizing someone as a great ape versus a monkey, and many people use great ape and monkey interchangeably. Also, no one to our knowledge, has made a connection between great apes and the uncanny valley. To fully understand the significance of simianization as a monkey versus a great ape, it is necessary to understand the history of the relationship of slavery and the discovery of great apes. And to understand why simianization of this kind relates to real world harm, it is necessary to understand the difference between a monkey and a great ape.

Trouble at the border

Monkeys include both new and old world primates, of which they are at least 178 species. Monkeys have tails, usually live in trees, and although they can be large, like mandrills, are usually less than 50 pounds and several feet in height. Great apes include humans, chimpanzees, bonobos, gorillas, and orangutans. They can reach 5-6ft in height and, in the case of a gorilla, weigh up to 400 pounds. For good measure, some great apes (humans and chimpanzees) occasionally hunt and eat monkeys.

Europeans have been familiar with monkeys since the eleventh century, kept at royal courts as status symbols to amuse and entertain the nobility. The earliest written record is of Count William of Italy who kept a monkey called "maimo" (Buquet, 2013) in his castle. In the fifteenth century, Queen Isabeau of France also kept a monkey (Buquet, 2013). Even clerics kept monkeys "in their windows so as to impress the passing rabble with the glory of their possessions" (Janson, 1952). According to representations in medieval artwork, these monkeys seem to be mostly Barbary macaques (*Macaca sylvanus*), black and white colobus (*Colobus guereza*) and baboons (*Papio sp*), brought to Western Europe from Africa via the trade routes in the Near East (Vadillo, 2011). Many monkeys were pampered by their owners, dressed in richly decorated collars and leashes, although they were also tethered to heavy bowls or blocks to prevent them destroying valuable possessions, like the Duke of Burgundy's monkey who tore up some important documents in 1288 (Buquet, 2011). Although monkeys were occasionally associated with the devil, they were more often seen as a parody of humans, full of harmless mischief and their sexuality amusingly on display, to be petted and indulged by their elite owners, who knew better (Vadillo, 2011).

The great apes who arrived in Europe were different. From the first sightings, great apes were known as "wild men" or "half humans." As early as the sixteenth century, there were rumors from Portuguese sailors of "monsters... in all proportion like a man; but that he is more like a giant in stature than a man; for he is very tall, and hath a man's face, hollow-eyed, with long haire upon his browes" (Huxley, 1863). These first accounts describe great apes with a mix of fascination and horror; "they goe many together and kill many negroes... [they] are never taken alive because they are so strong, that ten men cannot hold one of them". One of the first written records of an encounter with a gorilla recalls "some hellish dream creature – being of that hideous order, half-man half-beast" (Du Chaillu, 1868). Other mythical monsters of the African continent, like fire-breathing dragons or beasts that fed on human corpses, were eventually dismissed as folklore. But with the expansion of European colonies into Africa and Southeast Asia, the existence of great apes became startlingly real when their bodies began to arrive in Europe.

The first record of an adult great ape skeleton to arrive in Europe was an orangutan in 1781. Europeans had already seen live great apes on display in freak shows and menageries, but these were all juveniles, and they all died before they could reach adulthood. Compared to these young apes, this adult orangutan was "a monster almost twice their size, of vast strength and fierceness, and very brutal in expression" (Huxley, 1863). One of the first adult chimpanzee skeletons was described in 1835 (Huxley, 1863, p. 20), and the gorilla from a skull in 1847. Bonobos were not recognized by Western science as a species until 1933.

There were several centuries between when great apes were first seen by Europeans explorers in Africa and when great apes arrived in Europe to be gawked at, dissected, and puzzled over, which made their taxonomy a mess. To the untrained eye, juveniles of different great apes species can look more similar to each other than they do to the adults. As a result, all great apes were called orangutans. For a time, great apes were also confused with human pygmies, who were not officially described by Europeans until after gorillas, in 1870 (Du Chaillu, 1870).

While monkeys were mocked as ridiculous parodies of humans, much like the puppets in the flourishing puppet shows in Victorian England, as the uncanny valley would predict, the similarity of great apes to humans, in both size and behavior, made people uncomfortable. When Queen Victoria was introduced to a young orangutan at the London zoo, she described the orangutan as "painfully and disagreeably human" (van Wyhe & Kjærgaard, 2015). When Carolus Linneaus attempted to put great apes in the same class as humans, other scientists revolted, and were compelled "to defend the rights of mankind and to contest the ridiculous association with the true ape" (van Wyhe & Kjærgaard, 2015). This debate was reignited continuously through the nineteenth century, including at the publication of Darwin's *Descent of Man* (van Wyhe & Kjærgaard, 2015). With the strict social hierarchy of the day, "a discrete ladder of God's creations with the orangutan supplying the step before humans" (van Wyhe & Kjærgaard, 2015) was too difficult for many to embrace. But the similarity of great apes to humans was undeniable, and the classification of great apes with humans was inevitable. Darwin wrote that humans are "but one of several exceptional forms of Primates," while Theodor Bischoff wrote "that the apes, and especially the orang, chimpanzee and gorilla, come very close to man in their organization, much nearer than to any other animal, is a well-known fact, disputed by nobody" (Darwin, 1880). In an attempt to overcome the uncanny valley, early anthropologists broadened the distance between humans and great apes by inserting another rung into the metaphorical ladder.

The next rung down

In the same two centuries that rumors of great apes, then great apes themselves, were shipped from Africa to Europe, millions of slaves were shipped from Africa to the American colonies. If great apes were the intermediary between humans and the animal kingdom, Europeans decided that black people could be the intermediary between themselves and great apes. Respected members of British society emphasized a similarity between black people and great apes that bordered on interchangeability. British Historian Edward Long wrote "[t]hat the oran-outang and some races of black men are very nearly allied" (Long, 1774). Anthropologist James Hunt wrote in 1864, "the analogies are far more numerous between the ape and the Negro than between the ape and the European" (Hunt, 1863). At the core of this interchangeability was sexuality. It was taken as common knowledge that great apes had sexual relations with black women. For instance, naturalist George Buffon remarked on "the compulsive or voluntary intermixture of the negresses with the apes, the produce of which has united into both species" (comte de Buffon, 1797) and Thomas Jefferson wrote of "the preference of the Oran-utan for the black women over those of his own species" (Jefferson, 1781).

Inserting an extra rung into the discrete ladder had the advantage of solving another dilemma, which was how to reconcile the horrors of the slave trade with the morality of upper-class elites in the Age of Enlightenment. In England, by the eighteenth century, the abolitionist movement was already well underway. In America, the irony was not lost on the founding fathers that the freedom they demanded from the British did not extend to the slaves who worked their plantations. British and American advocates of lucrative slave labor may have pointed to the similarities they saw between black people and great apes to justify moral exclusion from the life, liberty and happiness that were the inherent rights of all *humans.*

Simianization did not end with the slave trade, nor was it only solely directed at Africans. Darwin was often caricatured as an ape, inviting another ape to contemplate its reflection in a hand mirror, and escorting an ape down a wedding aisle (Browne, 2001). The Irish were simianized in both Britain and America in the nineteenth century (Kenny, 2006), while the Japanese were simianized during the Second World War. Eurasians frequently simianized one another in the twentieth century. In the build-up to major conflicts Germans, Chinese, Prussians, and Jews were all simianized.

But while the simianization of Irish, Japanese, and other groups eventually faded, simianization of black people continued. And while black people were frequently simianized as monkeys, simianizing as great apes lent a more sinister, threatening quality. In fact, after the abolition of slavery in America, simianization became even worse. Black people (especially black men) were portrayed as lust-crazed apes, driven by insatiable sexual appetites, who would rape white women at every opportunity (Livingstone Smith & Panaitui, 2015). Simianization was a propaganda tool used to whip up local fury, which fueled the lynching epidemic on either side of the twentieth century. For example, in 1893, a mentally handicapped ex-slave called Henry Smith was accused of raping and dismembering the daughter of a policeman in Texas. Smith was innocent, but fabricated reports were spread by the local clergyman of how the girl was "torn asunder in the mad wantonness of gorilla ferocity" (Livingstone Smith & Panaitui, 2015). In the 1905 Ku Klux Klan novel *The Clansman*, Gus, a former slave who rapes a white girl, is described as "apelike" (Dixon, 2015). Before the Second World War, Hitler described black people as "half-apes" (Staples, 2009).

One of the most effective and enduring popular manifestations of simianization was the 1933 movie *King Kong*. In retrospect, the racist undertones are obvious: a white woman goes to a jungle island, where she is met by "black savages" under the

FIGURE 7.1   Image of *King Kong* movie poster from 1933, RKO Radio Pictures, and a linocut of the Scottsboro trials (Shi Khan, L. & Perez, T. 1935).

thrall of an enormous, threatening gorilla who takes an unnatural sexual interest in the white woman. The white woman brings the black gorilla back to white civilization, which he is unable to appreciate. White men destroy the black gorilla before he destroys white civilization, and the white woman falls helplessly into the arms of the main white man and "natural order" is restored (Affeldt, 2015)

Even if this racial narrative was not the intention of the film makers, it was a reflection of public sentiment towards black people at the height of the Jim Crow era. In 1933, the case of the Scottsboro boys was national news. Nine black youths were arrested for raping two white women on a train in Alabama. The accusation was false, but on barely any evidence, eight of the nine boys were sentenced to death by electric chair. The youngest, who was 12, had a hung jury between the death sentence and life imprisonment. In a linocut from the period, one of the boys clutches the limp body of a naked white woman in a clear reference to King Kong (see Figure 7.1).

During the civil rights movement, cartoons of apelike black men making advances on white women were widespread. "No Negro or Ape Allowed Inside Building" read one sign outside a country store in Calhoun, South Carolina in 1959 (Williams, 1995).

A new dawn

The second half of the twentieth century saw much-awaited progress in equality and civil rights. Baseball, the military, and schools were desegregated. Nonviolent protests in the form of sit-ins, bus boycotts, and freedom rides gradually over-came segregation in most of the USA. In 1964, President Lyndon Johnson signed the Civil Rights Act, which was followed by voting rights, the Fair Housing Act, and affirmative action. While each of these milestones was fraught with conflict, setbacks, and in some cases, violence, the vast majority agreed that the general dir-ection was forward.

By 2000, many social scientists declared racism dead in America. Or at least the kind of racism that lead to lynchings, segregation, and internment camps. The rationale was that after the Second World War, America's new role as a superpower and the moral compass of the world led to an "absence of intentional discrim-ination" (Roithmayr, 2014), a "decline in overt and covert racism" in the USA (Fleegler, 2006; Fredrickson, 2015). Some argued that racism was defeated so easily because it is relatively new "and did not infect Europe itself prior to the period between the late medieval and early modern periods" (Fredrickson, 2015). The civil rights movement "smashed the legal apparatus of segregation and political exclusion." "Old racism," or negative feelings towards black people and the belief that black people are inferior to white people "has declined substantially over time" and instead is replaced with new racism, "a simmering resentment over preferential treatment and programs for black people" (Huddy & Feldman, 2009). New racism is "subtle" (Ponds, 2013), "diffuse" (Clair & Denis, 2015), "covert" (Hogan & Mallott, 2005), and "path dependent" (Roithmayr, 2014). This new racism can also include negative stereotypes about black people, and feelings of insecurity that black people pose a threat to white people's position in the racial hierarchy (Bobo & Hutchings, 1996; Sniderman & Carmines, 1997). With the election of Barack Obama, some political scientists argued that racism no longer factors at all in political decisions towards black people (Huddy & Feldman, 2009). "Whites refusing to vote for black candidates has finally gone the way of segregated water fountains" (Thernstrom, 2008). Representative and civil rights icon John Lewis said in 2002: "It's a different world we live in. ... We have changed. We've come a great distance. ... It's not just in Georgia but in the American South. I think people are preparing to lay down the burden of race" (Thernstrom, 2008).

If this is true, then perhaps simianization is no longer relevant, and we should turn our attention to more implicit forms of bias. The widespread celebration of the great apes as our intelligent cousins who show signs of empathy, love and grief, may mean that simianization is no longer a form of dehumanization, but even a sort of compliment. On the contrary, we hypothesize that due to the uncanny valley, simianization will always be a particularly pernicious form of dehumanization that frequently results in real-world harm.

The uncanny valley

In 1970, the Japanese roboticist, Masahiro Mori, published an essay in an obscure journal on how people's reaction to robots would suddenly shift from empathy to revulsion as the robots looked and behaved almost, but not quite, human. As an

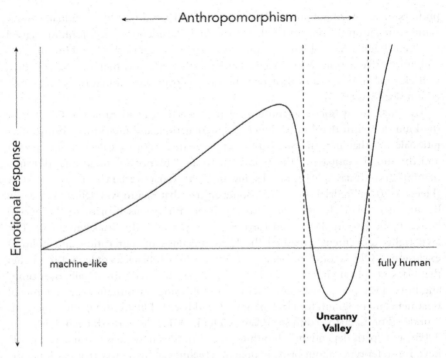

FIGURE 7.2 Mori's uncanny valley. Where the Y axis is the valence of emotional response, from negative to positive, and the X axis is the perception of someone as human, from machine like to fully human (based on Mori, 1970).

example, Mori invoked a crowd of mannequins that suddenly began to move in "the dead of night," suddenly becoming "a horror story" (Mori, 1970).

In Mori's view (see Figure 7.2), the positive emotional valence increases with increasing humanness until it reaches "the first peak." Following is a sharp dip towards a negative emotional response – the uncanny valley, rising again to the "second peak" which is as someone is perceived as fully human.

The uncanny valley has gained attention in recent years, especially with the rise of artificial intelligence, and the technology that has allowed the development of robots who increasingly look and behave like real humans. We predict that great apes, who differ from monkeys, not just in their size and strength, but also in intelligence and emotional capacities, are close enough to humans to fall into this uncanny valley.

As evidence, in 2009, a cartoon in the *New York Post* showed a dead chimpanzee with three bullet holes in front of two policemen. The speech bubble above the policemen says, "They'll have to find someone else to write the next stimulus bill." With the presidential campaign of Barack Obama, simianization was out in full force. There were chimpanzee T-shirts and chimpanzee dolls. A bar owner in Georgia sold T-shirts of Curious George eating a banana with "Obama 08" written

FIGURE 7.3 Adaption of March of Progress by Garde (2007). The drawing reflects the discrete ladder of the seventeenth and eighteenth centuries – the outdated idea that evolution is linear and humans are the pinnacle. Although the graphic is a terrible illustration of evolution, it has recently become a useful tool in measuring dehumanization, specifically simianization.

underneath. This persisted throughout the presidency. George W. Bush was also simianized, in an infamous website "Bush vs Chimp" that juxtaposed photos of Bush's facial expressions to chimpanzees. But this simianization did not pre-date his presidency, and was not extended to his entire family. In 2016 a Clay county employee called Michelle Obama "an ape in heels." Fox News viewers called Obama's daughter an "ape" (D'Onofrio, 2016). The Obamas were also frequently simianized as monkeys, but visual simianization – photoshopped pictures of the Obamas – almost always used great ape faces (chimpanzee and gorilla).

Recent research has illustrated how this type of simianization is related to real-world harm. Ashley Jardina used the March of Progress scale (see Figure 7.3) to ask 2000 white people how evolved black people were compared to white people (Jardina & Piston, 2016).

Overall, white people responded that black people were less evolved than white people, or closer to nonhuman apes, on the evolutionary scale. When Jardina broke down the responses by demographic, simianization was not limited to older, rural Republican men. Instead, Democrats and Republicans, conservatives and liberals, men and women, high income and low income, Southern and non-Southern, Young and Old – all demographics had a significant proportion who rated black people as less evolved, and more ape-like, than white people. Survey work by Jardina also found that people who rated blacks as more ape-like than white people were more likely to support the death penalty (Jardina & Piston, 2016).

Simianization also seems to occur on a subconscious level. Psychologist Philip Goff and colleagues primed people with subliminal words associated with apes or big cats, then showed them a video in which a police officer beats a suspect who is either black or white. When the person was primed with words associated with great apes, they were significantly more likely to condone police violence against a black suspect (Goff,

Eberhardt, Williams, & Jackson, 2008). Goff and colleagues also measured prejudice in their participants, and anti-black prejudice was unrelated to whether or not they condoned police violence towards black people – at least in this case, simianization, rather than prejudice was a more effective predictor. In a review of criminal cases, Goff found that articles about black people accused of capital crimes were more likely to contain references to apes than their white contemporaries, and that cases with ape-like language of this kind were more likely to result in the death penalty. Goff and colleagues (2008, p. 304) write that "these apelike representations were associated with the most profound outcome of intergroup dehumanization: death."

Finally, we propose that simianization might explain other instances of real-world harm, that are as yet unaccounted for. For instance, Kelly Hoffman and colleagues found that 40 percent of second-year medical students believed that the skin of black people is thicker than that of white people (Hoffman, Trawalter, Axt, & Oliver, 2016). Medical students who think black people have thicker skin are less likely to correctly treat black people for pain (Hoffman et al., 2016). Doctors are more likely to underestimate the pain of black patients who visit an emergency room. Black people with fractured limbs are less likely to receive pain medication, as are black cancer patients, or black people with migraines or back pain (Cintron & Morrison, 2006). Even black children with appendicitis are less likely to receive pain medication (Cintron & Morrison, 2006).

Too much admired

In terms of real-world harm, the interchangeability between humans and great apes might even be affecting the great apes themselves. The similarities between humans and great apes led to chimpanzees becoming the favored model for behavioral and biomedical testing, and for their frequent capture for zoos, circuses, and the pet trade. Thousands of chimpanzees were captured and exported to the USA, Europe, and the UK between the 1950s and 1980s. Initially, chimpanzees were used in biomedical research with great success. They were used to develop a vaccine for hepatitis A and B, and antibodies that are used to treat rheumatoid arthritis, non-Hodgkin's lymphoma, and psoriasis.

In the 1980s, the US government poured millions of dollars into AIDS research and biomedical labs began a massive breeding program to find a cure (Grimm, 2015). But although chimpanzees could be infected with HIV, they did not develop AIDS, and so were an ineffective research model. By 1997, the US government owned over 1000 chimps at a cost of $7.3 million a year. The same year, the National Institute of Health (NIH) banned the breeding of chimpanzees in federally funded biomedical laboratories. Funding for biomedical testing on chimpanzees has dried up, and as a result, many of these chimpanzees are left to spend the rest of their 50–70 year lifespan in cages.

The media industry that made great apes the darlings of the general public has also participated in their exploitation. Great ape infants are taken from their mothers and raised to perform in circuses, dressed up in clothes and filmed or photographed for commercials. Chimpanzees, for example, reach puberty as early as seven, and can live for 50 or 70 years. At puberty, they become too strong for their owners to control. They are often savagely beaten in attempts to intimidate them into submission. When this fails, they are often abandoned as they become a financial burden. Many of them end up in solitary confinement in unregulated backyard zoos, without adequate space or shelter. Even if they are rescued, these chimpanzees, having been raised by humans, often have difficulty integrating into a chimpanzee group at a sanctuary or zoo. This kind of abuse not only harms individual chimpanzees, but the entire species. Our research group has shown that seeing these entertainment chimpanzees leads people to conclude that chimpanzees are not endangered and make good pets – the opposite message that conservationists are trying so hard to convey (Schroepfer, Rosati, Chartrand, & Hare, 2011).

Great apes are not faring much better in the wild. The United Nations Environment Program (UNEP) estimates that between 2005–2011 over 22,000 great apes were either killed or captured from the wild (Nellemann, 2013). These great apes are mainly destined for Europe, the Middle East, China, and Southeast Asia, where they will be sold to unregulated backyard zoos, breeding centers, entertainment companies, circuses, and exotic pet collectors.

According to the same report, trafficking in live great apes has become part of organized environmental crime which is worth several billion dollars. Dealers target great apes for a sophisticated trade that has smooth passageways into major destinations around the world. With price tags from US$20,000 to US$300,000, the illegal trade in live apes is fast becoming an industry. In Shanghai Zoo we observed several infant chimpanzees living in isolation. In the wild, infants of this age would be completely dependent on their mothers for several more years. Each of them had been captured from the wild, shot off their mothers' backs, to be imported for the purpose of filling a zoo exhibit in China. To perform for the amusement of the audience they are trained using force (we observed an animal trainer holding an iron bar while leading the animals). With the market established, conservationists fear the worst. The population of every great ape species is in decline (Junker et al., 2012; Köndgen et al., 2008). In these cases, their similarity to us has not helped them. In fact, it may nudge them towards extinction.

The first peak

Dehumanization is a relatively recent topic of study, and received only scattered attention before 2000 (Haslam, 2006; Haslam & Loughnan, 2014). However, in recent years, systematic research on dehumanization has been growing, adding to our knowledge of the underlying mechanisms of dehumanization, and hopefully leading to methods of manipulations.

A promising area of study is the human-animal divide. Studies by Costello and Hodson (2010; see also Hodson, Dhont, & Earle, this volume) looked at the distance people perceived between animals and humans, also called the human-animal divide. The authors found that people who believed animals were more similar to humans were less likely to dehumanize immigrants. Those who saw a greater difference between animals and humans were more likely to dehumanize immigrants. This result also seemed to be amenable to manipulation, in that when people read an article that emphasized animals were similar to humans they became less likely to dehumanize immigrants (Costello & Hodson, 2010).

In another study (Costello & Hodson, 2014), white children were asked to place pictures of a white person and an animal on a horizontal board, with the space between them signaling how different the children believed them to be. Then, they were asked to do the same thing with a picture of a black person and an animal. Children who put a greater distance between the white person and the animal on the board tended to put less distance between black children and animals. White children who put a greater distance between the white person and the animal, also tended to attribute less "uniquely human" emotions to black children, like love, sympathy and guilt. They also tended to attribute fewer "uniquely human" traits to black children, like curiosity and creativity. Instead, these children attributed more basic emotions and traits that we share with animals, like happiness, fear, and nervousness. Encouragingly, however, when children watched a video showing the similarities between animals and humans, they reported a narrower difference in the human–animal divide (Costello & Hodson, 2014).

Further research is needed to test how flexible the human–animal divide is, how it relates to humanizing or dehumanizing other humans, and whether certain animals are more or less effective in decreasing this divide. Because of the uncanny valley, we would predict that humanizing great apes produces positive affect, but only up to a certain point. We have found initial support for this proposition when asking people in two conditions whether chimpanzees were endangered, and whether they made good pets. In the first condition, people saw photos of chimpanzees dressed up in human clothes; and in the second condition, people saw a video of chimpanzees in the wild. It was the video where chimpanzees were shown in the wild, with narration about their natural history and threats to their survival that made people aware of their endangered status (Schroepfer et al., 2011). In another study at the Lola ya Bonobo in the Democratic Republic of Congo, education paired with the experience of seeing bonobos playing in the natural habitat of the sanctuary was the most powerful method of conveying the conservation value of bonobos to Congolese children (André, Kamate, Mbonzo, Morel, & Hare, 2008).

Mori suggested that to avoid the uncanny valley, which falls just below the second peak of fully human likeness, robot designers should instead "take the first peak as their goal, which results in a moderate degree of human likeness and a considerable sense of affinity." Perhaps, instead of trying to fully humanize great apes, where they will always fall short of ourselves, we should appreciate the likeness that takes them to the first peak, and then celebrate the differences that make them unique.

In conclusion, great apes have been celebrated in popular culture since the 1970s, and their physical, cognitive, and behavioral similarity to us facilitates an interchangeability between our species. At the same time, civil and social rights movements have made huge strides in overcoming equality disparity, leading some to declare overt racism firmly in the past. However, we argue that simianization is one of the most pernicious forms of dehumanization because of the uncanny valley – the point at which something or someone becomes almost human. The uncanny valley can lead to feelings of revulsion and disgust. We propose that great apes can invoke the uncanny valley, and that when people are simianized, especially as great apes, it can lead to real-world harm. This form of simianization has been targeted towards black people since the transatlantic slave trade and persists today. Initial research on the human-animal divide has been an encouraging start in understanding the psychological underpinnings of dehumanization, but further research is needed to understand the role of the human-animal divide in simianization, and dehumanization more generally, and the extent to which it can be manipulated to improve the welfare of both humans and great apes.

Note

1 Humans belong to the great ape family, along with chimpanzees, bonobos, orangutans and gorillas. In this article, we will refer to nonhuman great apes simply as "great apes."

References

Affeldt, S. (2015). Exterminating the Brute. In W. D. Hund, C. W. Mills, & S. Sebastiani (Eds.), *Simianization: Apes, Gender, Class, and Race* (pp. 139–169). Zurich: Lit Verlag.

André, C. (2006). Une tendresse sauvage. Paris: Calmann-Lévy.

André, C., Kamate, C., Mbonzo, P., Morel, D., & Hare, B. (2008). The conservation value of Lola ya Bonobo Sanctuary. In T., Furuichi & J. Thompson (Eds.), *The Bonobos* (pp. 303–322). New York: Springer.

Bobo, L., & Hutchings, V. L. (1996). Perceptions of racial group competition: Extending Blumer's theory of group position to a multiracial social context. *American Sociological Review*, 951–972. http://dx.doi.org/10.2307/2096302

Browne, J. (2001). Darwin in caricature: A study in the popularisation and dissemination of evolution. *Proceedings of the American Philosophical Society, 145*(4), 496–509.

Buffon, G. L. L., comte de (1797). *Buffon's natural history: Containing a theory of the Earth, a general history of man, of the brute creation, and of vegetables, minerals, &C, &C* (Vol. 1). London: HD Symonds.

Buquet, T. (2013, June 23). Preventing "Monkey Business." Fettered apes in the Middle Ages. *Medieval Animal Data-Network*. Retrieved from http://mad.hypotheses.org/37

Carpenter, M., Nagell, K., & Tomasello, M. (1998). Social cognition, joint attention, and communicative competence from 9 to 15 months of age. *Monographs of the Society for Research in Child Development, 63*, 1–174. http://dx.doi.org/10.2307/1166214

Cintron, A., & Morrison, R. S. (2006). Pain and ethnicity in the United States: A systematic review. *Journal of Palliative Medicine, 9*(6), 1454–1473. https://doi.org/10.1089/jpm.2006.9.1454

Clair, M., & Denis, J. (2015). Sociology of racism. In J. D. Wright (Ed.), *The International Encyclopedia of the Social and Behavioral Sciences*, Vol 19, 2nd ed. (pp. 857–863). Oxford: Elsevier. https://doi.org/10.1016/B978-0-08-097086-8.32122-5

Costello, K., & Hodson, G. (2010). Exploring the roots of dehumanization: The role of animal—human similarity in promoting immigrant humanization. *Group Processes & Intergroup Relations, 13*(1), 3–22. https://doi.org/10.1177/1368430209347725

Costello, K., & Hodson, G. (2014). Explaining dehumanization among children: The interspecies model of prejudice. *British Journal of Social Psychology, 53*(1), 175–197. http://dx.doi.org/10.1111/bjso.12016

D'Onofrio, K. (2016, May 3). Fox News readers bash Obama's daughter with racial slurs, "ape," "monkey," *Diversity Inc.* Retrieved from www.diversityinc.com/news/malia-obama-fox-news

Darwin, C. (1880). *Descent of Man, and Selection in Relation to Sex, New edition, revised and augmented.* New York: D. Appleton & Company.

Dixon, T. (1905/2015). *The Clansman: An Historical Romance of the Ku Klux Klan.* In C. D. Wintz (Ed.). New York: Routledge.

Du Chaillu, P. B. (1868). *Explorations and adventures in equatorial Africa: With accounts of the manners and customs of the people, and of the chase of the gorilla, the crocodile, leopard, elephant, hippopotamus and other animals.* New York: Harper and Brothers.

Du Chaillu, P. B. (1870). Equatorial Africa, with an account of the race of Pigmies. *Journal of the American Geographical and Statistical Society, 2*(2), 99–112.

Fleegler, R. L. (2006). Theodore G. Bilbo and the Decline of Public Racism, 1938–1947. *Journal of Mississippi History, 68*(1), 1–27.

Fredrickson, G. M. (2015). *Racism: A Short History.* Princeton, NJ: Princeton University Press.

Goff, P. A., Eberhardt, J. L., Williams, M. J., & Jackson, M. C. (2008). Not yet human: implicit knowledge, historical dehumanization, and contemporary consequences. *Journal of Personality and Social Psychology, 94*(2), 292–306. doi: 10.1037/0022-3514.94.2.292

Grimm, D. (2015). New rules may end US chimpanzee research. *Science, 394,* 777. doi: 10.1126/science.349.6250.777

Hare, B. (2011). From hominoid to hominid mind: What changed and why? *Annual Review of Anthropology, 40,* 293–309. https://doi.org/10.1146/annurev-anthro-081309-145726

Hare, B., Call, J., Agnetta, B., & Tomasello, M. (2000). Chimpanzees know what conspecifics do and do not see. *Animal Behaviour, 59*(4), 771–785. https://doi.org/10.1006/anbe.1999.1377

Hare, B., & Kwetuenda, S. (2010). Bonobos voluntarily share their own food with others. *Current Biology, 20*(5), R230–R231. https://doi.org/10.1016/j.cub.2009.12.038

Haslam, N. (2006). Dehumanization: An integrative review. *Personality and Social Psychology Review, 10*(3), 252–264. https://doi.org/10.1207/s15327957pspr1003_4

Haslam, N., & Loughnan, S. (2014). Dehumanization and infrahumanization. *Annual Review of Psychology, 65,* 399–423. https://doi.org/10.1146/annurev-psych-010213-115045

Hobolth, A., Christensen, O. F., Mailund, T., & Schierup, M. H. (2007). Genomic relationships and speciation times of human, chimpanzee, and gorilla inferred from a coalescent hidden Markov model. *PloS Genetics, 3*(2), e7. https://doi.org/10.1371/journal.pgen.0030007

Hodson, G., Dhont, K., & Earle, M. (this volume). Devaluing animals, "animalistic" humans, and people who protect animals. In K. Dhont & G. Hodson (Eds.), *Why We Love and Exploit Animals: Bridging Insights from Academia and Advocacy.* Abingdon: Routledge.

Hodson, G., MacInnis, C. C., & Costello, K. (2014). (Over) valuing" humanness" as an aggravator of intergroup prejudices and discrimination. In P. Bain, J. Vaes, & J. P. Leyens (Eds.), *Are We All Human? Advances in Understanding Humanness and Dehumanization* (pp. 86–110). London: Psychology Press.

Hoffman, K. M., Trawalter, S., Axt, J. R., & Oliver, M. N. (2016). Racial bias in pain assessment and treatment recommendations, and false beliefs about biological differences between blacks and whites. *Proceedings of the National Academy of Sciences, 113*(16), 4296–4301. https://doi.org/10.1073/pnas.1516047113

Hogan, D. E., & Mallott, M. (2005). Changing racial prejudice through diversity education. *Journal of College Student Development, 46*(2), 115–125. doi: 10.1353/csd.2005.0015

Huddy, L., & Feldman, S. (2009). On assessing the political effects of racial prejudice. *Annual Review of Political Science, 12,* 423–447. https://doi.org/10.1146/annurev.polisci.11.062906.070752

Hund, W. D., Mills, C. W., & Sebastiani, S. (Eds.). (2015). *Simianization: Apes, Gender, Class, and Race.* Zurich: LitVerlag.

Hunt, J. (1863). On the Negro's place in nature. *Journal of the Anthropological Society of London, 2,* pp. xv–lvi. doi: 10.2307/3025197

Huxley, T. H. (1863). *Evidence as to Man's Place in Nature.* New York: D. Appleton and Company.

Janson, H. W. (1952). *Apes and Ape Lore: In the Middle Ages and the Renaissance* (Vol. 20). Warburg Institute, University of London.

Jardina, A., Piston, S. (2016). *Dehumanization of Black People Motivates White Support for Punitive Criminal Justice Policies.* Paper presented at the 2016 Annual Meeting of the American Political Science Association, Philadelphia, PA.

Jefferson, T. (1781). Notes on Virginia. *The Life and Selected Writings of Thomas Jefferson.* New York: Penguin.

Junker, J., Blake, S., Boesch, C., Campbell, G., Toit, L. d., Duvall, C., … Kuehl, H. S. (2012). Recent decline in suitable environmental conditions for African great apes. *18*(11), 1077–1091. https://doi.org/10.1111/ddi.12005

Kano, F., & Call, J. (2014). Great apes generate goal-based action predictions: an eye-tracking study. *Psychological science, 25*(9), 1691–1698. https://doi.org/10.1177/0956797614536402

Kenny, K. (2006). Race, violence, and anti-Irish sentiment in the nineteenth century. In Lee, J. J., & Casey, M. R. (Eds.). *Making the Irish American: History and Heritage of the Irish in the United States* (pp. 364–378). New York: New York University Press.

Koblin, J. (2018, May 29). After racist tweet, Roseanne Barr's show is Canceled by ABC, *New York Times.* Retrieved from www.nytimes.com/2018/05/29/business/media/roseanne-barr-offensive-tweets.html

Köndgen, S., Kühl, H., N'Goran, P. K., Walsh, P. D., Schenk, S., Ernst, N., … Leendertz, F. H.. (2008). Pandemic human viruses cause decline of endangered great apes. *Current Biology 18*(4), 260–264. https://doi.org/10.1016/j.cub.2008.01.012

Livingstone Smith, D., & Panaitui, I. (2015). Aping the Human Essence. In W. D. Hund, C. W. Mills, & S. Sebastiani (Eds.), *Simianization: Apes, Gender, Class, and Race* (pp. 77–103). Zurich: LitVerlag.

Long, E. (1774). *The History of Jamaica: Or, general survey of the antient and modern state of the island: with reflections on its situation settlements, inhabitants, climate, products, commerce, laws, and government.* London: T. Lowndes.

Melis, A. P., Hare, B., & Tomasello, M. (2006). Chimpanzees recruit the best collaborators. *Science, 311*(5765), 1297–1300. doi: 10.1126/science.1123007

Melis, A. P., Hare, B., & Tomasello, M. (2009). Chimpanzees coordinate in a negotiation game. *Evolution and Human Behavior, 30*(6), 381–392. https://doi.org/10.1016/j.evolhumbehav.2009.05.003

Mori, M. (1970). The uncanny valley. *Energy, 7*(4), 33–35.

Nellemann, C. (2013). Stolen apes. A UNEP rapid response assessment: UNEP.

Ponds, K. T. (2013). The trauma of racism: America's original sin. *Reclaiming Children and Youth, 22*(2), 22–24.

Roithmayr, D. (2014). *Reproducing Racism: How Everyday Choices Lock in White Advantage.* New York: New York University Press

Savage-Rumbaugh, E. S., Murphy, J., Sevcik, R. A., Brakke, K. E., Williams, S. L., Rumbaugh, D. M., & Bates, E. (1993). Language comprehension in ape and child. *Monographs of the society for research in child development,* i-252.

Schroepfer, K. K., Rosati, A. G., Chartrand, T., & Hare, B. (2011). Use of "entertainment" chimpanzees in commercials distorts public perception regarding their conservation status. *PLoS One, 6*(10), e26048. https://doi.org/10.1371/journal.pone.0026048

Sniderman, P. M., & Carmines, E. G. (1997). Reaching beyond race. *Political Science & Politics, 30,* 466–471. https://doi.org/10.2307/420124

Staples, B. (2009, February 27). The ape in American bigotry, From Thomas Jefferson to 2009, *The New York Times.* Retrieved from www.nytimes.com/2009/02/28/opinion/28sat4.html

Thernstrom, A. T. S. (2008, March 02). Taking race out of the race. *LA Times.* Retrieved from http://articles.latimes.com/2008/mar/02/opinion/op-thernstrom2

Tomasello, M., & Call, J. (1997). *Primate Cognition.* New York: Oxford University Press.

Vadillo, M. (2011). *Apes in Medieval Art.* Paper presented at the International Medieval Congress, Leeds, UK.

van Wyhe, J., & Kjærgaard, P. C. (2015). Going the whole orang: Darwin, Wallace and the natural history of orangutans. *Studies in History and Philosophy of Science Part C: Studies in History and Philosophy of Biological and Biomedical Sciences, 51,* 53–63. doi: 10.1016/j.shpsc.2015.02.006

Williams, C. J. (1995). *Freedom & justice: Four decades of the civil rights struggle as seen by a Black photographer of the deep South.* Macon, GA: Mercer University Press.

Yerkes, R. M. (1925). *Almost Human.* New York: Century.

ASKHAM BRYAN
COLLEGE
LEARNING RESOURCES

8

WHY PEOPLE LOVE ANIMALS YET CONTINUE TO EAT THEM

Jared Piazza

Abstract

We, as humans, are deeply fascinated by animals from an early age yet we also in many ways exploit animals. Eating animals is a particular "blind spot" in our ethical thinking about animals. Recent work in psychology illuminates why it is so difficult to think straight ethically about our use of animals as food. We are often mindless when we eat; rarely do we connect animal products with their animal origins. When we eat animals, we are motivated to disregard information we would otherwise find relevant to our ethical concerns. We think we have good reasons for eating animals, which overrides our empathy for their suffering. This chapter reviews the psychological literature on our biased thinking about animals we eat and concludes with some thoughts on what might be done about it.

Biases in our thinking about animals

Our relationships with animals are riddled with contradictions and prejudice. From a young age, we are fascinated with animals. Research shows that children as young as one year of age naturally gravitate towards animals. When given the choice of playing with a wide range of attractive toys, from dolls to toy airplanes, and living animals, one- and two-year-olds decidedly prefer to interact with a fish or hamster (LoBue, Pickard, Sherman, Axford, & DeLoache, 2013). Children are even interested in animals that for many adults would be potentially threatening – spiders, snakes; they enjoy watching and talking about these animals even if their parents do not. Indeed, adolescent children seem to know more about the lives of certain animals (e.g., invertebrate) than most adults do (Kellert, 1985).

Yet, also from an early age, we can observe the seeds of our prejudice in the way we think about animals. Children feel a greater emotional attachment for animals they have as pets than animals farmed and raised for food (Daly & Morton, 2006); children attribute greater intelligence and sentience, for example, to dogs than cows (Hawkins & Williams, 2016). The use of animals for food does not seem to bother most children, especially those who grow up in rural areas and are accustomed to connecting animals to food products (Bray, Zambrano, Chur-Hansen, & Ankeny, 2016). Children's concern for animals, like adults, is anthropocentric, that is, we use our own experiences in our efforts to understand and relate to animals, and we give preference to those animals we believe are most like us (Plous, 1993; Serpell, 2004). We, as humans, like to think of ourselves as special and superior in many ways to animals, and research by Leite, Dhont, and Hodson (2019) shows the more we endorse our superiority to animals, the less we believe we need to care for or show concern for them.

A study by Borgi and Cirulli (2015) presented kindergarten children with 48 different animal species. The animals were presented in pairs, and the children were asked which of the two they liked more. The results were largely consistent with those observed in adults (e.g., Tisdell, Wilson, & Nantha, 2006) – the children exhibited a preference for animals phylogenetically similar to humans. Mammals were preferred to birds, which were preferred to herptiles (e.g., lizards, frogs) and invertebrates (e.g., insects, worms). There were a few exceptions – butterflies were liked quite a lot due to their aesthetic qualities, and snails and frogs got the same amount of "likes" as monkeys – but by and large children exhibited preferences for animals that they believed to be similar to them. Furthermore, mammals commonly used as food (e.g., cows, pigs, sheep) fared much lower in children's regard compared to their non-farmed counterparts (e.g., elephants, dolphins, kangaroos).

As we enter adulthood, our prejudiced thinking about animals only solidifies further. Our decisions about which animals should be protected and conserved are guided not by which animals are most vulnerable or in the greatest need of our help, but by our emotional preferences – which animals we like and that are most popular (Colléony, Clayton, Couvet, Jalme, & Prévot, 2017; Tisdell et al., 2006). We become comfortable with the fact that we offer legal protections and moral status to some animals (e.g., dogs), while denying the same protections to other species (e.g., pigs) with little concern for their comparable capacity for thought and emotion (Bilewicz, Imhoff, & Drogosz, 2011; Joy, 2010). When we see animals caged and in distress our bodies react most strongly (e.g., we sweat more in empathic concern) when the animal is phylogenetically similar to us; for example, the suffering of a monkey elicits a stronger empathic response than the suffering of a chicken (Westbury & Neumann, 2008). We afford less standing and concern to animals that appear threatening and dangerous (Piazza, Landy, & Goodwin, 2014), and give greater standing to animals that are cute and evoke positive emotions (Piazza, McLatchie, & Olesen, 2018). Indeed, research shows that cute dogs, with baby-like features, tend to be preferred as pets (Weiss, Miller, Mohan-Gibbons, & Vela, 2012) and are treated more humanely by their owners (Thorn, Howell, Brown, & Bennett, 2015). Likewise, when meat is known to be sourced from a baby animal it is found less appetizing than the very same meat sourced from a mature animal of the same species (Piazza et al., 2018).

Eating animals

Of all the ways we use and exploit animals, probably nowhere else are human biases on such conspicuous display as in the domain of food. When we think of an animal as a food source our moral attitudes about them drastically shift (Loughnan, Bastian, & Haslam, 2014; Piazza & Loughnan, 2016). This is true when thinking about different animal species – i.e., those we use for food and those we do not (Herzog, 2010; Joy, 2010) – but also when thinking about the same animal differently. In one study, participants read a story about the Bennett's Tree Kangaroo living in Papua New Guinea (Bratanova, Loughnan, & Bastian, 2011). Some participants learned some interesting biological facts about the animal, while other participants learned these facts and they also read that a group of people indigenous to Papua New Guinea hunt the tree kangaroo for meat. Others learned that the indigenous people simply collected dead kangaroos who had been knocked out of trees in a violent storm to use for meat. Regardless of whether the indigenous group was responsible for the tree kangaroos' death, when participants were thinking about the kangaroo as meat to be eaten they attributed less suffering to the kangaroos and less moral concern for them than when they were simply thinking of the kangaroos as animals.

Something appears to happen to our moral concern for animals when we stop thinking about them as living beings and start thinking of them as food. In ongoing research from my lab (Piazza, Gardiner, & Stones, unpublished), we manipulated people's perceptions of whether an exotic culture from South America had a traditional meal that revolved around meat from a particular animal – tapirs in one study, and wild boars in another. Participants read basic facts about the animal and either learned that the exotic culture had a tradition of using meat from the animal in their cooking, or they learned that the traditional meal did not include meat from the animal, however, a member of the group proposed to incorporate meat from the animal into the meal. Participants judged the acceptability of killing the tapir or wild boar to use as food. In both studies, we observed that our British participants (a total of 173 across studies) found it more acceptable for the foreign people to kill and eat the animal when there already existed a tradition around eating the animal than when it was newly proposed – as if the tradition was reason enough to accept the use of the animal. This was true despite not eating the animal themselves, and independent of other characteristics of the animal, including how harmless or young was the animal.

Thus, thinking about animals as food appears to alter our moral judgments of animals, and this seems to be particularly true when there is a history of animal use or a social consensus about its acceptability. When culture or tradition dictates certain animals have a particular use, it can be difficult to think critically about our treatment of that animal.

Animals and food: An ethical blind spot

There are many reasons why people have trouble thinking consistently about animals. But probably the biggest reason is the same reason why people find themselves often behaving morally inconsistent in other domains of life: we have ethical blind spots (Bazerman & Tenbrunsel, 2011; Sezer, Gino, & Bazerman, 2015). That is, we often fail to realize our actions have ethical implications or should be framed in moral terms. This is essentially the problem facing us in the realm of animals and food. Most people simply fail to see eating animals as a moral issue, or if they have some doubts about its ethicality, they easily convince themselves they have good reasons for doing it (Piazza, Ruby, Loughnan, et al., 2015). This is not to say we do not face similar challenges in other domains of animal use: animal testing, entertainment, population management, and so on (see Knight, Vrij, Cherryman, & Nunkoosing, 2004). However, the social consensus about using animals for food is overwhelmingly affirmative, thus, the ethical blind spot with regards to eating animals is incredibly pervasive.

Because eating animals is socially normative in most places and circles, people rarely find themselves in a compelling environment that forces them to question its ethicality. Thus, the solutions that moral psychologists commonly prescribe for counteracting ethical blind spots do not easily apply. One popular psychological perspective on why people fail to act morally claims that people fail not because they lack the motivation to be moral, but because they are situationally inattentive to their internal standards when facing a temptation (Mazar, Amir, & Ariely, 2008). *Reminding* people of their standards, making these standards more *visible* within the decision-making context, and *engaging* people's moral self-concepts prior to acting have been recommended as three ways for overcoming moral temptations (Ayal, Gino, Barkan, & Ariely, 2015). This advice, as helpful as it is, can only be applied when people are willing to see an action in moral terms. Reminding people of their motivation to be good or drawing their attention to the questionable aspects of eating animals will have little effect if people continue to see themselves on the right (or popular) side of the issue. Indeed, when people believe it is their right to use animals as food they respond only with reactance and hostility (e.g., attacking the motivations of vegetarians) when they believe this right is under threat (Dhont & Hodson, 2014). It seems obvious, but the principal challenge in convincing people that using animals for food is morally problematic is its widespread acceptance.

Mindless eating: Dissociating meat from animal

When a behavior is socially normative and widespread, routines often form around that behavior, leading to yet another roadblock to social change: mindless behavior. *Mindlessness* is the automatization of behavior, such that one does not have to think about what one is doing. One merely enacts pre-set scripts, routines or procedures that have been used successfully in the past (Neal, Wood, & Quinn, 2006). In the context of food, this means thoughtlessly purchasing and eating foods that one has always eaten or is already comfortable with. Mindlessness about food can be problematic because it prevents us from carefully considering the ethicality of the choices we make (see Bastian & Loughnan, 2017).

Interrupting routines can be useful in correcting unethical behavior (Sezer et al., 2015). When people are given more time to reflect on their decisions they are often able to overcome biases in their thinking. For example, reductions in gender-based prejudice have been observed when people are asked to carefully consider job candidates on the basis of their qualifications and not their gender (Bohnet, Van Geen, & Bazerman, 2015). Eating meat is normative and routine for many people, and the mindlessness of this behavior manifests in the *dissociation* of meat from its animal source (Joy, 2010). Dissociation enables people to mindlessly order and devour a cheeseburger without any thought of the animals that suffered to produce it.

Kunst and Hohle (2016) demonstrated in a series of studies how dissociation is often the default for meat eaters. When cooking meat or selecting a meat dish, little thought is given to the animal source. Yet when efforts are made to draw people's attention to the animal it came from, people's appetite for meat declines. For example, in one study, the researchers compared participants' judgments of chicken presented as either: a whole body without its head, cut into parts (legs, wings, and breast), or minced into a meat purée. They found that meat eaters were more likely to think about the animal source when the whole body of the animal was presented in the meat dish than when the meat was presented to look less like an animal. Reminding meat eaters of the animal origins had the consequence of increasing participants' empathy towards the animal, and decreasing their appetite for the meat product.

One conclusion you might draw from this research is that drawing people's attention to the living animal origins of their meat might be sufficient to overcome the psychological barrier posed by mindless eating. If only things were that simple. Although research suggests that connecting people's thoughts about meat and its animal source can make meat less appetizing, meat eaters are still motivated to buck the animal welfare implications of this connection, either because it presents a challenge to their moral self-concept or they simply prefer to believe in the correctness of what they are doing.

The meat-motivated mind

Mindless dissociation can keep us from thinking about the moral implications of what we eat. Yet, when the thought arises that eating meat may not be fully ethical, there are other psychological strategies available to help us rebuff these irksome thoughts. Indeed, such thoughts may be increasingly common in today's society. The horrors of factory farming have been tirelessly documented and exposed by animal advocates, such that the public is becoming more and more aware of the ethical controversy surrounding meat production, particularly in the West. Though vegetarians and vegans remain minorities in every country (Ruby, 2012), and face criticism for taking such an unpopular stance towards meat (MacInnis & Hodson, 2017), their presence in society makes it increasingly difficult for people to remain entirely guileless about their meat consumption (throw in concerns about health and the environment and the potential for a guilty conscience intensifies further). At the same time, animal products, and meat in particular, are a source of great pleasure for many people (Zaraska, 2016). The pleasure of cooked meat remains a fundamental barrier to reductions in consumption (Lea & Worsley, 2003), and, indeed, a barrier to our moral thinking about animals as well (Bastian & Loughnan, 2017; Piazza & Loughnan, 2016; Piazza et al., 2015).

Research in psychology has shown that when people desire something to be true, we are quite adept at fitting our beliefs to support our preferences (Baron, 1995; Kunda, 1990). This seems to be true even when we are confronted with information that might suggest our beliefs may be incorrect and need to be updated. As a general rule, consumers will go out of their way to avoid exposing themselves to information that could interfere with their purchasing choices (Ehrich & Irwin, 2005). But when such information cannot be avoided, rather than accommodating our beliefs to fit this new information we often disregard it, particularly when the information has implications for cherished practices or beliefs that we hold about the world or ourselves (Piazza & Loughnan, 2016).

Denial and disregard: Actively avoiding the ethical implications of meat

To accept that meat comes from a once living, feeling, thinking being is ethically disorienting for many people (Loughnan et al., 2014). Believing that one is in some way responsible for the injury and death of innocent animals is a distressing thought. How do most consumers who have such thoughts deal with them?

One strategy seems to be to convince oneself that animals are not really harmed or mistreated when they are processed for food (Bastian, Loughnan, Haslam, & Radke, 2012; Rothgerber, 2012). In one study, my colleagues and I had meat eaters from the United States tell us why it is okay to eat meat (Piazza et al., 2015). One line of argumentation that was quite common was to contend that animals killed for meat are treated and slaughtered humanely, or that the participant made efforts to purchase exclusively "humane" meat products. In a similar line of research, Dowsett, Semmler, Bray, Ankeny, and Chur-Hensen (2018) found that roughly 7 percent of Australians in their sample asserted that they only ate humane or free-range animal products. Setting aside whether there is such thing as "humane meat" (see McWilliams, 2015), the *belief* in humane meat appears to be quite a useful psychological strategy for neutralizing the ethical dissonance one might otherwise experience when considering the production of meat.

Another vehicle for avoiding ethical implications of meat is to deny that animals used for food are capable of suffering in any meaningful way. Loughnan, Haslam, and Bastian (2010) found that animals used for food are often denied the same cognitive and emotional capabilities as animals exempt from such treatment. Simply considering an animal (e.g., cow or sheep) as food, or in the process of becoming food (e.g., being transported to the slaughterhouse), is enough to motivate meat eaters to reduce their opinion about the mental capabilities of an animal (Bastian et al., 2012). This *denial of mind* is argued to reduce the perceived unethicality of meat consumption by removing the perceived harm done to animals.

Even more challenging to address is the tactic of *acknowledging, yet disregarding*, information that could bear on one's moral thinking. Steve Loughnan and I recently found that people often fail to utilize relevant information about animals in their moral thinking when the animal poses an ethical dilemma for them (Piazza & Loughnan, 2016). This seems to be true even when relevant information is made readily available to consumers. In one study, omnivores read about the intellectual capabilities of one of three animals: pigs (animals eaten in their society), tapirs (animals not eaten in their society), and trablans (a fictional animal). Participants either learned that the animal was quite smart and could outperform dogs on certain problem-solving tasks, or they learned that the animal was not so smart and that dogs outperformed them. All participants further read about how the animal was currently being confined, mistreated and slaughtered for meat: pigs in North America and Europe, tapirs in South America and Asia, and trablans on a fictional planet. They then made a moral judgment about the treatment of the animal. A distinct, but unsurprising, pattern of results emerged: when the animal was described as highly intelligent, participants found it morally worse to treat them in the manner described. However, this was only true for tapirs and trablans. Pigs were the clear exception.

The most surprising finding was not that pigs were treated morally different from the other animals – we expected this to be the case given the normative use of pigs as food in many cultures. The most surprising finding was that all three animals were rated equally intelligent by our omnivorous participants when such information was made available to them. In other words, our participants acknowledged the great intelligence of pigs, yet this information made no impact whatsoever upon their moral judgment. It was as though the information no longer mattered, while this was not the case when they were judging the moral worth of the other animals that were not being used for food in their own culture.

In short, it was only when the animal source had direct implications for the participants themselves that they disregarded intelligence information as morally relevant input into their judgment of how the animal should be treated. This study highlights probably the most exasperating psychological maneuver adopted by meat eaters: when problematic information cannot be reframed or contested, one can simply disregard the information or treat it as irrelevant.

Rationalizing: Thinking we have good reasons

Because eating meat is practiced in just about every country and community on the planet, it is generally assumed that we have good reasons for doing it (Joy,

2010; Piazza et al., 2015). This is often how our moral reasoning works – rather than arriving at conclusions through careful analysis of the facts, we often have a conclusion in mind that we would prefer and we search for facts and arguments that support it (Haidt, 2001; Nickerson, 1998). Indeed, when we think we have the right position we tend to believe most of the facts, or at least the most *legitimate* facts, are on our side – what psychologists call "my-side bias" (Baron, 1995; Liu & Ditto, 2013). This is another way in which our thinking about the use of animals for food can be difficult to overturn. Because meat eating is so pervasive people assume there are good reasons for it, otherwise this would mean billions of people – including our caring and sensible family members, friends, and neighbors – have it wrong.

To probe people's reasons for eating meat, my colleagues and I had American meat eaters tell us "why it is okay to eat meat" (Piazza et al., 2015). Our team was inspired by the writing of psychologist and activist, Melanie Joy, founder of the international organization *Beyond Carnism*. Joy (2010) theorizes in her book, *Why We Love Dogs, Eat Pigs, and Wear Cows: An Introduction to Carnism*, that there are three principal arguments people historically have used to justify widespread yet ethically questionable practices, whether it's slavery, gender inequality, or animal use. These justifications entail asserting the *normalness* of an action (i.e., that it is traditionally or widely practiced); its *necessity* (i.e., that its undoing would lead to the disruption of health, safety, or human flourishing); and its *naturalness* (i.e., it arises due to a natural law governing human behavior). Joy termed this set of beliefs the "3Ns of Justification" and speculated about their application to meat consumption.

Inspired by Joy's theorizing, we sought to determine whether the 3Ns of Justification might account for the lion's share of justifications people offer for eating meat. Confirming Joy's view and adding to it, we found that the vast majority of justifications offered fell into four categories: that eating meat is Natural (we are naturally designed to have dominion over animals and to eat meat), Necessary (our bodies need meat to be completely healthy and strong), Normal (most people eat meat and have done so for millennia), and Nice (yum, bacon) – what we labelled "the 4Ns." Based on these responses we developed a 4N Scale for assessing people's endorsement of the 4Ns. What we observed was no surprise: meat eaters endorsed the 4Ns more strongly than meat reducers, who endorsed the 4Ns more strongly than meat avoiders. This is consistent with the notion of "my-side bias" – people's understanding of the *facts* about meat (e.g., whether we need meat to be healthy) depended on their *stance* towards eating meat. But, interestingly, most meat eaters did not fully endorse the 4Ns. On a 1–7 scale, 7 = "completely agree," the average 4N score for meat eaters fell in the 3–5 range (with 4 = "neither agree nor disagree").

These middling mean scores have at least two potential implications. First, they seem to imply that some, possibly many, people continue to eat meat despite lacking strong justifications for doing so. For these people, other forces – such as habit, convenience, coordinating meals with others, or lack of knowledge of meat-free cuisine – may be the principal barriers to effective dietary change. Second, these neutral endorsements could imply that some meat eaters have ambivalent attitudes towards meat. This provides animal advocates with a potential ray of hope: meat eaters may believe there are good reasons for using animals for food, but their endorsement of these beliefs is not at ceiling – there is plenty of room for doubt. Given the right circumstances, meat eaters might be persuaded otherwise. But what are the right circumstances?

Confronting our rationalizations

We know from research on social influence that people often resist obvious attempts at persuasion, largely because we like to feel that we are autonomous agents, acting of our own accord, and not modelling clay in the hands of others (Fransen, Smit, & Verlegh, 2015). Because meat is a source of great pleasure for many, and the thought of losing something one enjoys is aversive, meat eaters are likely to be on guard for such deliberate persuasion attempts. Less direct approaches to social influence are likely to be more effective in moving people in a more compassionate direction.

One potential strategy that my lab has been recently investigating is *self-persuasion* (Aronson, 1999), which entails engaging people in a process whereby they freely generate counterarguments on a perspective or behavior – and the perspective happens to be their own position or a behavior they may perform (for an example of self-persuasion applied to condom use, see Aronson, Fried, & Stone, 1991). Having participants persuasively argue for a position they might otherwise disagree with, or that conflicts with their behavior, creates a state of tension or dissonance that demands resolution (Aronson, 1999; Aronson et al., 1991). One potential solution is to shift one beliefs or practices in the direction of the argued position.

Self-persuasion is a subtle form of persuasion and it might be put to some use on behalf of animals. To test this possibility, in one study, we had 114 British omnivores consider pro-N (i.e., pro-meat) arguments, such as, "It is *necessary* to eat meat to be strong and healthy" or "Meat is *delicious*." In one condition, we had them imagine that they were in an argument with a friend and were trying to convince the friend that these pro-N statements were true. In the self-persuasion condition, participants argued the counter-attitudinal or anti-N position, such as, "It is *not* necessary to eat meat to be strong and healthy" or "Meat is *disgusting*." After spending 5 or so minutes developing and writing down their arguments, we had them answer a few questions assessing their concern for farmed animals ("pigs, cows, sheep, chickens, etc."), with items such as, "Farmed animals deserve to be treated with care and compassion" and "Farmed animals do not really suffer when being raised and killed for meat" (reverse scored). We also asked them if they would like to take home with them some vegetarian recipes, and we left a stack of recipes by the door for them to take on their way out.

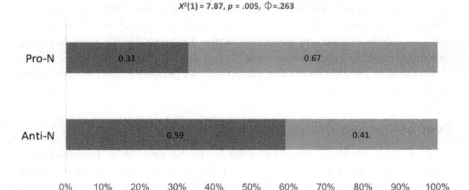

FIGURE 8.1   Percentage of participants taking vs not taking vegetarian recipes home with them after writing Pro-N (pro-meat) or Anti-N (anti-meat) arguments.

The results offered some preliminary positive support for the power of self-persuasion, at least in the short term. Participants who had written counter-attitudinal arguments (e.g., why it's *not* necessary to eat meat) reported having greater moral concern for farmed animals (mean scores on a 1–9 scale ranged from 6.35 to 6.62) than those who argued for their own position (mean scores ranged from 5.87 to 5.88). Furthermore, as can be seen in Figure 8.1, participants in the anti-N condition were more likely to take vegetarian recipes home with them afterwards (59%) compared to those in the pro-N condition (33%).

Of course, we cannot know for sure if any of our participants prepared and ate the recipes they brought home with them, and it's certainly far from clear whether self-persuasion could be harnessed to bring about long-term changes in behavior. Although these findings are preliminary, they suggest that one potential way to encourage people to rethink their attitude towards meat is not to try to persuade them out of it, but to give them opportunities to persuade themselves. The trick of course is figuring out natural ways of doing this without being too obvious about the aim of the method itself.

Empathizing with animals that are eaten

As we have seen, there are many psychological processes that serve to reinforce the belief that meat eating is an acceptable practice. In her book, *Why We Love Dogs, Eat Pigs, and Wear Cows*, Melanie Joy (2010) argues that our collective dissociation of meat from animals blinds us to their suffering, which prevents us from acting upon our universal impulses for compassion. The solution to this collective state of dissociation, she argues, is for more and more people to "bear witness" to the suffering of farmed animals. She writes, "collective witnessing closes the gap in social consciousness. Collective witnessing leads to an informed public and a system in which values and practices are more aligned …. The goal of all justice movements is to activate collective witnessing so that social practices reflect social values" (pp. 138–139).

Joy shrewdly highlights the immense hurdle animal advocates face in convincing a meat-loving world that there is a disconnect between our concern for those who suffer and our love of meat (and other animal products). Indeed, research shows that Joy's portrait of our current state as a state of collective dissociation is an accurate one. While meat eaters love meat, they aren't insensitive to the pain of farmed animals. One study found that relative to meat eaters, when vegetarians and vegans viewed images of meat dishes their brains produce heightened electrical activity occurring 300 to 700 ms after the images appeared (Stockburger, Renner, Weike, Hamm, & Schupp, 2009). Previous work shows that this late positive potential (or LPP) occurs when individuals are presented highly affective or aversive stimuli (Cuthbert, Schupp, Bradley, Birbaumer, & Lang, 2000). This brain response correlated reliably with meat avoiders' negative ratings of the meat dishes, highlighting their disgust towards meat.

Yet, despite differences in the affective reactions vegetarians and meat eaters have towards meat, these groups show little divergence in their affective reactions to the suffering of animals used to produce meat. In a recent unpublished study, my colleagues and I recorded brain activity (event-related potentials) from 20 meat eaters and 18 ethically motivated vegans, while they viewed images of farmed animals (cows, sheep, and pigs) being jabbed, prodded, and stabbed with sharp instruments (these were static images we constructed – no animals were harmed for this research, and ethical approval was obtained before collecting data; Piazza, Crespo Llado, Linnert, & McLatchie, in preparation). Both groups of participants produced a pattern of brain waves typically observed in studies of pain perception with human targets (see for example Suzuki, Galli, Ikeda, Itakura, & Kitazaki, 2015). Critically, compared to neutral trials where the animals were not depicted in painful situations, both groups showed comparably heightened reactions to the painful stimuli 200 to 700 ms following the onset of the images (vegans showed a heightened orienting response to animal pain as early as 100 ms, and this was the only reliable difference found between groups). Thus, we observed little evidence suggesting there is an empathic divide between meat avoiders and meat eaters in their attentional and emotional response to the pain of farmed animals.

In other words, while meat eaters and meat avoiders have different reactions to meat, their empathic response to animals looks quite similar. Although these findings are preliminary, they suggest that where these groups differ is not in their level of empathy, or potential for or **when** not thinking in their motivation to overcome the dissociation and denial we all face when thinking about animals as food, or not thinking about animals at all when eating them.

What to do?

Given that our failure to act on behalf of animals eaten is *not* due to a deficiency in empathy, but a failure to allow our empathy to guide our thinking as a result of other psychological processes related to mindlessness, motivated cognition, and rationalization, how might we make progress in our moral thinking about animals that we eat? This is one of the big questions researchers like myself wrestle with. We know that people do not like to be told that their actions, such as eating a delicious hamburger, is rife with moral implications. As we have seen, this causes defensive reactions, such as disparaging meat avoiders as moral "extremists" (Hodson, Dhont, & Early, this volume; MacInnis & Hodson, 2017), and drives motivated thinking (Piazza & Loughnan, 2016) and rationalization (e.g., asserting the nutritional benefits of animal products or our supremacy over animals; Piazza et al., 2015). As discussed above, there may be some value in having people confront their own reasons for eating animals. But this strategy is limited in that it relies on people being willing to engage in such counter-attitudinal thinking, and there must be convincing arguments available for consumers to be persuaded by, including the belief that eating meat is *not* necessary, natural, normal, or nice. Furthermore, people must have desirable alternatives to turn to, if they are going to give up something so enjoyable and habitual as eating meat.

The desire for meat is unlikely to abate any time soon. While the demand for tasty plant-based alternatives appears to be on the rise, particularly in Europe and the United States (Webber, 2017), the demand for meat also continues to soar (Waite, 2018). This makes the development of affordable, high-quality meat alternatives, such as the Beyond Burger and clean meat, key to confronting our motivated thinking about eating animals (see Leenaert, 2017). Because our desire for meat often guides our moral thinking about animals, rather than the other way around (Piazza & Loughnan, 2016), progress in this area will require the advancement, promotion, and uptake of desirable meat alternatives. As people come to see that they do not need to eat animals to be contented consumers, they will find the moral arguments against eating animals more palatable, and their motivations will cease to get in the way of extending moral courtesies to animals that we eat.

Waking up

In the case of unprovoked violence or neglect, we all agree animal cruelty should end. The problem is that eating animals is not a case of clear cruelty for many people. Through various processes, reviewed in this chapter, we have seen that people have many safeguards in place that help close themselves off from otherwise worrisome thoughts that they might experience when confronting the suffering of animals slaughtered for food. Thus, before any effort can be made to draw on the emotions and sympathies of caring people, we must first wake people up to the reality of our psychological predicament: when it comes to thinking about farmed animals, our love of meat shapes what we believe is true about the lives and moral worth of these animals. In our efforts to aid animals, particularly farmed animals, we must not only consider the quality of arguments we present to people, but also address the underlying appetites and motivations that determine whether an audience will be at all receptive to listen.

References

Aronson, E. (1999). The power of self-persuasion. *American Psychologist, 54*(11), 875–884.

Aronson, E., Fried, C., & Stone, J. (1991). Overcoming denial: Increasing the intention to use condoms through the induction of hypocrisy. *American Journal of Public Health, 18,* 1636–1640. doi: 10.2105/AJPH.81.12.1636

Ayal, S., Gino, F., Barkan, R., & Ariely, D. (2015). Three principles to REVISE people's unethical behavior. *Perspectives on Psychological Science, 10*(6), 738–741. doi: 10.1177/1745691615598512

Bastian, B., & Loughnan, S. (2017). Resolving the meat-paradox: A motivational account of morally troublesome behavior and its maintenance. *Personality and Social Psychology Review, 21*(3), 278–299. doi: 10.1177/1088868316647562

Baron, J. (1995). Myside bias in thinking about abortion. *Thinking & Reasoning, 1,* 221–235. doi: 10.1080/13546789508256909

Bastian, B., Loughnan, S., Haslam, N., & Radke, H. R. M. (2012). Don't mind meat? The denial of mind to animals used for human consumption. *Personality and Social Psychology Bulletin, 38*(2), 247–256. doi: 10.1177/0146167211424291

Bazerman, M. H., & Tenbrusel, A. E. (2011). *Blind Spots: Why We Fail to Do What's Right and What To Do About It.* Princeton, NJ: Princeton University Press.

Bilewicz, M., Imhoff, R., & Drogosz, M. (2011). The humanity of what we eat: Conceptions of human uniqueness among vegetarians and omnivores. *European Journal of Social Psychology, 41,* 201–209. doi: 10.1002/ejsp.766

Bohnet, I., van Green, A., & Bazerman, M. H. (2015). When performance trumps gender bias: Joint versus separate evaluation. *Management Science, 62*(5), 1225–1234. doi: 10.1287/mnsc.2015.2186

Borgi, M., & Cirulli, F. (2015). Attitudes toward animals among kindergarten children: Species preferences. *Anthrozoös, 28*(1), 45–59. doi: 10.2752/089279315X14129350721939

Bratanova, B., Loughnan, S., & Bastian, B. (2011). The effect of categorization as food on the perceived moral standing of animals. *Appetite, 57,* 193–196. doi: 10.1016/j.appet.2011.04.020

Bray, H. J., Zambrano, S. C., Chur-Hansen, A., & Ankeny, R. A. (2016). Not appropriate dinner table conversation? Talking to children about meat production. *Appetite, 100,* 1–9. doi: 10.1016/j.appet.2016.01.029

Colléony, A., Clayton, S., Couvet, D., Jalme, M. S., & Prévot, A.-C. (2017). Human preferences for species conservation: Animal charisma trumps endangered status. *Biological Conservation, 206,* 263–269. doi: 10.1016/j.biocon.2016.11.035

Cuthbert, B. N., Schupp, H. T., Bradley, M. M., Birbaumer, N., & Lang, P. J. (2000). Brain potentials in affective picture processing: Covariation with autonomic arousal and affective report. *Biological Psychology, 52,* 95–111. doi: 10.1016/S0301-0511(99)00044-7

Daly, B., & Morton, L. L. (2006). An investigation of human-animal interactions and empathy as related to pet preference, ownership, attachment, and attitudes in children. *Anthrozoös, 19*(2), 113–127. doi: 10.2752/089279306785593801

Dhont, K., & Hodson, G. (2014). Why do right-wing adherents engage in more animal exploitation and meat consumption? *Personality and Individual Differences, 64*, 12–17. doi: 10.1016/j.paid.2014.02.002

Dowsett, E., Semmler, C., Bray, H., Ankeny, R. A., & Chur-Hansen, A. (2018). Neutralising the meat paradox: Cognitive dissonance, gender, and eating animals. *Appetite, 123*, 280–288. doi: 10.1016/j.appet.2018.01.005

Ehrich, K. R., & Irwin, J. R. (2005). Willful ignorance in the request for product attribute information. *Journal of Marketing Research, 42*, 266–277. doi: 10.1509/jmkr.2005.42.3.266

Fransen, M. L, Smit, E. G., & Verlegh, P. W. (2015). Strategies and motives for resistance to persuasion: An integrative framework. *Frontiers in Psychology, 6*(1201), 1–12. doi: 10.3389/fpsyg.2015.01201

Haidt, J. (2001). The emotional dog and its rational tail. A social intuitionist approach to moral judgment. *Psychological Review, 108*, 814–834. doi: 10.1037/0033-295X.108.4.814

Hawkins, R. D., & Williams, J. M. (2016). Children's beliefs about animal minds (child-BAM): Associations with positive and negative child-animal interactions. *Anthrozoös, 29*(3), 503–519. doi: 10.1080/08927936.2016.1189749

Herzog, H. (2010). *Some We Love, Some We Hate, Some We Eat: Why It's So Hard to Think Straight About Animals.* New York: Harper Perennial.

Hodson, G., Dhont, K., & Earle, M. (this volume). Devaluing animals, "animalistic" humans, and people who protect animals. In K. Dhont & G. Hodson (Eds.), *Why We Love and Exploit Animals: Bridging Insights from Academia and Advocacy.* Abingdon: Routledge.

Joy, M. (2010). *Why We Love Dogs, Eat Pigs, and Wear Cows: An Introduction to Carnism.* San Francisco, CA: Red Wheel/Weiser.

Kellert, S. R. (1985). Attitudes toward animals: Age-related development among children. *The Journal of Environmental Education, 16*, 29–39. doi: 10.1080/00958964.1985.9942709

Knight, S., Vrij, A., Cherryman, J., & Nunkoosing, K. (2004). Attitudes toward animal use and belief in animal mind. *Anthrozoös, 17*, 43–62. doi: 10.2752/089279304786991945

Kunda, Z. (1990). The case for motivated reasoning. *Psychological Bulletin, 108*, 480–498. doi: 10.1037/0033-2909.108.3.480

Kunst, J. R., & Hohle, S. M. (2016). Meat eaters by dissociation: How we present, prepare and talk about meat increases willingness to eat meat by reducing empathy and disgust. *Appetite, 105*, 758–774. doi: 10.1016/j.appet.2016.07.009

Lea, E., & Worsley, A. (2003). Benefits and barriers to the consumption of a vegetarian diet in Australia. *Public Health Nutrition, 6*, 505–511.

Leenaert, T. (2017). *How to Create a Vegan World: A Pragmatic Approach.* Herndon, VA: Lantern Books.

Leite, A. C., Dhont, K., & Hodson, G. (2019). Longitudinal effects of human supremacy beliefs and vegetarianism threat on moral exclusion (vs. inclusion) of animals. *European Journal of Social Psychology, 49*, 179–189. doi: 10.1002/ejsp.2497

Liu, B. S., & Ditto, P. H. (2013). What dilemma? Moral evaluation shapes factual beliefs. *Social Psychological & Personality Science, 4*, 316–323. doi: 10.1177/1948550612456045

LoBue, V., Pickard, M. B., Sherman, K., Axford, C., & DeLoache, J. S. (2013). Young children's interest in live animals. *British Journal of Developmental Psychology, 31*, 57–69. doi: 10.1111/j.2044-835X.2012.02078.x

Loughnan, S., Bastian, B., & Haslam, N. (2014). The psychology of eating animals. *Current Directions in Psychological Science, 23*(2), 104–108. doi: 10.1177/0963721414525781

Loughnan, S., Haslam, N., & Bastian, B. (2010). The role of meat consumption in the denial of moral status and mind to meat animals. *Appetite, 55*, 156–159. doi: 10.1016/j.appet.2010.05.043

MacInnis, C. C., & Hodson, G. (2017). It ain't easy eating greens: Evidence of bias toward vegetarians and vegans from both source and target. *Group Processes & Intergroup Relations, 20*(6), 721–744. doi: 10.1177/1368430215618253

Mazar, N., Amir, O., & Ariely, D. (2008). The dishonesty of honest people: A theory of self-concept maintenance. *Journal of Market Research, 45*, 633–644. doi:10.1509/jmkr.45.6.633

McWilliams, J. (2015). *The Modern Savage: Our Unthinking Decision to Eat Animals.* New York: St. Martin's Press.

Neal, D.T., Wood, W., & Quinn, J. M. (2006). Habits – A repeat performance. *Current Directions in Psychological Science, 15*, 198–202. doi: 10.1111/j.1467-8721.2006.00435.x

Nickerson, R. S. (1998). Confirmation bias: A ubiquitous phenomena in many guises. *Review of General Psychology, 2*, 175–220.

Piazza, J. Gardiner, L., & Stones, K. (2018). [Moral judgments of exotic animal killing as a function of food tradition and animal characteristics]. Unpublished raw data.

Piazza, J., Landy, J. F., & Goodwin, G. P. (2014). Cruel nature: Harmfulness as an important, overlooked dimension in judgments of moral standing. *Cognition, 131*, 108–124. doi: 10.1016/j.cognition.2013.12.013

Piazza, J., & Loughnan, S. (2016). When meat gets personal, animals' minds matter less: Motivated use of intelligence information in judgments of moral standing. *Social Psychological and Personality Science, 7*(8), 867–874. doi: 10.1177/1948550616660159

Piazza, J., McLatchie, N., & Olesen, C. (2018). Are baby animals less appetizing? Tenderness towards baby animals and appetite for meat. *Anthrozoös, 31*, 319–335. doi: 10.1080/08927936.2018.1455456

Piazza, J., Ruby, M. B., Loughnan, S., Luong, M., Kulik, J., Watkins, H. M., & Seigerman, M. (2015). Rationalizing meat consumption: The 4Ns. *Appetite, 91*, 114–128. doi: 10.1016/j.appet.2015.04.011

Plous, S. (1993). Psychological mechanisms in the human use of animals. *Journal of Social Issues, 49*, 11–52. doi: 10.1111/j.1540–4560.1993.tb00907.x

Rothgerber, H. (2012). Real men don't eat (vegetable) quiche: Masculinity and the justification of meat consumption. *Psychology of Men & Masculinity, 14*, 363–375. doi: 10.1037/a0030379

Ruby, M. B. (2012). Vegetarianism: A blossoming field of study. *Appetite, 58*, 141–150. doi: 10.1016/j.appet.2011.09.019

Serpell, J. (2004). Factors influencing human attitudes to animals and their welfare. *Animal Welfare, 13*, 145–151.

Sezer, O., Gino, F., & Bazerman, M. H. (2015). Ethical blind spots: Explaining unintentional unethical behavior. *Current Opinion in Psychology, 6*, 77–81.

Stockburger, J., Renner, B., Weike, A. I., Hamm, A. O., & Schupp, H. T. (2009). Vegetarianism and food perception: Selective visual attention to meat pictures. *Appetite, 52*(2), 513–516. doi: 10.1016/j.appet.2008.10.001

Suzuki, Y., Galli, L., Ikeda, A., Itakura, S., & Kitazaki, M. (2015). Measuring empathy for human and robot hand pain using electroencephalography. *Scientific Reports, 5*, 15924. doi: 10.1038/srep15924

Thorn, P., Howell, T. J., Brown, C., & Bennett, P. C. (2015). The canine cuteness effect: Owner perceived cuteness as a predictor of human-dog relationship quality. *Anthrozoös, 28*, 569–585. doi: 10.1080/08927936.2015.1069992

Tisdell, C., Wilson, C., & Nantha, H. S. (2006). Public choice of species for the 'Ark': Phylogenetic similarity and preferred wildlife species for survival. *Journal for Nature Conservation, 14*, 97–105. doi: 10.1016/j.jnc.2005.11.001

Waite, R. (2018, Jan 24). 2018 will see high meat consumption in the U.S., but the American diet is shifting. *World Resources Institute.* Retrieved from www.wri.org/blog/2018/01/2018-will-see-high-meat-consumption-us-american-diet-shifting

Webber, J. (2017, Dec 29). Demand for vegan and vegetarian food increased by 987% in 2017. *LiveKindly.* Retrieved from www.livekindly.co/just-eat-veganism/

Weiss, E., Miller, K., Mohan-Gibbons, H., & Vela, C. (2012). Why did you choose this pet? Adopters and pet selection preferences in five animals shelters in the United States. *Animals, 2,* 144–159. doi: 10.3390/ani2020144

Westbury, H. R., & Neumann, D. L. (2008). Empathy-related responses to moving film stimuli depicting human and non-human animal targets in negative circumstances. *Biological Psychology, 78,* 66–74. doi: 10.1016/j.biopsycho.2007.12.009

Zaraska, M. (2016). *Meathooked: The History and Science of Our 2.5-Million-Year Obsession With Meat.* New York: Basic Books.

9

FEATHERLESS CHICKENS AND PUPPIES THAT GLOW IN THE DARK

Moral heuristics and the concept of animal "naturalness"

Christopher J. Holden and Harold Herzog

Abstract

Heuristics are unconscious strategies that we all use to speed up our thinking. In some cases, these are effective, and in others, they can lead to inconsistencies in judgments. Further, it has been suggested that there is a set of moral heuristics that functions in the same way when we are making moral judgements. One way in which we often make moral judgments using heuristics is when we are deciding on what is natural. Put another way, some things just seem "more natural" than others. For example, we might see eating poultry as natural, but genetically engineering the chickens used in factory farming as unnatural, even if the genetic engineering was done to lessen their suffering. In this chapter, we explore how these moral heuristics function in the case of determining animal naturalness and explore their implications in terms of how we think about and interact with animals. In a broad sense, it appears that different people make different decisions regarding what is and is not natural.

Some animals just seem more natural than others. Take chickens. In 1954, one male and four female baby chicks were born with no feathers in the poultry sciences department at the University of California at Davis. A spontaneous genetic mutation caused this strange skin disorder. The researchers named the mutation "Scaleless," and the five animals were used to develop a new genetic line of chickens. No one paid much attention to this research project until 2000 when an Israeli geneticist named Avigdor Cahaner began interbreeding featherless chickens with commercial broiler stock (i.e., chickens bred for the production of meat; see Figure 9.1). Cahaner realized that overheating was a serious problem which inhibits poultry production in hot climates, and his goal was to produce broilers that would thrive under tropical conditions. Initially, the featherless birds were produced by

FIGURE 9.1   Featherless chicken photograph.

Source: Photograph by Tim Flach.

selective breeding. In recent years, however, researchers have used genetic engineering in an attempt to produce improved lines of featherless birds (Wells et al.,
2012). Their efforts have been successful. In a subsequent report, Cahaner's team
wrote: "It appears that broiler meat production in hot regions and climates can
be substantially improved by introducing the featherless gene into contemporary
broiler stocks" (Hadad, Halevy, & Cahaner, 2014, p. 1067).

As you might expect, reactions to the prospect of large-scale production of featherless chickens have been mixed. A few poultry scientists have argued that not
having feathers actually enhances animal welfare because the birds are more comfortable in tropical conditions (Davis, 2014). In addition, Cahaner claims featherless
birds are environmentally friendly. He says they save energy by eliminating the need
for air conditioning in broiler houses and the need to pluck feathers from carcasses.
Animal protectionists disagree. As one activist wrote: "Genetically engineering any
creature for human convenience, and the "bottom line" is UNNATURAL!" (van
der Sluis, 2007).

When compared to their progenitor the red jungle fowl of Southeast Asia, however, all poultry sold in supermarkets are decidedly unnatural. The Cobb 500, for instance, is touted by the Cobb-Vantress Corporation as the world's most popular broiler. In 1925, it took four months and ten pounds of feed to produce a scrawny two -and-a-half pound bird. Thanks to advances in poultry science, today a Cobb 500 chick can grow to nearly five pounds in six weeks. More remarkably, now it takes only a pound and a half of feed to produce a pound of chicken.

Disputes about what is natural frequently crop up in discussions of ethics and chickens. Animal protectionists denounce the unnatural lighting regimes of broiler houses, the unnatural intensive crowding conditions, and the unnaturally large size and overly rapid growth rates of modern chickens. Ironically, participants in a brutal blood sport of cockfighting also invoke claims about what is natural when defending the practice. One of us (Harold Herzog) spent several years investigating the clandestine rooster fighting subculture in the southern Appalachian Mountains of the United States. A focus of this research was how individuals justified their involvement in an activity that is patently cruel and nearly always ends in death (Herzog, 2010). Perhaps the most frequent defense of cockfighting among participants was their conviction that game roosters are natural fighters. As one claimed, "What we do is an act of nature. We don't make these roosters fight. That's what they were put here for." Another argued, "It would be cruel not to let them fight."

Mental shortcuts affect how we think about animals

Our revulsion at the prospect of a genetically engineered featherless chicken, the decision to pay more for a broiler labeled "all natural," and the "it's only natural" justification for cockfighting, all reflect the operations of processes cognitive psychologists refer to as *heuristics*. Heuristics are unconscious intuitive rules of the thumb that facilitate rapid judgments and decision making. We use these mental short cuts to solve the Sunday crossword puzzle, and emergency room doctors use heuristics in deciding if a patient is suffering from a heart attack or indigestion. A study by the Nobel Prize-winning psychologist Daniel Kahneman illustrates how heuristics influence our thinking about other species. *Anchoring* is a common heuristic in which our cognitive processes are biased by an initial reference point[1]. In his book, *Thinking, Fast and Slow,* Kahneman (2011) described how he once asked visitors to the San Francisco Emporium how much they would pay to save sea birds from oil spills. Some of the visitors were given an initial anchor point by asking "Would you be willing to pay X dollars to save 50,000 birds?" and then they were asked what was the *most* they would be willing to pay. In some cases X was $5 and in others it was $500. Participants who were given no anchor said they would be willing to donate, on average, $64. But respondents given the $5 anchor said they would be willing to pay an average of $20. In contrast, the people who were given the $500 anchor indicated they would be willing to pay $142 to save 50,000 birds. In this case, simply changing the anchor made for a seven-fold difference in the amount of money people would be willing to donate in order to save the same number of birds. This also suggests that participants were using the anchor to determine the appropriate amount to donate, without relying on other sources of information. For example, they may not have considered their budget, or their concerns for animal welfare may not have factored into this decision. Thus, the use of an anchor is a heuristic in the sense that it may help to facilitate faster decisions, but may not lead to the most accurate or sound decisions.

"It's natural" is a moral heuristic

The legal scholar Cass Sunstein (2005) used the term *moral heuristics* to refer to unconscious rules of thumb that influence our judgments about ethical issues. The psychologist Lewis Petrinovich and his colleagues used hypothetical ethical quandaries to explore how our thinking about animals is influenced by the quick and dirty rules of thumb (Petrinovich, O'Neill, & Jorgenson, 1993). Take, for example, the Trolley Problem, a well-known moral dilemma. Here is the most common version: *An out-of-control trolley car is careening down the tracks toward a group of five people. But you can save them by pulling a switch that will send the trolley toward a spur where one person is standing. Should you kill one person if it will save five people?* Faced with this decision, most people say that sacrificing one person to save five people would be the right thing to do. However, what if you had to *personally push* one person onto the tracks in order to save five lives? In this case, most people change their mind and say no, even though the utilitarian calculus is the same in both situations.

Petrinovich and colleagues' (1993) trolley car scenarios pitted the lives of people against the lives of animals. Suppose the trolley was headed for a group of the world's last five remaining mountain gorillas. Should you save them by pushing the switch and sending the trolley car towards a 25-year-old man? In this case, most participants said they would not push the switch and kill one man so the gorillas could live. Likewise, in a hypothetical lifeboat situation, nearly all of Petrinovich and colleagues' (1993) participants said a dog should get tossed overboard rather than a person. After testing hundreds of people using dozens of scenarios, the researchers reported the single most powerful heuristic they uncovered was, "Save people over animals." Note, however, that in a more recent study, 30% of men and 45% women said they would sacrifice the life of a foreign stranger in the path of a runaway bus if it would save the life of their own dog (Topolski, Weaver, Martin, & McCoy, 2013).

A similarly powerful moral heuristic is the belief that "natural is good." Claims about natural and unnatural behaviors frequently come up in discussions about the rights of animals. For example, Melanie Joy (2011) points out in her book, *Why We Love Dogs, Eat Pigs, and Wear Cows*, the "it's only natural" argument is one of the most common rationalizations for eating animals. Jared Piazza of Lancaster University and his colleagues asked college students to write down three reasons why they eat meat. Among the most common statements were comments like, "It's natural for humans to eat meat," "Evolutionarily, hominids have always eaten meat," "Organisms consuming each other is something that is prevalent in nature," and "Humans were meant to have dominion over animals" (Piazza, et al. 2015; see also Loughan & Davies, this volume; Piazza, this volume).

Variations of "it's natural" are sometimes used by philosophers who make the case *for* animal rights. Bernard Rollin's argument for giving moral status to non-human species hinges on the Greek concept of *telos*. Rollin believes that animals have an essential nature that deserves respect. Rollin writes, "As ordinary people know well, animals too have natures, genetically based, physically and psychologic-ally expressed which determine how they live in their environments. Following Aristotle, I call this the *telos* of an animal, the pigness of the pig, the dogness of the dog – "fish gotta swim, birds gotta fly"" (Rollin, 1995, p. 157). Rollin's ideas about animal naturalness and animal welfare have been influential. His notion of *telos* was incorporated into a bill introduced into the Scottish Parliament in 2017 to ban traveling circus animal acts. Indeed, one of the often-cited Five Freedoms of Animal Welfare is "the freedom to express normal and natural behavior (e.g., accommodating for a chicken's instinct to roost)" (American Humane Association, 2016; Pierce, this volume).

"Natural" and the problem of pets

Pet-keeping raises a host of ethical issues related to "natural" and "unnatural." Cats are particularly problematic. Unlike dogs, cats require meat in order to thrive (Bradshaw, 2013). But the fact that cats are natural meat-eaters poses a problem for cat owners who believe the consumption of meat is immoral. As reported by the psychologist Hank Rothgerber, they can experience *pet food guilt* (Rothgerber, 2013). He found that cat-owning vegans suffered more pet food guilt than did vegetarians who had pet cats. Further, people who avoided meat for ethical reasons were more likely to experience pet food guilt than people who had given up meat because of health concerns.

Then there is the problem of living with a predator. Not all cats are habitual killers, but pet cats do take a heavy toll on wildlife. This raises the issue of whether pet cats should be allowed to spend part of their day outdoors where they can express their natural hunting proclivities. Rollin's argument for animal welfare would seem to support the idea that confining a cat 24/7 inside a house or an apartment violates its *telos*. If, as Rollin claims, "fish gotta swim, and birds gotta fly," it seems to follow that "cats gotta kill." The bioethicist Jessica Pierce also believes that it is unethical to force cats to spend their lives essentially caged inside a house. In her 2016 book *Run, Spot, Run: The Ethics of Keeping Pets* she writes, "Many cat owners, myself included, feel that cats lose something important when denied access to the outside world and that the risks posed by the outside world are ones that at least some cats would happily trade for freedom" (p. 109). Advocates on different sides of the cat predation issue disagree on the role cats play in nature.

Researchers at North Carolina State University found that 60% of feral cat colony managers, but only 5% of bird conservationists, agreed with the statement "Feral cats fill a natural role as predators" (Peterson, et al., 2012).

Cats and dogs are also different in ways other than their dietary needs. Dogs are the oldest domestic animal. Experts argue about the exact time(s), place(s), and reason(s) wild canids hooked up with our ancestors, however, there is general agreement that dogs evolved from grey wolves between 15,000 and 25,000 years ago. For nearly all of this time, dogs were the not the pampered pets seen in developed countries today but free-ranging creatures living a semi-wild existence in and around human habitations. Even now, 85% of the roughly one billion dogs on Earth are not owned by a human nor do they live in houses or eat commercial pet food (Lord, Feinstein, Smith, & Coppinger, 2013). Rather they are feral creatures referred to variously as pariah dogs, village dogs, or street dogs. They hang around rural villages and scavenge in urban garbage dumps, and they control their own reproductive destinies. Over the past 1,000 years, a few functional body types emerged in dogs used for, say, hunting or guarding. Humans, however, only really began to intentionally mold canine morphology about 200 years ago with the rise of kennel clubs and the development of formal breeds selected on the basis of arbitrary standards of beauty. There are now over 400 recognized breeds, and the domestic dog has become the most variable animal on Earth. The size difference between a two-pound Yorkshire terrier and a 200-pound Mastiff, for instance, is proportionately the same as the difference in size between an adult human male and an African bull elephant.

In recent years, modern technology has enabled the direct modification of dog genes. Snuppy, the world's first cloned dog, was born in Korea in 2005, and over the last decade research on dog cloning has expanded. Researchers have even produced a litter of cloned drug detector dogs, writing, "breeding cloned dogs with excellent abilities will be an optimal method to produce more candidate dogs for unique purposes (Lee et. al., 2016 p. 411). In 2009, the same research team that produced Snuppy inserted a gene into a beagle named Ruppy which made the dog glow in the dark (Callaway, 2009). And recently Chinese researchers used CRISPR, a new gene editing tool, to produce dogs with twice the normal amount of muscle mass. Their goal is to develop dogs with disorders such as Parkinson's disease and muscular dystrophy. These genetically modified animals would then be used in the search for treatments of these afflictions in humans (Griffin, 2015).

Studying perceptions of animal naturalness

From preferences for free-range beef to the logical foundations of animal protectionism, beliefs about of what is "natural" profoundly affect how we think about members of other species. As illustrated by the reactions to featherless chickens, some animals are perceived as less natural than others. Yet, as far as we know, no attempts have been made to systematically investigate how people view the naturalness of different types of animals or the factors that reduce perceptions of animal naturalness. Here we describe our efforts to develop a scale to assess differences in the perceived naturalness of animals in a variety of circumstances. We believe it is important to understand how the "natural is good" heuristic influences human attitudes towards other species. For instance, the perception that meat grown in laboratories (also called *in vitro meat* or "clean meat") is unnatural is likely to be a serious deterrent to the acceptance of technological advances that have the potential to greatly reduce animal suffering.

To explore perceptions of animal naturalness, we adapted methods developed by University of Pennsylvania psychologist Paul Rozin (2005) to study the meaning of "naturalness" primarily in the contexts of foods and medicines. He asked people to rate the naturalness of foods and other entities on a scale of 0 (not natural at all) to 6 (completely natural). The technique allowed Rozin to assess decrements in perceived naturalness associated with manipulations such as freezing and genetic modification. For example, if a subject were to rate organic milk as a 6 and skim milk as a 1 on the naturalness scale, the process of removing the fat from milk would have produced a 83% decrement in perceived naturalness. Using this method, Rozin found, for instance, that chemical changes such as removing fat from milk produced larger decrements in perceived naturalness than did physical processes such as grinding peanuts into peanut butter.

Rozin's research focused primarily on the naturalness of foods and water. But to examine the impact of domestication, his participants also rated the naturalness of several breeds of dogs. Intrigued by Rozin's findings, we developed the Animal Naturalness Scale specifically to investigate factors that affect human perceptions of animals in different situations. We were interested in three questions. First, can human subjective perceptions of the "naturalness" of animals be reliably measured? Second, what factors impact the perceived naturalness of animal species? And finally, do those who are more sensitive to changes in animal naturalness have different attitudes towards animal welfare, trust in science, and the naturalness of various foods?

To study how people perceive the naturalness of animals in different situations, we recruited 137 participants through Amazon's Mechanical Turk (MTurk). MTurk is an Internet participant pool which is now widely accepted as a source of participants for online psychological studies (Buhrmester, Kwang, & Gosling, 2011). The participants were equally divided between men and women, and they were, on average, 32 years old. Each participant was paid $1 for taking the survey. The Animal Naturalness Scale included 32 items, each describing an animal in a specific situation.[2] We used these items to compare evaluations of the naturalness of the same species in different settings. We were particularly interested in effects on perceived naturalness caused by three processes. They were domestication (e.g., a feral dog versus a mixed-breed dog), captivity (e.g., a dolphin in the Atlantic Ocean versus a dolphin in a tank at Sea World), and changes caused by biotechnology (e.g., a dairy cow versus a dairy cow genetically modified to produce insulin in its milk). In addition, we included comparisons of special interest such as differences in the perceived naturalness of pedigree dogs versus mixed breed dogs and between indoor cats and pet cats allowed to go outside.

TABLE 9.1 Average perceived animal "naturalness" scores

Animal	"Naturalness" Score
An elephant living in the wild of Africa.	5.80
A wolf living in the wild.	5.80
A monkey living in a jungle in the Congo.	5.78
A fish living in a tropical reef.	5.77
A dolphin living in the Atlantic Ocean.	5.75
A horse living on the plains in Montana.	5.51
A feral (unowned) dog living in a village.	4.54
A rat living in a garbage dump.	4.5
A mixed breed cat.	4.42
A mixed breed dog.	4.23
A family cat which is allowed to spend part of the day outside.	4.01
A free range chicken raised on a small farm.	3.95
A horse on a family farm.	3.82
A dairy cow.	3.66
A purebred Siamese cat.	3.55
A purebred pedigree poodle.	3.08
A family cat which is always kept indoors.	2.95
A thoroughbred race horse.	2.54
A tropical fish living in an aquarium.	2.22
A dairy cow produced by artificial insemination.	2.18
A monkey living in a zoo in the United States.	2.01
An elephant in a zoo in the United States.	1.85
A dolphin living in a tank at Sea World.	1.47
A broiler chicken raised in intensive housing conditions on a "factory farm."	1.38
A rat in a psychology laboratory.	1.32
A genetically-modified dairy cow that produces insulin in its milk.	1.30
A dairy cow produced by artificial insemination.	1.26
A pet dog that is the product of reproductive technology (cloning).	1.15
A monkey living in a biomedical research facility.	1.11
A genetically-modified rat used for research.	0.99
A genetically-engineered aquarium fish with a gene inserted that makes it glow in the dark.	0.95
A chicken which has been genetically engineered so it does not experience pain.	0.93

How natural is purebred poodle or an indoor cat?

We included items related to three dogs and a wolf in our version of the Animal Naturalness Scale (see Table 9.1). With an average naturalness score of 5.8, a wolf living in the wild had the highest naturalness rating on the 0 to 6 scale. A cloned dog, in contrast, only scored 1.2. At 3.1, poodles were in the middle. Our participants rated a mixed breed dog as 27% less natural than a wolf, and a purebred poodle was viewed as 13% less natural than a mixed breed dog. Cloning caused the biggest decline; a cloned dog was perceived as 69% less natural than a mixed breed dog.

Because of their predatory inclinations and inherent wildness, we thought that cats might be perceived as more natural than dogs. While the oldest skeletal remains of domestic cats date back 7000 years, some researchers argue that cats have only really become domesticated in the last two centuries (Driscoll, Macdonald, & O'Brian, 2009). But contrary to our expectations, cats were not perceived as being more natural than dogs. A mixed breed cat had, on average, a naturalness rating of 4.4 compared to 4.2 for a mixed breed dog, a negligible difference. On the other hand, a purebred Siamese cat, was viewed as 12% less natural than a mixed breed cat. Perhaps our most interesting feline finding was that being kept indoors reduced a cat's perceived naturalness by 28%.

Domestication, captivity, and biotechnology

The average naturalness scores of all 32 animals are listed in order from most "natural" to least "natural" in Table 9.1. Note that the five animals rated as most natural were wild creatures living in their natural habitats. In contrast, manipulations causing the five largest decrements in perceived naturalness involved science and technology. To assess the relative power of domestication, captivity, biotechnology to reduce the perceived naturalness of animals, we selected five comparisons involving each of the three factors. For instance, to represent the impact of captivity we used: a fish in an aquarium versus a fish on a tropical reef, a monkey in zoo versus a monkey living in an African jungle, a factory farm chicken versus a free-range chicken, a dolphin living in a tank at Sea World versus a dolphin in the ocean, and an elephant in a zoo versus an elephant in a jungle in Africa.

The results were striking. Captive animals were perceived as being, on average, 66% less natural than their wild counterparts. Biotechnological manipulations had a similar effect, producing, on average, a 64% reduction in naturalness. Domestication, on the other hand, only reduced average naturalness rating by 15%. The difference we found between the impact of domestication and biotechnology mirrors the findings of Rozin (2005).

People differ in their perceptions of animal naturalness

We also examined individual differences among our participants' susceptibility to circumstances that decreased animal naturalness. Researchers have found that beliefs about the treatment of animals are associated with an array of psychological traits and constellations of attitudes. For example, people who are concerned about animal welfare tend to score low on measures of social dominance, high on empathy, and be sensitive to moral disgust (Dhont, Hodson, Costello & MacInnis, 2014; Herzog & Golden, 2009; Taylor & Signal, 2005). Likewise, opposition to GMO foods and genetic engineering is associated with an absolutist moral orientation and intuitive styles of thinking, viewing nature as sacred, and belief in God

(Ragojlovic & Einsiedel, 2013; Scott, Inbar, & Rozin, 2016; Saher, Lineman, & Hursti (2006).

In light of these previous findings, we developed three hypotheses related to individual differences. We anticipated that participants who perceived greater declines in the naturalness of animals across the categories (i.e., domestication, captivity, and genetic modification) would (a) be more concerned with animal welfare, (b) be more sensitive to technological manipulations of foods, (c) have more negative attitudes toward science. Our rationale for these hypotheses was, first, that practices like captivity and genetic modification can be, and are often seen as, harmful to the animals. Some of this perception likely stems from the fact that it is considered unnatural to do so. Therefore, individuals who are concerned with animal welfare might be particularly attuned to changes that detract from the naturalness of animals. Additionally, we thought that this same sensitivity to changes that detract from an animal's naturalness would carry over to perceptions of food, particularly if that food was genetically modified. Thus, we incorporated the scale assessing perceived naturalness of alternative foods. Finally, as many of the genetic modifications being made to animals are relatively novel scientific developments, we thought that the perceptions people have about the application of science would play a role in how natural people found things such as a genetically modified cow. For example, someone who feels as though it is appropriate to change a baby's genetic characteristics to avoid the risk of disease might also feel that it is natural and appropriate to genetically modify animals. To test these ideas, our participants completed the four following scales.

- The *Attitudes toward Animals Scale (10 Item Version)* is a measure of general attitudes and beliefs about the treatment of animals (Herzog, Grayson, & McCord, 2015). ("It is morally wrong to hunt wild animals just for sport.")
- The *Alternative Food Sources Naturalness Scale* was developed for the present study. It included 12 items in which the participants rated the naturalness of foods such as GMO vegetables and lab grown meat on a 0 (not at all natural) to 6 (completely natural) scale.
- The *Belief about Science Scale* contains ten items pertaining to how individuals value science as way of knowing, for example, "Science is the most efficient means of attaining truth" (Farias, Newheiser, Kahane & de Toledo, 2013).
- The *Science Application Scale* is a five-item measure of beliefs taken from the Pew Science Survey designed to assess how people view science in terms of understanding human origins and interactions with our environment, and how scientific discoveries can be applied to health and the production of food. For example one item assessing general beliefs about science reads: "Humans and other living things have evolved due to natural processes such as natural selection"; whereas an item assessing broader applications of science reads: "All children should be required to be vaccinated against diseases such as measles, mumps, rubella, and polio" (Pew Research Center, 2009).

We were surprised by the large individual differences in how susceptible our participants were to decreases in the perceived naturalness of animals. Some people, for example rated an elephant in a zoo as 20% less natural than an elephant living in an African jungle, while others rated the zoo elephant as 100% less natural than its wild counterpart. To explore these individual differences, we calculated a score reflecting how susceptible our participants were to the effects of domestication, captivity, and biotechnology. More specifically, we computed a "change in naturalness score" for each of our participants by averaging their percent change in perceived naturalness over the 15 comparisons we used to assess the impacts of domestication, captivity, and biotechnology.

Our participants reported an average change in naturalness of 52%. However, some participants had average scores of less than 10% while others had scores over 85%. We predicted that individuals who perceived greater change in the naturalness of animals would be more concerned with animal protection as measured by the Animal Attitudes Scale. We also thought they would have less favorable attitudes toward science as measured by the Beliefs about Science Scale and the Science Application Scale. Finally, we reasoned that people who perceived greater change in naturalness would tend to give *lower* scores when it came to rating the naturalness technologically manipulated foods as measured by the Alternative Food Sources Scale.

Two of the hypotheses were supported. We found a statistically significant relationship between individual differences in concern for animal welfare and the amount of perceived change in animal naturalness ($r = .30$). This finding indicates that individuals with more concern for animal welfare perceived more of a reduction in naturalness across the categories of domestication, captivity, and genetic modification. There was also a strong inverse relationship between animal naturalness decrement scores and perceptions of the naturalness of foods ($r = -.52$). That is, people who perceived greater declines in animal naturalness across the categories were more inclined to perceive foods such as lab grown meat and GMO vegetables as unnatural. Contrary to our expectations, however, neither of the two measures of attitudes toward science were related to perceptions of decrements of animal naturalness. This would suggest that the scientific perspective that people have on the world does not influence their perception of naturalness; or that they may perceive the changes in naturalness as justified.

Where lies the future? The blind hens' challenge

In her book *Frankenstein's Cat: Cuddling Up To Biotech's New Beast,* Emily Anthes (2013) asks, "What does biotechnology really mean for the world's wild things? And what do our brave new beasts say about us?" (p. 11). It is a good question. Developments in biotechnology present new challenges in how we think about ethical treatment of animals. Take, for example, "the blind hens' challenge."

In a 1985 article published in the journal *Poultry Science,* Ahmed Ali and Kimberly Cheng reported the results of a study of egg production in a genetic line of blind hens. Like the first featherless chickens, blindness originally appeared in a flock of chickens as the result of a random mutation. Researchers then used these animals to develop a strain of sightless birds. Surprisingly, Ali and Cheng found that some of the problems that plague hens used in industrial egg production such as cannibalism and feather-picking were notably absent in their blind chickens. In fact, they reported that the hens coped with the crowded conditions of large scale poultry production amazingly well. The sightless birds produced 13 percent more eggs and consumed less feed than sighted hens. And physiological indicators suggested that the blind chickens suffered less stress. Encouraged by these results, Ali and Cheng proposed that blind hens be used for commercial egg production.

Subsequent studies, however, found that blind chickens do have their share of problems. For example, they show abnormal patterns of preening and tend to walk around in circles (Collins, Forkman, Kristensen, Sandøe, & Hocking, 2011). But as Peter Sandøe and his colleagues (2014) point out, either way, the specter of sightless chickens which are oblivious to the indignities of factory farms raises ethical issues. The philosopher Adam Shriver (2009) agrees. In his article, "Knocking Out Pain in Livestock: Can Technology Succeed Where Morality Fails?" Shriver writes, "we may be very close to, if not already, at the point where we can genetically modify factory-farmed livestock with a reduced or completely eliminated capacity to suffer." Shriver then constructs a logical argument that leads him to the conclusion that at some point we need to seriously consider replacing current farm animals with cattle and pigs genetically engineered for diminished abilities to suffer.

The natural heuristics and the ick factor

We found Shriver's line of thinking convincing, but we were not sure how his ideas would resonate with the public. So we ran it by our friend Barbara one afternoon during happy hour at Maggie B's, our local watering hole. First, we told Barbara how the creation of a line of blind hens might dramatically reduce the cruelty inherent in large-scale egg production. Then we told her about Adam Shriver's contention that advances in molecular neurobiology might soon enable researchers to produce cattle and hogs incapable of suffering.

"Think about it, Barbara. Cruelty-free meat. Wouldn't that be great?"

With a horrified look on her face, Barbara said, "That's disgusting."

We were curious about her lack of enthusiasm for cruelty-free meat and eggs, so taking a page from the playbook of the psychologist Johnathan Haidt, we decided to push her reasoning. To study the role of emotions such as disgust in moral judgments, Haidt confronted people with ethically problematic scenarios that did not involve any harm. He found that even though no one was hurt, most people objected to acts like cleaning a toilet with an American flag or eating a family dog which had just been killed by a car. Often, however, Haidt's participants became flustered when they could not come up with any logical justifications for their decisions, and they would often say something like "I don't know, I can't explain it, I just know it's wrong" (Haidt, 1995).

Much like Haidt's participants, Barbara became flustered when we asked why she was so adamantly opposed to genetically engineered, cruelty-free meat. After hemming and hawing a few minutes, she finally blurted out, "It's the *ick* factor. Blind chickens and cows that don't feel pain? That's just…unnatural!"

Why the study of perceptions of animal naturalness is important

Barbara's opposition to sightless hens and cows with re-jiggered brains illustrates the powerful influence of the natural heuristic on how people think about the treatment of animals. Research on attitudes towards lab grown *in vitro* meat by Matti Wilks and Clive Phillips (2017) supports the view that animal protectionists need to understand public concerns about animal naturalness and unnaturalness. Wilks and Phillips found that although participants reported a number of practical concerns regarding *in vitro* meat such as taste and price of the new product, many also saw the product as being unnatural. These findings are interesting considering that participants also reported viewing in vitro meat as a more ethical alternative to factory farming (Wilks & Phillips, 2017). Overall, this leads to the conclusion that concerns about naturalness are likely to be a serious deterrent to the acceptance of technological advances that have the potential to greatly reduce animal suffering (Wilks & Phillips, 2017).

It was our intent to build on these lines of research to better understand how people make decisions about what is moral and what is natural. The research we have described is an initial attempt to investigate psychological aspects of beliefs about animal naturalness. We found that animals vary greatly in their perceived naturalness. We also found that captivity and biotechnology cause substantial decrements in the perceived naturalness of animals while domestication has much smaller effects. Finally, our studies suggested individuals who are concerned about animal welfare tend to be particularly sensitive to changes in animal naturalness.

As is true of nearly all research, our attempt to measure perceptions of animal naturalness has limitations. We obtained our sample from the MTurk subject pool. Thus the degree that the results are generalizable to other populations is unclear. MTurk is an international platform, but we chose to limit our pool to participants within the United States. Although the MTurk population is more diverse than other participant pools (e.g., college undergraduate samples), it may not be entirely representative of the US population (Ross, Zaldivar, Irani, & Tomlinson, 2010). A potential problem stems from our estimates of the relative importance of domestication, captivity, and biotechnology in causing declines in the perceived naturalness of animals. These estimates were calculated by comparing the naturalness ratings of a more "natural" animal (e.g., a monkey in a jungle) with a less natural counterpart (e.g., a monkey in a zoo). In some cases, however, the "more versus less natural" comparisons involved could have involved multiple factors. For example, the use of animals in science involves captivity as well as biotechnology.

Despite these limitations, we believe this study demonstrates that Rozin's method of rating the perceived naturalness of foods can be fruitfully applied to investigations of the perceptions of naturalness of animals. We hope our efforts will spur further research. Variations based on the animal naturalness scale could be adapted to a wide variety of topics. These include studies of

- cultural differences in the perceived naturalness of animals;
- in-depth investigations of perceived naturalness of specific groups, for example, breeds of dogs or animals in various habitats in zoos;
- the influence of perceived animal naturalness on human behavior such as food choices and donations to animal protection organizations and environmental causes;
- factors that are associated with sensitivity to decreased in perceived naturalness;
- and, finally, assessing the acceptability of types of robotic pets.

We are now studying additional factors that affect individual differences in perceptions of animal naturalness including individual differences in fundamental ethical values and political ideology. This is an important addition to this line of research because it can better elucidate how people come to their moral decisions. More specifically, it has been proposed that people make decisions about what is right and wrong based on different values, and that these values differ across cultures and groups (Graham, Haidt, & Nosek, 2009; Haidt & Joseph, 2004). For example, some individuals may consider whether a harm was done in determining whether an action was moral or not, and others may be more influenced by factors such as purity, loyalty, or fairness (Graham et al., 2013).

We also plan on conducting studies that will better assess how this naturalness heuristic operates. More specifically, we will have participants make decisions about certain foods and products based on their description of being natural or not. For example, if when given a limited budget, are participants more likely to buy a food that is labeled as natural, despite it costing more? If so, this would be fairly strong evidence for the naturalness heuristic in action.

We hope experiments like these can bring some clarity to the messy ways by which humans decide what is natural and what is not. We have already seen that different individuals make somewhat different decisions regarding what is natural, and this could be important as the technology surrounding genetic modification of humans and animals continues to develop. In some cases, this genetic modification could prevent suffering, or could lead to higher yields of food, which both have global consequences. However, these issues are likely to be contested as it appears there is no universal means by which people determine what is natural.

Notes

1 It should be noted that there has been recent discussion suggesting that anchoring effects may not truly be a heuristic (Frederick & Mochon, 2012), however some argue that

anchoring is a heuristic (Shah & Oppenheimer, 2008). It has also been suggested that people may adjust from the initial anchor, but that these adjustments may not always be accurate (Epley & Gilovich, 2006) and may not always have adequate information to make those adjustments (Simmons, LeBoeuf, & Nelson, 2010)

2 The full-length items are included in Table 9.1. In the reported study, the internal consistency (i.e., Cronbach's alpha) of this measure was $\alpha = .89$.

References

American Humane Association (2016, October 17) Five Freedoms: The Gold Standard of Animal Welfare [Weblog post]. Retrieved from www.americanhumane.org/blog/five-freedoms-the-gold-standard-of-animal-welfare/

Anthes, E. (2013). *Frankenstein's Cat: Cuddling up to Biotech's Brave New Beasts.* New York: Farrar, Straus & Giroux.

Bradshaw, J. (2013). *Cat Sense: How the New Feline Science Can Make You a Better Friend to Your Pet.* New York: Basic Books.

Callaway, E. (2009). Fluorescent puppy is world's first transgenic dog. *New Scientist, 2706.* Retrieved from www.newscientist.com/article/dn17003-fluorescent-puppy-is-worlds-first-transgenic-dog/

Collins, S., Forkman, B., Kristensen, H. H., Sandøe, P., & Hocking, P. M. (2011). Investigating the importance of vision in poultry: Comparing the behaviour of blind and sighted chickens. *Applied Animal Behaviour Science, 133*(1–2), 60–69.

Buhrmester, M., Kwang, T., & Gosling, S. D. (2011). Amazon's Mechanical Turk: A new source of inexpensive, yet high-quality, data? *Perspectives on Psychological Science, 6,* 3–5.

Davis, K. (2014) Anthropmoporphic visions of chickens bred for human consumption. In J. Sorenson (Ed.) *Critical Animal Studies: Thinking the Unthinkable.* (pp. 169–187). Toronto: Canadian Scholars Press.

Dragojlovic, N., & Einsiedel, E. (2013). Playing God or just unnatural? Religious beliefs and approval of synthetic biology. *Public Understanding of Science, 22,* 869–885.

Dhont, K., Hodson, G., Costello, K., & MacInnis, C. C. (2014). Social dominance orientation connects prejudicial human–human and human–animal relations. *Personality and Individual Differences, 61,* 105–108.

Driscoll, C. A., Macdonald, D. W., & O'Brien, S. J. (2009). From wild animals to domestic pets, an evolutionary view of domestication. *Proceedings of the National Academy of Sciences, 106,* 9971–9978.

Epley, N., & Gilovich, T. (2006). The anchoring-and-adjustment heuristic: Why the adjustments are insufficient. *Psychological Science, 17,* 311–318.

Farias, M., Newheiser, A. K., Kahane, G., & de Toledo, Z. (2013). Scientific faith: Belief in science increases in the face of stress and existential anxiety. *Journal of Experimental Social Psychology, 49,* 1210–1213.

Frederick, S. W., & Mochon, D. (2012). A scale distortion theory of anchoring. *Journal of Experimental Psychology: General, 141,* 124–133.

Graham, J., Haidt, J., Koleva, S., Motyl, M., Iyer, R., Wojcik, S. P., & Ditto, P. H. (2013). Moral foundations theory: The pragmatic validity of moral pluralism. *Advances in Experimental Social Psychology, 47,* 55–130.

Griffin, A. (2015, October). Genetically-engineered, extra-muscular dogs created by Chinese scientists. *The Independent.* Retrieved from www.independent.co.uk/news/science/mutant-extra-muscular-dogs-created-by-chinese-scientists-a6701156.html

Hadad, Y., Halevy, O., & Cahaner, A. (2014). Featherless and feathered broilers under control versus hot conditions. 1. Breast meat yield and quality. *Poultry Science, 93*, 1067–1075.

Haidt, J. (1995). The emotional dog and its rational tail: A social intuitionist approach to moral judgment. *Psychological Review, 108*, 814–834.

Haidt, J., & Joseph, C. (2004). Intuitive ethics: How innately prepared intuitions generate culturally variable virtues. *Daedalus: Special Issue on Human Nature, 133*(4), 55–66.

Herzog, H. A. (2007). Gender differences in human–animal interactions: A review. *Anthrozoös, 20*, 7–21.

Herzog, H., (2010). *Some We Love, Some We Hate, Some We Eat: Why It's So Hard to Think Straight About Animals.* New York: Harper.

Herzog, H. A., & Golden, L. L. (2009). Moral emotions and social activism: The case of animal rights. *Journal of Social Issues, 65*, 485–498.

Herzog, H., Grayson, S., & McCord, D. (2015). Brief measures of the Animal Attitude Scale. *Anthrozoös, 28*, 145–152.

Joy, M. (2011). *Why We Love Dogs, Eat Pigs, and Wear Cows: An Introduction to Carnism.* Newburyport: Conari Press.

Kahneman, D. (2011). *Thinking, Fast and Slow.* New York: Farrar, Straus and Giroux.

Kim, M. J., Oh, H. J., Kim, G. A., Setyawan, E. M. N., Choi, Y. B., Lee, S. H., … Lee, B. C. (2017). Birth of clones of the world's first cloned dog. *Scientific Reports, 7*, 15235.

Lee, J. H., Kim, G. A., Kim, R. S., Lee, J. S., Oh, H. J., Kim, M. J., … Lee, B. C. (2016). Reproductive ability of a cloned male detector dog and behavioral traits of its offspring. *Journal of Veterinary Science, 17*, 407–411.

Li, M., & Chapman, G. B. (2012). Why do people like natural? Instrumental and ideational bases for the naturalness preference. *Journal of Applied Social Psychology, 42*, 2859–2878.

Lord, K., Feinstein, M., Smith, B., & Coppinger, R. (2013). Variation in reproductive traits of members of the genus Canis with special attention to the domestic dog (*Canis familiaris*). *Behavioural Processes, 92*, 131–142.

Loughnan, S. & Davies, T. (this volume). The meat paradox. In K. Dhont & G. Hodson (Eds.), *Why We Love and Exploit Animals: Bridging Insights from Academia and Advocacy.* Abingdon: Routledge.

Macdonald, B. N., & Vivalt, E. (2017). Effective strategies for overcoming the naturalistic heuristic. Retrieved from https://osf.io/ndtr2/

Pew Research Center (2009) Scientific Achievements less prominent than a decade ago. Retrieved from https://ir.stonybrook.edu/xmlui/bitstream/handle/11401/8070/pewreport709.pdf?sequence=1

Peterson, M. N., Hartis, B., Rodriguez, S., Green, M., & Lepczyk, C. A. (2012). Opinions from the front lines of cat colony management conflict. *PLoS One, 7*, e44616.

Petrinovich, L., O'Neill, P., & Jorgensen, M. (1993). An empirical study of moral intuitions: Toward an evolutionary ethics. *Journal of Personality and Social Psychology, 64*, 467–468.

Piazza, J. (this volume). Why people love animals yet continue to eat them. In K. Dhont & G. Hodson (Eds.), *Why We Love and Exploit Animals: Bridging Insights from Academia and Advocacy.* Abingdon: Routledge.

Piazza, J., Ruby, M. B., Loughnan, S., Luong, M., Kulik, J., Watkins, H. M., & Seigerman, M. (2015). Rationalizing meat consumption. The 4Ns. *Appetite, 91*, 114–128.

Pierce, J. (this volume). Putting the "Free" back in freedom: The failure and future of animal welfare science. In K. Dhont & G. Hodson (Eds.), *Why We Love and Exploit Animals: Bridging Insights from Academia and Advocacy.* Abingdon: Routledge.

Pierce, J. (2016). *Run, Spot, Run: The Ethics of Keeping Pets.* Chicago, IL: University of Chicago Press.

Rollin, B. (1995). *The Frankenstein Syndrome. Ethical and Social Issues in the Genetic Engineering of Animals.* Cambridge, MA: Cambridge University Press.

Roman, S., Sanchez-Siles, L. M., & Siegrist, M. (2017). The importance of food naturalness for consumers: Results of a systematic review. *Trends in Food Science & Technology, 67,* 44–57.

Ross, J., Irani, I., Silberman, M. Six, Zaldivar, A., & Tomlinson, B. (2010). "Who are the Crowdworkers? Shifting Demographics in Amazon Mechanical Turk." In: CHI EA 2010. (pp. 2863–2872).

Rothgerber, H. (2014). Carnivorous cats, vegetarian dogs, and the resolution of the vegetarian's dilemma. *Anthrozoös, 27,* 485–498.

Rothgerber, H. (2013). A meaty matter. Pet diet and the vegetarian's dilemma. *Appetite, 68,* 76–82.

Rozin, P. (2005). The meaning of "natural" process more important than content. *Psychological Science, 16,* 652–658.

Saher, M., Lindeman, M., & Hursti, U. K. K. (2006). Attitudes towards genetically modified and organic foods. *Appetite, 46,* 324–331.

Sandøe, P., Hocking, P. M., Förkman, B., Haldane, K., Kristensen, H. H., & Palmer, C. (2014). The Blind Hens' Challenge: Does it undermine the view that only welfare matters in our dealings with animals? *Environmental Values, 23,* 727–742.

Scott, S. E., Inbar, Y., & Rozin, P. (2016). Evidence for absolute moral opposition to genetically modified food in the United States. *Perspectives on Psychological Science, 11,* 315–324.

Shah, A. K., & Oppenheimer, D. M. (2008). Heuristics made easy: An effort-reduction framework. *Psychological Bulletin, 134,* 207–222.

Shriver, A. (2009). Knocking out pain in livestock: Can technology succeed where morality has stalled? *Neuroethics, 2,* 115–124.

Simmons, J. P., LeBoeuf, R. A., & Nelson, L. D. (2010). The effect of accuracy motivation on anchoring and adjustment: Do people adjust from provided anchors? *Journal of Personality and Social Psychology, 99,* 917–932.

Solon, O. (2012, February) Food projects proposes matrix-style vertical chicken farms. *Wired.* Retrieved from www.wired.com/2012/02/headless-chicken-solution/

Sunstein, C. R. (2005). Moral heuristics. *Behavioral and Brain Sciences, 28,* 531–541.

Taylor, N., & Signal, T. D. (2005). Empathy and attitudes to animals. *Anthrozoös, 18*(1), 18–27.

Terry, J. (2000). "Unnatural acts" in nature: The scientific fascination with queer animals. *GLQ: A Journal of Lesbian and Gay Studies, 6,* 151–193.

Thompson, P. B. (2008). The opposite of human enhancement: Nanotechnology and the blind chicken problem. *Nanoethics, 2,* 305–316.

Topolski, R., Weaver, J. N., Martin, Z., & McCoy, J. (2013). Choosing between the emotional dog and the rational pal: A moral dilemma with a tail. *Anthrozoös, 26,* 253–263.

van der Sluis. (2007, April 4). Featherless: The future or an unsalable concept? [Web log post]. Retrieved from www.poultryworld.net/Breeders/General/2007/4/Featherless-the-future-or-an-unsaleable-concept-WP002971W/

Wells, K. L., Hadad, Y., Ben-Avraham, D., Hillel, J., Cahaner, A., & Headon, D. J. (2012). Genome-wide SNP scan of pooled DNA reveals nonsense mutation in FGF20 in the scaleless line of featherless chickens. *BMC Genomics, 13,* 257.

Wilks, M., & Phillips, C. J. (2017). Attitudes to in vitro meat: A survey of potential consumers in the United States. *PloS one, 12,* e0171904.

Young, E. (2002, May). Featherless chicken creates a flap. *New Scientist.* Retrieved from www.newscientist.com/article/dn2307-featherless-chicken-creates-a-flap/

10

ACCOMPLISHING THE MOST GOOD FOR ANIMALS

Jon Bockman

Abstract

Animals are suffering in countless ways, whether it be through starvation on the streets, product testing in laboratories, or abuse on farms. Many of us want to help animals, but we do not know where we can have the most impact. Using a lens of effective altruism – a philosophy examining the ways that we can use our time and money to accomplish the greatest good in the world – we start to understand the ways in which we can help the largest number of animals. By viewing different cause areas in terms of scale, neglect, and tractability, we can identify areas that have the greatest potential for reducing and eliminating animal suffering. A close examination of these factors shows that if we seek the most effective ways to help animals, we should focus on animals raised for food. However, our research must not end there. Even within that cause area, we must continue using evidence and reason to investigate the most impactful interventions to help farmed animals.

What if you were told that you could be *twice as effective* as you currently are with your efforts to help animals? Of course, you would be interested to hear how to double your outcome. After all, the goal of charity – through donations or through work – is to accomplish good in the world. If we could accomplish twice as much good, we would want to do that.

What if you were told that you could *be ten times as effective* with your efforts to help animals? Again, you would likely be interested, but you might start to experience some skepticism. Doubling your effectiveness might feel plausible through strategic planning. But multiplying it by a factor of ten? Highly unlikely. Still, though, it seems intriguing, so you are willing to hear out the argument.

Now what if you were told that you could be *100 times as effective* with your efforts to help animals? At this point, your skepticism meter has probably skyrocketed, and you are ready to move on from the discussion. One hundred times more effective? That cannot be real. Anyone who has heard the old adage "If it is too good to be true, then it probably is" will be inclined to dismiss this claim as impossible. But it is not. Not only is it not impossible, it is true, and it is backed up by data (Animal Charity Evaluators, 2018a). Most animal advocates are directing their donor dollars to animal shelters (Animal Charity Evaluators, 2016), which, while a noble effort, simply pales in comparison to other methods of helping animals. For those individuals, being 100 times more effective with their time and money is a completely achievable goal. This chapter will provide guidance to all readers on how they can optimize their efforts to help animals.

Difficult choices

Even after we have made the choice to help animals, there are still many decisions that will need to be made. After all, there are a tremendous number of animals who need our help. Companion animals are starving in the street, animals used in circuses are housed in tiny cages and forced to perform, wild animals die by numbers too big to imagine from starvation, disease, and predation, farmed animals are intensively confined and slaughtered at an early age, and animals used in research are subject to harmful experiments. How do we decide where to focus our efforts?

It is a difficult question to answer, and one that will inevitably involve frustration. After all, when we choose to do one particular thing, this often means that we are choosing *not* to do another. When thinking about the most impactful way to advance any particular cause, we should use evidence and reason to determine how we can use our time and money effectively. This philosophy, referred to as effective altruism, enables us to maximize our contributions to the world.

Effective altruism is a rising social movement that encourages using research and data to efficiently accomplish our charitable goals. William MacAskill, a philosophy professor at the University of Oxford, explains the concept in a simple and earnest way: "Of all the ways that we could make the world a better place, which will do the *most* good? Which problems should we tackle immediately, and which should we leave for another time? [...] In order to make comparisons between actions, we need to ask: How many people benefit, and by how much?" (MacAskill, 2015, p. 37). These simple questions can make a tremendous difference in the amount of good that we are able to accomplish in the world. We can use them at each stage of our process, including selecting a cause area, selecting which interventions to use, and selecting which charities to support.

All too often, we as a society tend to overly rely on our hearts when making decisions about how to contribute to the world, and unfortunately this can stunt our ability to create the change we wish to see. Instead of solely relying on our intuition, we should use effective altruism to help us identify the most impactful ways to help animals. To begin with, we need to determine where exactly we should focus our efforts, and for that purpose, it is useful to consider the following three factors:

- Scale: How many animals are affected in this area? How extreme is their suffering?
- Neglectedness: How much time and money are already given to this area?
- Tractability: How likely is it that we can make substantive progress in this area?

By considering these three factors, we can use objective criteria to focus on the core goal of accomplishing the most good.

The big picture

Over the course of 2016 in the United States, only 3% of donations went to *either* animal causes *or* environmental causes (Giving USA, 2017). The numbers are so small that analysis of the data has led to lumping the two categories together. To simplify this discussion, let's assume that half of that (1.5%) went to animal charities. That means that we are only directing about 1.5% of our donations to helping what is arguably the biggest collective body of suffering in the world, seeing as how nonhuman animals *significantly* outnumber humans. Because we want to stretch that money into helping as many animals as possible, it becomes clear that we need to think critically about how to prioritize our efforts.

We can begin to do this by first taking a closer look at available data. *At any one time*, the following animals exist in the world:[1]

- >10 trillion wild animals (Tomasik, 2017)[2]
- >32 billion farmed animals (Food and Agriculture Organization of the United Nations Statistics Division FAOSTAT, 2017)[3]
- >7 billion humans (US Census Bureau, 2015)[4]
- 1 billion companion animals (Animal Charity Evaluators, 2018a)
- 100 million animals used in scientific and medical research (Hunter, 2014)

At first glance, it seems like we should focus all of our efforts on alleviating the suffering of wild animals. After all, their number completely dwarfs the next highest category of farmed animals. Since there are at least 300 times as many wild animals as animals used in farms – this is likely a conservative estimate – we should invest all our resources in that area, right? The number of wild animals is indeed important, but we should not end our thinking there. We also need to consider other important questions. How significant and widespread is their suffering, and how many resources are dedicated to solving that problem? Is there another area that is so underserved that we should consider supporting it even though it does not involve the largest number of animals? What about the overall problem of suffering in the wild – do we know how to solve it, and do we even have a good understanding of it in the first place? In the coming sections, I will examine how tractability and neglect complement a consideration of scale to help define the most impactful methods to help animals. To do that, I will begin by exploring wild animals as a cause area.

Animals in the wild

At any one time, roughly ten trillion wild animals exist in the wild. The magnitude of this number can be unfathomable, especially because each of these ten trillion animals is a living being who seeks to avoid suffering. In order to figure out what to do with such a large number, we need to consider this information alongside other factors. Doing so will help us optimize our efforts.

Conventional wild animal charities target low-hanging fruit

Many conventional charities that work on behalf of wild animals attempt to help them through conservation efforts. It is much easier for people to grasp the concept of saving a specific animal or even a particular species, especially when the species in question is at risk due to human exploitation of them or their environment. Campaigns such as "Save the whales" or "Save the tigers" also have the benefit of advocating for a particularly majestic animal. At the very least, those campaigns are more attractive than advocating for, say, a dung beetle. Is this a problem? After all, conservation campaigns are targeted at saving animals who need our help. It seems even more sensible that we intervene in those situations – since, in many cases, the animals involved are suffering as a result of human activity. At the very least, we owe them our efforts to save their species, right?

Yes and no. Yes, I agree we should consider helping those animals, but no, I am not convinced that we should do so solely because a species is in decline, or because we are the cause of their suffering. Instead, we need to consider a variety of factors. Do the animals in question require rescue, and thus food, lodging, and potentially medical assistance? If they do, how much will it cost for each animal involved? There are even deeper questions that we need to ask. How does one value an elephant when compared with, say, a squirrel, or a snake? We might joke that elephants are majestic and snakes are slimy, but how do they compare with regard to their ability to feel pain and to suffer? And how should that difference factor into our decision about who to help? Should we give additional consideration to animals with more complex cognitive abilities? Or should we focus more on those who are closer to extinction? We do not have a clear consensus on the answers to these questions, so we need to examine other factors to increase our understanding of the situation. Apart from conservation efforts, what else is being done to help wild animals?

Wildlife rehabilitation

I spent six years as a wildlife rehabilitator, which means I took in injured and orphaned wildlife, rehabilitated the ones I could, and released them back into the wild. I rescued and released about 2000 wild animals during this time, in addition to running educational programs. Compared with conservation efforts, it seems like money directed to wildlife rehabilitation has more potential to do good. After all, many wild animals who come into these centers require only a small amount of assistance before release, and in that way rehabilitation can help a large number of animals at a minimal cost. Unfortunately, this is not always the case: sometimes animals come in with more substantial injuries that require large investments of time and money from staff, and certain species simply require more resources for their care regardless of injury due to expensive food and habitat requirements. For example, it might cost a total of $10 to rehabilitate an individual songbird fledgling, whereas it might cost $1,000 to rehabilitate a single injured owl – a significant difference.

Those of us who are driven by effectiveness will need to ask ourselves a difficult question: at what point do we decide that treatment is not cost-effective? If surgery to repair one owl's broken wing will cost $1,000, but we are able to rescue and release 100 songbirds for that same price, we may have to make a difficult choice with our limited resources. The dilemma becomes even more complicated when you consider the severity of suffering involved. The above example is an easier decision to make, because the number of animals is so significantly different that one would not think that the suffering of the one owl could approach the suffering of 100 orphaned songbirds. If the numbers were such that we could help *one* owl's broken wing, or rehabilitate *two* orphaned songbirds, then we would also want to think about quantifying the amount of suffering that each animal is experiencing when making our decision.

For those motivated by accomplishing the greatest good for animals, in most cases like the above example we should choose to help the many over the few. When possible, we should also factor in the severity of suffering. We need to extend this strategic consideration to all efforts to help wild animals – and by extension, to all efforts to help all animals.

A natural life

> The blind forces of evolution do not optimize for happiness. The pain endured by a fish afflicted with parasites or a rat swallowed alive by a snake is no more tolerable than the "natural" suffering of humans due to malaria, cancer, or starvation. Both deserve our attention.
>
> *Tomasik, 2009*

Many of us think of the lives that wild animals experience as "natural." After all, they live in nature! They have the freedom to do whatever they want whenever they want – to seek out experiences they enjoy and avoid those they do not, to sleep when they want to, to fight when they want to, to flee when they need to, and so on. We accept all of these behaviors as natural – and indeed as perks of living in the wild. But what we do not see in conventional footage of pandas munching on bamboo or a lioness playing with her cubs is the suffering that accompanies such an existence. For many wild animals, their freedom comes at a high cost: they must constantly survey the environment to avoid predators, find food to feed themselves and their families, and identify sheltered locations for sleep. Living in nature is no easy feat.

The sad reality is that vast numbers of animals live with these problems and much more. Consider the life of an American Bullfrog. A mother frog might lay 20,000 eggs in a single season (Bruening, n.d.). The majority of those that metamorphose into tadpoles will die from either starvation, predation, or disease in very early stages of their life – after all, we do not see 20,000 bullfrogs for every mother frog. That means that *thousands upon thousands of animals are suffering and dying for every single clutch of eggs*. If that number sounds high, that is because it is; in fact it is gargantuan. It is difficult to estimate how many animals in the wild are suffering and dying at any given moment, but the numbers must be incredibly high. If there was a way to meaningfully reduce this suffering, we should consider intervening, but the decision of whether or not to do so is ultimately a complicated one. First, we do not have a good idea of an effective way to combat this high level of suffering on a large scale – its tractability is not well understood. When we ask ourselves how well we understand the problem and how well we understand potential solutions, the answer with regard to wild animals is – unfortunately – not very well at all. Even if we did, and even if we were able to, for example, prevent the tadpoles from the previous example from dying a premature death, what would the world do with millions of extra frogs?

A particularly important consideration with regard to helping animals in these situations is the ways in which our assistance might affect our ecosystem. It is possible that saving large numbers of wild animals might upset the natural balance of predator populations and result in increased suffering via hunger and starvation. Our well-intentioned efforts could conceivably create an even bigger mess. At one point we thought that exterminating wolves would provide safety to our society – but we later found that reintroducing them brought substantial benefits (Lister, 2006). It is entirely possible that we might be able to develop some interventions that could mitigate these consequences. For example, perhaps we could find a way to use a contraceptive to limit the amount of offspring, and implement controls on the frequency of breeding. However, even if this did drastically reduce the suffering of that species, it is still difficult to understand the potential ecological consequences of such actions. This is an area we should care about, but we are not prepared to intervene at this stage. Further research is needed, and encouraged, but that is not an easily accessible form of advocacy for the average person looking to help animals. Thus, we need to consider other factors to make an informed judgment on how we can maximize the good that we create in the world today, which means we need to investigate other cause areas.

Animals used in research

Looking at the least populated category on our initial list – animals used for research – we know that at any one time around 100 million animals are likely living out their lives in small cages while they are used for medical and cosmetic research. The total number of animals might be on the lower end of the scale when compared with other cause areas, but there are still significant numbers of animals suffering because of their use in research.

I spent four years on the Institutional Animal Care and Use Committee (IACUC) at Northern Illinois University, and have some first-hand knowledge of the experiences that many of these animals endure. Our university did not deal with primates, instead mostly housing mice, rats, rabbits, and chickens. Having personally conducted inspections of these facilities, it was clear that their very minimal physiological needs were being met; they had proper access to food and water, quality air, clean cages, and regular monitoring from a veterinarian genuinely concerned with their well-being. Of course, that is not particularly surprising, as such facilities need animals to be reasonably healthy in order to conduct research that will produce usable data. However, this care was generally skewed specifically toward the animals' short-term needs. Long-term care was not given priority unless a study was examining something that would require it. That meant that the animals were not provided with ways to regularly exercise and move around, and that their psychological needs were often not properly addressed. This last point is particularly important. Clean cages, food and water, and good air quality are all vital for good health. But they do not truly address the *needs* of the animals. How would any of us feel to be trapped in a tiny cage for our entire life? Or perhaps more relatable, how would we feel if any of our cats or dogs were trapped in a tiny box barely bigger than three body-lengths without any sort of stimulation? Caring for our companion animals in this way would be considered criminal in many countries.

To be fair, I never thought that the researchers themselves were necessarily bad people. In fact, many of them were indeed trying to accomplish good for humans. They simply lacked an understanding of the needs of these animals, a situation that I would attribute to their involvement in a field that uses animals as tools for science and therefore conditions researchers to believe that research involving animals is necessary and normal. It is striking how quickly humans can acclimate to a particular viewpoint when they are surrounded by people behaving a certain way (Wolford, 2014); with most researchers laser-focused on accomplishing gains in science, it is not surprising that the needs of animals being used are not fully considered. Regardless of the reason, it is clear that many animals used in research are not given proper consideration when it comes to their well-being.

Growing public interest in the welfare of animals has helped lead to improved consideration for the treatment of animals used in research, which has resulted in relatively significant updates in the Guide for the Care and Use of Laboratory Animals – an instructive book that provides guidance on how to care for animals used in research (Committee, 2011). The latest version of the guide calls for increased cage sizes and even has a section on enrichment, which was previously rarely discussed in these circles. Some researchers found themselves needing to update their outlook on research involving animals based on these new requirements. It signified a positive collective step forward.

While raising quality standards is a way to improve the treatment of animals used in research across the board, there are other ethical questions that we must consider. Even if using animals for research *did* produce a cure for humans who are afflicted with a disease, then we need to ask ourselves what trade-offs would be considered acceptable. Those with a less favorable view of animals might feel that any benefit to humans is worth any cost to animals. Others might feel that there needs to be an equitable exchange, but would err on the side of benefitting humans – those individuals might think that using 10,000 animals for research was acceptable if it saved even a modest number of 20 humans. Still others might think that the benefit needs to exceed the cost – perhaps those individuals would think that using 10,000 animals for research was permissible if it saved 10,001 humans. And finally, some individuals would feel that no use of animals is ever permissible, since they are unable to give us their consent to be used in the first place.

These considerations affect our understanding of the problem, as it is not clear what level of justification of the use of animals (if any) is appropriate. Critics of animal testing will point to the usefulness (or lack thereof) of research conducted on animals, as it is not uncommon for research findings to be deemed not generalizable to humans (Pound, Ebrahim, Sandercock, Bracken, & Roberts, 2004). For example, thalidomide does not cause birth defects in rodent populations, but it causes severe birth abnormalities when given to pregnant women (Greek, Shanks, & Rice, 2011). Furthermore, research involving animals can go beyond simply failing to provide generalizable findings; it can be harmful to key stakeholders. As Kramer and Greek (2018) note: "Overall, the use of animal models in medical research has poor predictive value in terms of its ability to distinguish between treatments that will be helpful versus harmful to humans …. Reliance on animal-based research causes much more harm to humans than the accidental good that arises from using animals in the context of drug development." (p. 5)

Because of these complications, we do not have a good understanding of how to make a large impact in this area. However, that does not mean that nothing can be done. Writing letters to the editor of a newspaper raises awareness about the issues involved with using animals for research. Joining an Institutional Animal Care and Use Committee could enable us to affect the treatment of animals at a local university (though these positions are very rarely available). Pouring over documents might unearth some examples of mistreatment of animals in individual circumstances and potentially lead to improvements or even closures of labs. Avoiding purchasing products tested on animals helps show that the public does not approve of such testing. People who engage in these efforts should be commended for the animals that they help. But how many animals does each of these actions affect? Do any of them truly address the systemic problem? Theoretically, efforts to prove the problems with using animal models in research might result in changes, but this seems like a particularly difficult type of advocacy to undertake given that very few of us would be qualified to offer that proof. Additionally, the research community has a vested interest in allowing the use of animals so as to continue their current research. Advocating for animals used in research is a worthwhile thing to do, and advocates have achieved meaningful victories through the rescue of those being used and the cessation of certain kinds of testing. However, given the relatively small numbers involved, those concerned with affecting the greatest number of animals might consider other causes.

Companion animals

At any one time, there are approximately one billion companion animals in the world. The scale of this area certainly deserves consideration. However, just like with wild animals and animals used in research, we need to consider other factors when determining how to allocate resources to advocacy efforts on their behalf. For example, companion animals are given the most attention out of all the groups of animals that make up this list of cause areas. We pamper our cats and dogs, and fuss over every little problem in their life. I have given many presentations at colleges, and upon asking students to indicate if they or a family member had a pet at one point in their life, virtually everyone raises their hand. When asked if they loved that animal, again, near universal participation. Simply put, we truly care about our cats and dogs.

To be sure, some cats and dogs truly need our help. Homeless animals reside on the street without access to clean food, water, or shelter. This situation is exacerbated by pet owners' reluctance to spay and neuter their animals, which can result in repeated accidental pregnancies and subsequent litters. There is a large body of animal suffering here that we should care about. That said, when you look at the numbers, they show that we actually *do* care about it. In fact, we care about it disproportionately relative to other animal cause areas. In 2015 alone, Americans spent over $60 billion on their pets (American Pet Product Association, 2017). A huge 66% of all donations that go to animal charities go to companion animal charities (Animal Charity Evaluators, 2018a), despite the fact that the majority of those 1 billion animals are extremely well-cared for and do not need anywhere near the same amount of assistance as do animals used in research or on farms. The reasoning behind this distribution is not difficult to infer. After all, it is easiest for us to relate to companion animals, seeing as how so many of us live with them and develop strong relationships with them – we tend to donate more to individuals and causes that are familiar to us (Macaulay, 1975). Another factor that drives many people's philanthropy is proximity, as 54% of donors prefer to give to local or regional nonprofits (Camber Collective, n.d.). This means that more than half of all donors interested in helping animals may end up donating to support companion animals because companion animal shelters and humane societies are by far the most ubiquitous establishments helping animals. Additionally, donors are motivated by being able to see results – donating locally is a way we can put our efforts toward something that directly helps the communities where we live, where we can more often see the effects of our contributions.

Unfortunately, supporting this area does not lead to the greatest good for animals. Companion animals are by far the least neglected group of animals, and severely suffering companion animals are few in number relative to other cause areas. We invest the majority of our donor dollars that are reserved for animals on companion animals, and caring for these animals is a costly endeavor – Animal Charity Evaluators estimates that donors only save about 2.45 animals for a $1,000 donation to a conventional animal shelter (Animal Charity Evaluators, 2018a). As such, advocacy for companion animals is simply not the best option for someone wanting to help as many animals as possible.

One positive note worth mentioning is that companion animal advocacy is highly tractable, as advocates do understand the problem and potential solutions. We recognize that rehoming efforts at shelters are a band-aid to fix a bigger problem, which is the overpopulation of companion animals. Because of this, many organizations develop methods to address the root causes of the problem by setting up TNR (Trap, Neuter, Release) programs – attempting to reduce new offspring – and by focusing on education in order to encourage the public to adopt from shelters instead of buying from pet stores (which may have purchased their dogs from puppy mills, an inhumane method of breeding dogs that involves intensive confinement). These efforts increase the impact of work on behalf of companion animals, and should be commended. Still, scale and neglectedness indicate that we should focus our efforts elsewhere.

Animals raised for food

Thus far, we have discussed three important factors that we should use when determining which animal cause areas to support: Scale, tractability, and neglect. While each of the previous causes that we have examined rank highly in one or two of these categories, none of them score highly in all three. So how do farmed animals stack up?

With regard to scale, it is clear that we should give attention to farmed animals. Worldwide, a shocking total of 32 billion animals are raised in farms at any one given moment in time, with the explicit goal of getting them to market weight as quickly as possible in order to be slaughtered for consumption (FAOSTAT, 2017). Even more surprising, this number only includes *land animals*, as the number of sea animals is simply measured in terms of metric tons rather than individuals. It is difficult to estimate the total number of fish killed for food, although some people, such as economist Harish Sethu, certainly try. Sethu (2015) points out that the total numbers should not just include the fish we directly eat – we also need to include the number of fish who are fed to other fish in aquaculture, which is the aquatic version of factory farming. He notes that "Between 144 and 293 wild sea animals are captured and killed annually to feed the aquacultured fish and shrimp eaten by the average American consumer," indicating that a far greater number of animals are affected by our choice of fish or shrimp at the dinner table than would seem apparent.

Exacerbating the problem is the fact that these numbers do not account for the way that commercial fishing also harms ecosystems through invasive methods of capturing their prey – disturbing not just terrain, but unintentionally capturing and killing large numbers of dolphins, sea turtles, and sea birds that are swept up in their nets. In fact, 40 percent of fish catch is categorized as this "by-catch," which results in undesired dead or dying animals being simply dumped back into the ocean (Fish Forward Project, n.d.). Another factor worth including in our consideration of scale is the lack of federal laws protecting all species of farmed animals in the United States. Not only are we caging and eating massive numbers of animals, but they have little to no protections in place on behalf of their welfare. This is an irony of the Animal Welfare Act – its name would suggest that it is protecting all animals, but in practice it *explicitly excludes farm animals* from consideration, essentially ensuring that the animals that actually need protection the most do not receive it (Favre, 2002).

It is true that farmed animal advocacy involves large numbers of individual animals, and that they are not adequately protected by legislation, but what about the amount of attention that they receive? It seems like undercover investigations come out regularly, and we are always hearing about the need to improve conditions for farmed animals. Are we not already adequately addressing this issue? Unfortunately, although animal advocates have enjoyed an increasing number of victories over the past ten years or so, there is still very far to go. The number of animals slaughtered for food each year remains astronomically high, and people in many areas of the world are increasing their consumption of animal products at an alarming rate.

FIGURE 10.1 Allocation of Donations. Graphs showing the percentage of animals killed and used annually in the United States compared to the distribution of donations between different cause areas.

Note: "Farm" refers to farmed animals, "Shelters" refers to animals in humane societies and shelters, "Clothing" refers to animals used to produce clothing materials, "Lab" refers to animals used in research, and "Other" encompasses areas not covered by the other terms, including equine charities, veterinary charities, and charities that work on multiple issues.

To understand the level of neglect in this area, it can be helpful to look at the amount of support directed to charities. Perhaps most telling is that of all donations that go to animal charities in the United States, only about 1% of these go to charities that specifically focus on farmed animal issues (Animal Charity Evaluators, 2016). Since only roughly 1.5%[5] of charitable donations go to animals overall in the first place, that means that about 0.015% of our total annual donations in the United States go toward what is arguably *the largest source of human-caused suffering in our planet's history* (see Figure 10.1). Farmed animals are severely neglected.

It is not difficult to understand why we prioritize our spending the way we do. As discussed earlier, most of us have developed a strong relationship with companion animals at some point in our life. It makes sense that we would gravitate toward helping them. Additionally, most of us were raised in environments where we consumed farmed animal products every day, often at every meal. I am no exception to this generality – for the first 21 years of my life, I had not considered a meal to be a meal unless there was a piece of meat on the plate (Note: I have been a vegan for 14 years). For many of us, the prospect of donating to help farmed animals might be seen as inconsistent with our dietary behavior, thus implying that a change is necessary. It is much easier to turn a blind eye to the suffering of these animals and instead prioritize other areas, allowing ourselves to be comforted by the fact that we are still helping animals to some extent. And as we have established, there is no shortage of other animals who need our help. But the reality is that farmed animals are suffering terrible abuses each and every day, and they deserve just as much consideration as other animals who suffer. Regardless of how uncomfortable it may make us, we should apply the framework of scale, neglect, and tractability to prioritize our efforts, and thereby focus more on farmed animals.

A note on tractability

The tractability of any particular cause area can be challenging to understand, as it can be difficult to truly determine the impact of our actions. For example, working in a soup kitchen is valuable in that it provides short-term assistance for homeless people who are without food. But how well does that action address the root causes of what displaced those people in the first place? It certainly helps the individual – they need to be fed – but the extent to which it makes an impact on the systemic problem is either extremely small or zero. That may not hold true for all situations. In fact, there is evidence that certain forms of direct giving to the poor – such as cash transfers – are actually extremely effective (Give Directly, 2017). However, these situations are complex. What seems like a good solution can sometimes counterintuitively harm the cause we intended to help. For example, welfare programs designed to help individuals with mental health problems can actually disincentivize a transition into paid work (Turton, 2001). It is therefore clear that we need to think critically about the effects of our efforts. Thankfully, some of the tactics commonly used to help farmed animals seem likely to have a positive impact.

Outreach on behalf of farmed animals

Families gave Christmas gifts to countless companion animals this past year, but how many of those families gave a single thought to the living conditions of the pig who served as the Christmas ham on their table? I know I am not alone when I say that I used to be in this camp myself, guilty of pampering my own pets without giving any consideration to how farmed animals are treated. But then, in college, I received a leaflet from an organization called Vegan Outreach that showed me how poorly farmed animals are treated. I decided that I did not want to support that exploitation, and I changed my behavior. More than that, I went on to help animals as a career, started donating to animal charities, and now seek to empower other advocates to be as effective as possible.

My transition was inspired by a common outreach tactic called *leafleting*, which basically involves handing out informational literature to people at public events or on public campuses. Outreach activities like leafleting or food demonstrations (where advocates sample relatively unknown products like vegan meat, dairy, and egg replacements) seem to be a reasonable method for creating change on a small scale, but it is not particularly efficient – you are only making appeals to individual people, and many will not be convinced by your arguments. We need to consider ways to create change on a broader level, and think about how we can reach those who are less likely to be influenced by a call to change their diet. There are other interventions that attempt to address the issue on a bigger scale, including tactics like corporate outreach (encouraging companies to reduce or eliminate animal products from their supply chains) and developing food technology (creating and promoting alternatives to animal products, including plant-based and cultured or "clean" meats). Through these efforts, we come closer to addressing the root of the problem, rather than hacking at the branches. Corporate campaigns and animal-free meat/dairy/egg alternatives offer tremendous potential, as using these strategies can empower us to help millions of animals through a single change. That said, even with the power of these types of efforts, there must be public demand for change to occur. Advocacy efforts such as leafleting and food demonstrations can be a powerful tool in creating that demand, so it is likely that these efforts work in concert with more scalable campaigns.

The above broad considerations about scale can be helpful in determining the ways we should prioritize helping animals, but they are far from exhaustive or conclusive. Sadly, the state of research on the effectiveness of farmed animal advocacy interventions leaves much to be desired. Advocates and donors simply do not have access to the data we need to make strong, confident recommendations about the most impactful interventions. To that end, an intervention that will significantly advance the cause of animal advocacy is research. Conducting research studies – particularly when done in collaboration with advocates, to ensure practicality and future buy-in, and with academics, to ensure strong quality controls – has the potential to shape the efforts of prominent advocates and organizations. For those interested, Animal Charity Evaluators offers grants for research on ways to help animals through their Animal Advocacy Research Fund, and they are more than willing to help guide applicants towards research that would be most valuable at any given point in time (Animal Charity Evaluators, 2018b).

Of note, not every type of advocacy will be a good fit for everyone. In fact, some may prefer to contribute donations to charities as a way to help animals effectively. The good news is that you can be an incredibly effective advocate by taking this approach, providing that you think carefully about which charities you support. Donations are important, regardless of size – widespread support is crucial in raising public awareness and in attaining campaign victories – but in particular, those with the ability to be major donors can create substantive change at an organization. For example, a donor who contributes \$100,000–\$150,000 to an organization who has met their funding goals for the year could feasibly help create two new positions at that organization, magnifying the good that the charity can do and creating more full-time equivalent employees than you would if you were to work for that charity directly. We should seek to empower charities when possible, as they are the groups that are regularly thinking about and working on the problems we hope to fix. In other words, they know what they are doing.

Except when they do not. Not all charities are equal, and sadly many charities suffer from low-quality leadership, poor prioritization, and lackluster execution. Thankfully, there are resources for donors who wish to identify strong groups to support. Animal Charity Evaluators provides updated charity recommendations every year on Giving Tuesday in November. Their recommendations include extensive comprehensive reports that provide donors with a wide range of information that they can use when deciding which groups to support. For those who simply wish to see the best of the best, their website includes a short list of Top Charities (Animal Charity Evaluators, 2018c).

Whatever you choose to do to help animals, it is important to think about all aspects of your efforts. Being an advocate and a donor can be incredibly rewarding, as we are taking active steps to combat a problem that we see in the world. In that way, we are thinking about the bigger picture, about people and animals other than ourselves, and it can be easy to forget that we need to also exercise self-care. In addition to objective considerations of impact, we also need to think about what will sustain animal advocates over time.

Sustainable advocacy

Many of us became interested in helping animals through some specific interaction. Maybe it was the experiences you had with your cats growing up. Or perhaps it was a trip to the zoo that introduced you to the majesty of a giraffe. Maybe your school class had a pet mouse, and you became determined to help other mice, such as those used in research. It is undeniable that these experiences shape our lives, and lead us to the path of becoming active for animals, either through our donations or through our work. Anyone who has a passion to help a specific group of animals – but also wants to be effective with their advocacy – is faced with a tough choice: How should they allocate their time and resources?

The good news is that the answer is not black or white. If you agree with the arguments presented in this chapter that identify farmed animal advocacy as the highest impact cause area but you still highly value a different set of animals, you do not have to work on one area exclusively. You could update your current efforts and invest a significant amount of your time or money in helping as many animals as possible by supporting effective farmed animal advocacy organizations, but still spend some resources on other groups of animals. This latter work would keep you motivated and excited for the work being done to help animals, and could ultimately result in a more sustainable career as a supporter of animal advocacy.

My wife and I live with three dogs. They are not the most cost-effective use of our resources to help animals; one of our dogs is older and requires a special diet and 6-month vet visits. But while I understand that I could save more animals by donating the money that I spend on their care elsewhere, I also recognize that they help fuel my fire to continue my work. They are a constant reminder of why I have spent my career working to help animals. They are indirectly making me a better advocate by inspiring me to continue the work that I do, and by enriching my day and recharging my batteries as I work to help the largest numbers of animals. One can focus on helping the maximum number of animals while also still enjoying personal relationships and passions. A person can have the best of both worlds – and in fact, for many of us, that may ultimately end up being the best way to maximize the impact of our advocacy.

Conclusion

We have seen that when thinking about how to help animals, we need to consider three factors: scale, neglect, and tractability. By examining these factors, it seems clear that we are not currently using our time and money to effectively help the largest numbers of animals, and that working to help farmed animals results in the biggest gains for animals overall. But your analysis should not end there. This chapter has identified farmed animal advocacy as the most effective cause area, but what exactly do we do to help farmed animals? Which actions should you take, and which charities should you support?

These are complex questions, and they do not have simple answers. Animal Charity Evaluators provides recommendations on their site (www.AnimalCharityEvaluators. org) based on rigorous comprehensive charity and intervention evaluations, along with a library of additional resources and general advice. While you may not find answers to all of the questions you have, you can still use some basic considerations to inform your decisions about how to help farmed animals. For example, although it is the first thing people tend to gravitate toward when they think of helping farmed animals, setting up a sanctuary is not an optimal way to do so – directly caring for animals costs substantially more money per animal helped than other, more scalable forms of advocacy such as corporate or institutional campaigns. You might still choose to support a sanctuary in some way because certain animals are particularly important to you, or because you see especially significant value in rescuing animals who are alive and in need. But it is vital that we invest efforts in highly impactful ways to help animals as well.

Most importantly, it is essential to continue asking questions. Performing cost-benefit analyses will help identify the most impactful ways to help animals, as it helps us avoid the pitfall of continuing to do something simply because that is the way it has been done in the past. We should try advocacy methods that are new to us, and be accepting of the occasional misstep – by being prepared to fail, we can use that failure to inform our subsequent actions. It is through the pursuit of knowledge that we will be able to truly determine the most effective ways to help animals. Whatever we do, we should act mindfully. If we have a choice to save one or save 100, we should almost always choose to save 100. By advocating strategically, we will maximize the good that we can do – and move our society toward a system that treats animals with respect.

Notes

1 Note that while these numbers refer to animals that exist at any one point in time in the world, there are other sections of the text where numbers refer to animals used and killed annually.
2 "Collectively, wild land vertebrates probably number between 1011 and 1014. Wild marine vertebrates number at least 1013 and perhaps a few orders of magnitude higher" (Tomasik, 2017).
3 This includes all categories of livestock tracked by the Food and Agriculture Organization of the United Nations Statistics Division (FAOSTAT), except for beehives. Their estimate for the total world livestock population in 2013 is 32,669,958,965.
4 According to the US Census Bureau's International Data Base, the world mid-year population for 2015 is estimated to be 7,256,490,000.
5 Note that 1.5% is an estimate. As noted in the earlier section of this chapter "The Big Picture," 3% of total donations in the United States go either to animals or environmental causes; 1.5% assumes an equal split between the two causes, but it is likely slightly higher or lower than that.

References

Animal Charity Evaluators (2016). Why farmed animals. Retrieved from https://animalcharityevaluators.org/donation-advice/why-farmed-animals/
American Pet Products Association (2017). Pet industry market size and ownership statistics. Retrieved from www.americanpetproducts.org/press_industrytrends.asp
Animal Charity Evaluators (2018a). Donation impact. Retrieved from https://animalcharityevaluators.org/donation-advice/donation-impact/
Animal Charity Evaluators (2018b). Animal advocacy research fund. Retrieved from https://researchfund.animalcharityevaluators.org
Animal Charity Evaluators (2018c). Recommended charities. Retrieved from https://animalcharityevaluators.org/donation-advice/recommended-charities/

Bruening, S. (n.d.). Lithobates catesbeianus. *Animal Diversity Web*. Retrieved from http://animaldiversity.org/accounts/Lithobates_catesbeianus/

Camber Collective (n.d.). Money for good 2015: Revealing the voice of the donor in philanthropic giving. Retrieved from http://static1.squarespace.com/static/55723b6be4b05ed81f077108/t/56957ee6df40f330ae018b81/1452637938035/$FG+2015_Final+Report_01122016.pdf

Committee for the Update of the Guide for the Care and Use of Laboratory Animals (2011). *The guide for the care and use of laboratory animals.* Washington, DC: The National Academies Press.

Favre, D. (2002). Overview of the Animal Welfare Act. Retrieved from www.animallaw.info/article/overview-us-animal-welfare-act

Fish Forward Project (n.d.). Bycatch – a sad topic. Retrieved from https://fishforward.eu/en/project/by-catch/

Food and Agriculture Organization of the United Nations Statistics Division (2017). Live animals. Retrieved from www.fao.org/faostat/en/#data/QA

Give Directly (2017). Research on cash transfers. Retrieved from www.givedirectly.org/research-on-cash-transfers

Giving USA (2017, June 12). See the Numbers – Giving USA 2017 infographic. Retrieved at https://givingusa.org/tag/giving-usa-2017

Greek, R., Shanks, N., & Rice, M. J. (2011). The history and implications of testing thalidomide on animals. *The Journal of Philosophy Science & Law*, *11*(3), 1–32. doi: 10.5840/jpsl20111133

Hunter, R. G. (2014). Alternatives to animal testing drive market. *Genetic Engineering and Biotechnology News*, *34*(1), 11. doi: 10.1089/gen.34.01.07

Kramer, L. A., & Greek, R. (2018). Human stakeholders and the use of animals in drug development. *Business and Society Review*, *123*(1), 3–58. doi: 10.1111/basr.12134

Lister/McDaniel (2006). The wolves of Yellowstone. *Ecology*, 7. Retrieved from www.bioinfo.rpi.edu/~bystrc/pub/artWolves.pdf

MacAskill, W. (2015). *Doing Good Better.* New York: Penguin.

Macaulay, J. (1975). Familiarity, attraction, and charity. *The Journal of Social Psychology*, *95*(1), 27–37. doi: 10.1080/00224545.1975.9923231

Pound, P., Ebrahim, S., Sandercock, P., Bracken, M. B., & Roberts, I. (2004). Where is the evidence that animal research benefits humans? *BMJ*, *328*, 514–517. doi: 10.1136/bmj.328.7438.514

Sethu, H. (2015, March 16). How many animals does a vegetarian save? Retrieved from www.countinganimals.com/how-many-animals-does-a-vegetarian-save/

Sethu, H. (2015, March 16). The fish we kill to feed the fish we eat. Retrieved from www.countinganimals.com/the-fish-we-kill-to-feed-the-fish-we-eat/

Tomasik, B. (2017). How many wild animals are there? *Reducing Suffering*. Retrieved from http://reducing-suffering.org/how-many-wild-animals-are-there/

Turton, N. (2001). Welfare benefits and work disincentives. *Journal of Mental Health*, *10*(3), 285–300. doi: 10.1080/09638230123664

US Census Bureau (2015). *International Data Base,* Version: Data: 12.0625 Code: 12.0321.

Wolford, B. (2014). Mob mentality: The brain suppresses personal moral code when in groups. *Medical Daily*. Retrieved from www.medicaldaily.com/mob-mentality-brain-suppresses-personal-moral-code-when-groups-288342

11

THE MEAT PARADOX

Steve Loughnan and Thomas Davies

Abstract

How are (most) people able to both eat animals and claim to love animals? This seeming paradox between harm and care is fundamentally psychological. In understanding the psychology behind this behavior, we can start to understand how it may be challenged and ultimately changed. In this chapter, we start by outlining the meat paradox as a form of cognitive dissonance, before mapping the psychology at the level of animals (eaten), consumers (eaters), and behavior (eating). We then tie these individual processes to a broader social system that serves to benefit the status quo and prevent change in how we treat animals. In the second section, we adopt a broader perspective, examining how these same psychological processes may relate to other forms of exploitation. Importantly, we argue that animal-human exploitation may be psychologically fundamental to other systems of oppression. We ultimately conclude that the psychology of eating animals tells us that hard work will need to be done to pierce the social and personal defenses people have arrayed to avoid feeling bad about their paradoxical treatment of animals. In doing so, we may unexpectedly benefit a wide range of struggles against systems of oppression, making the fight against animal mistreatment a fundamental concern.

Human exploitation of animals is perhaps the most longstanding, widespread, and harmful exploitation we have ever engaged in as a species. This book is testament to both the breadth of this harm and the emerging desire to see it corrected. Our contribution will be to hone in on one particularly damaging area of animal exploitation – meat eating. We have chosen meat eating as it reflects a voluntary and near ubiquitous form of exploitation, one which seems ripe to be challenged and changed by activists. Our aim will be to outline the psychology underlying eating animals and point to important implications both for animals and for other humans.

This chapter is comprised of three sections. In the first, we will outline the psychology of eating animals by dividing the psychological processes between those effecting how we view animals, how we view ourselves and others who both eat and refrain from eating animals, and how the physical process of eating meat changes how we think about meat. For brevity, we will present these as the eaten, the eaters, and the eating, in turn (Loughnan, Bastian, & Haslam, 2014). In the next section, we point to what we believe to be a major psychological process at the core of all three components; cognitive dissonance. We call this the "meat paradox" and it reflects the tension between exploiting animals and loving/caring for animals. We argue that activists interested in shifting how people treat animals – directly or, especially, indirectly – need to be mindful of this process. In the final section we adopt a broader approach and seek to examine how the psychology of human/ animal interaction undergirds the struggle for human freedom. In this way, we point to how understanding and addressing animal exploitation at a basic psychological level may unravel a range of additional exploitations.

Eating animals

The process of eating animals is deceptively simple at first glance – it just involves someone putting something into their mouth. Indeed, what they are being asked to eat is inherently pleasurable for a range of reasons. Meat is packed with fat, salt, and protein, all of which our ancestors found hard to extract from their evolutionary environment (Stanford & Bunn, 2001). In terms of sheer calorie density, until the modern age of corn syrup, sugar cane, and vegan ice-cream there was no better way to efficiently increase calorie intake than consuming animal products. These likely evolved tendencies remain with us today. In a study of people's justifications for eating meat, we found that the belief that meat is simply "too delicious to stop eating" was one of four major reasons given for eating animals (Piazza et al., 2015). Thus, on the surface eating meat seems simple; people are engaging in an inherently rewarding gustatory practice.

Dig a little deeper however, and the psychological processes supporting eating animals are far more complex. We argue they exist at three distinct but interlocking levels. We possess strong beliefs about the animals we eat which helps us eat them (the eaten). We possess beliefs about ourselves and others that serves to help us eat animals (the eaters). Finally, when the knife hits the plate, we possess beliefs about eating which further help us eat animals (the eating). We are not suggesting that all people engage in all processes simultaneously. It seems quite likely that whereas one person may rely heavily on beliefs about themselves and their identity (eaters), another still may rely on beliefs about the nature and psychology of animals (the eaten). Indeed, we can forward the novel hypothesis that as an omnivore becomes closer to the act of meat creation (farming, hunting, slaughtering, butchering), the more varied and robust their defenses become. We will dedicate the next three sections to exploring what psychology can tell us about the eaten, the eaters, and the eating.

The eaten

All (non-poisonous) animals are equally edible, but not all animals are equally eaten. A small part of the variability in which animals are eaten can be attributed to trivial factors like availability or toxicity. Even if we exclude these animals, there remains a huge number of available animals of which people eat a tiny fraction. For most Westerners just three animals – cows, pigs, and chickens – reflect the overwhelming majority of meat they consume. We suggest that this limited range of animals is not reducible to purely economic or historic factors, but rather includes a range of psychological beliefs about animals which serve to make eating them more morally palatable (see also Joy, 2010).

The difficulty in eating animals arises because animals are harmed by being eaten and this violates their moral rights and our desire to be non-violent. Thus, any beliefs about animals that can serve to undermine their moral standing, makes them easier to eat. A mind – or at least the capacity to suffer – has long been recognized as the key characteristic possessed by entities given moral rights (Bentham, 1789). Animals that are seen as more human-like (and thus more mindful) are seen as more capable of experiencing pain (Plous, 2003), and people who hurt such animals are deemed worthy of harsher punishment (Allen, Wilson, Ng, & Dunn, 2010). Therefore, it is unsurprising that people tend to avoid seeing meat animals as mindful. In an initial demonstration of this effect, we asked people to rate the edibility and capacity for mind of 32 different animals (Bastian, Loughnan, Haslam, & Radke, 2012). What we found was a robust negative relationship, such that people report animals with lots of "mind" to be less edible, and relatively "mindless" animals as more edible. This same effect has been shown by other research groups (Ruby & Heine, 2012). Of course, people are imperfect at determining the extent to which an animal possesses a complex mind or high intelligence. For example, Marino (2018) carefully maps the cognitive, social, and emotional lives of chickens in ways which show huge divergence from common expectation (see Bottomley & Loughnan, 2017). For instance, chickens understand that hidden objects continue to exist, they can count to at least five, understand the basics of time, can identify individual chickens and socially manipulate them. All of this is a far cry from the common understanding of a chicken. This variability and divergence from reality provides people considerable justifications for eating meat. In short, people are comfortable eating "mindless" animals.

The relationship between animals' minds and our willingness to eat them could cut either way; we might see animals as "mindless" – that is, stupid, lacking sentience or mental complexity – because we eat them, or eat them because we think they are mindless. Although both are possible, there is good psychological evidence that people engage in the former; they see meat-animals as mindless *because* they are considered meat-animals. A clear demonstration of this effect was provided in a study of food and animal categorization. Participants were presented with a novel animal (Bennetts Tree Kangaroo) and told that it was either used as food or simply lived in the wild (Bratanova, Loughnan, & Bastian, 2011). Compared to the animal living in the wild, being told that the tree kangaroo was a meat animal resulted in participants viewing the animal as possessing less mind and as less worthy of moral concern. Importantly, this occurred even though participants were not the eaters. If people do not like eating animals with lots of mind, might one strategy to reducing meat consumption be to highlight the mental life of meat animals? Several authors have suggested this (e.g., Marino, 2018; Loughnan et al., 2014). In an attempt to examine this possibility, we presented a group of participants with information about the intelligence of pigs (Piazza & Loughnan, 2016). Importantly, half of the participants learned that pigs were smart, whereas half learned that another student called "John" learned pigs were smart. For participants asked to imagine John, the information about pig intelligence resulted in them believing that John would think pigs were smarter and be less likely to eat them. In short, it worked as you might expect. However, when the information was presented to the participants themselves, while pigs were seen as more intelligent, they were not considered any more worthy of moral concern. Thus, simply telling people that animals are smart might increase their beliefs about the animal's intelligence, however at that point intelligence mysteriously decouples from concern (see also Dowsett, Semmler, Bray, Ankeny, & Chur-Hansen, 2018 for similar results). In summary, people see the minds of animals in a motivated fashion. They see meat-animals as particularly lacking in mind, however when faced with compelling evidence for animals' minds they fail to update their moral concern.

Factors beyond animal mind also matter for whether an animal is considered edible. One important emerging factor appears to be animal cuteness. Baby animals by virtue of their large eyes and heads relative to their bodies strike people as particularly cute and the evoked feelings of tenderness seem at odds with killing and eating them. Indeed, researchers have corroborated that people are less willing to eat cute animals, specifically because of increased moral concern (Piazza, McLatchie, & Olesen, in press).

In short, people are able to eat animals because they possess a series of beliefs about those animals. Central to this is the notion that some animals lack mind and therefore moral standing. By seeing 'meat-animals' as relatively mindless, people can undermine the animals moral standing and thus reduce negative feelings around consuming that animal. Importantly, once people learn an animal is used for meat they see it as lacking mind, indicating that they are using mind in a motivated fashion. Further evidence for this motivation comes from the finding that people resist using new knowledge that animals are intelligent to change their beliefs about the animals' rights.

The eaters

Of course, animals are only part of the story – the other half is the people who eat them. Eaters are not a dichotomous group; meat-eater *versus* non-meat eater. Here we will focus on meat-eaters, simply because they are the vast majority and to varying degrees vegetarians and vegans are not involved in the exploitation of animals (see Ruby, 2012 for an excellent review of the psychology of vegetarianism). Within meat-eaters, ambivalence toward eating animals is commonplace (Berndsen & van der Pligt, 2004). Studies that have employed nuanced measures of dietary practice often find large portions of their omnivore sample divided between hardened meat-eaters and conflicted omnivores (e.g., Piazza et al., 2015). A range of individual factors contribute to this variability in how people approach meat eating.

Identity plays an important role in people's tendency to eat animals. One of the most potent identities is to be masculine, frequently (but not exclusively) sought by men. The link between red meat and masculinity is robust. People think that men who eat meat are more masculine than men who do not, and men who highly value masculinity are more likely to choose to eat meat (e.g., Rothgerber, 2013; Ruby & Heine, 2011). Combined, meat serves to both signal and to exercise masculinity, making it particularly attractive for hyper-masculine men. The link between red meat and masculinity is so strong that meat itself has come to be masculine (Rozin, Hormes, Faith, & Wansink, 2012). In addition to masculinity, cultural identity is an important aspect of the self which can be exercised through eating animals. A moment's reflection on cultural celebrations reveals that they are often associated with specific foods, especially specific types of meat. This can occur at the level of regions (Texan BBQ, Memphis ribs), nations (Thanksgiving turkey, French coq-au-vin) and religions (fish on Fridays for Catholics, Eid-al-Adha for Muslim). The importance of meat for culture has been largely neglected by psychologists. People report meat eating to be a cherished practice independent of its costs (Graca, Calheiros, & Oliveira, 2014) indicative of an important role of culture. Where the prevailing culture opposes meat eating, people are less likely to eat meat especially if they are culturally conservative (Ruby, Heine, Kamble, Cheng, & Waddar, 2013). By contrast, when the prevailing culture endorses meat eating – as in the West today – conservatives eat more meat (Dhont & Hodson, 2014) and are more likely to lapse back into meat eating if they leave (Hodson & Earle, 2018). Although this work all points to an important role for culture, this topic awaits further exploration.

At the individual level eaters also possess a suite of specific meat-eating justifications which serve to sustain their meat consumption (see Joy, 2010). These "4Ns" of meat eating are commonly held beliefs that reflect over 80% of the justifications spontaneously generated for why omnivores eat meat (Piazza et al., 2015). The most common is Necessary, the belief that people need to eat meat to be healthy or fit. The second is Nice, which reflects a belief that meat is so pleasurable to consume that this enjoyment justifies its consumption. The third is Natural, that human beings evolved to eat animals and have done for millennia. Finally, eaters can believe that it is Normal to eat meat, that meat eating is just something that people do. While none of these are arguably strong moral arguments for eating animals, they are widely endorsed and do predict how much meat people eat (Dowsett et al., 2018; Piazza et al., 2015). Thus, meat eaters can defend their choice to eat animals by citing these 4Ns.

In addition to valued identities, meat eaters can possess a set of ideological perspectives which legitimate or valorize hierarchy and dominance. We call these people authoritarians, and this is often measured via Social Dominance Orientation (SDO; Pratto, Sidanius, Stallworth, & Malle, 1994) which captures the extent to which people are comfortable with hierarchy and inequality (see Dhont, Hodson, Costello, & MacInnis, 2014; Dhont, Hodson, Leite, & Salmen, this volume for its connection to animals). Since eaters are necessarily in a dominant position relative to the eaten, the more an individual values dominance the more likely they are to accept their role as dominators. We know that people who are high in SDO are more likely to eat animals and are more likely to eat more meat than omnivores who are low in SDO (Allen et al., 2010; Dhont & Hodson, 2014). Further, authoritarian omnivores are also more likely to be speciesists, and generally endorse the domination of animals by people even outside the domain of meat eating (Dhont et al., 2014, 2016; Caviola, Everett, & Faber, 2019). Thus, by accepting or even celebrating hierarchy and inequality, one can help alleviate concerns about eating animals; I can love animals, and still think that they are beneath me and that their exploitation is justifiable.

Finally, eaters tend to see animals as not only beneath but importantly different from humans. The tendency to see animals as human-like is called anthropomorphism, and it is not surprising that people who highly anthropomorphize animals do not support eating them (Niemyjska, Cantarero, Byrka, & Bilewicz, 2018). Importantly, the key factor linking anthropomorphism to eating animals is empathy. People who see animals as more human-like care about them more, and this increased care reduces their desire to eat meat (Niemyjska et al., 2018). Alongside chronic, individual differences in the tendency to anthropomorphize, there is evidence that artificially widening or shrinking the perceived gap between humans and animals alters how people want to see them treated. In a series of experiments, we artificially manipulated the perceived distance between humans and animals (Bastian, Costello, Hodson, & Loughnan, 2012). What we found was that when the distance between humans and animals was emphasized, people cared less about animals and were more accepting of their exploitation. Thus, by positioning themselves as not only superior but as importantly different to animals, people are able to justify their consumption.

So, who eats the most meat? It is a person who is comfortable with or even enjoys authority and dominance (SDO), dislikes deviance from the norm, values identities which are reinforced or celebrated by eating or even killing animals (e.g., masculinity), who views animals as very different to people, and believes in a set of justifications for eating animals, most notably that it is necessary for their health and wellbeing.

The eating

Ultimately meat eating is a behavior and does not occur until the eaten and the eater come into direct contact. The act of eating – or of consuming more broadly, for instance drinking milk – is one of the critical points in the psychology of eating animals because it reflects the moment when all the beliefs about animals and the self are most needed. Here we outline some of the psychological processes at work when people sit to eat meat.

Relatively few studies have examined the impact of actually eating meat on people's psychology. In one early study, we had a group of participants come into the lab to sample either cashew nuts or dried beef jerky (Loughnan, Bastian, & Haslam, 2010). Following this, we asked them to complete ratings of the mental abilities and capacity to suffer of a cow. We found that when people had just consumed meat they rated cows less mindful and less capable of suffering. In a follow-up study, we had participants rate the mind and moral standing of a cow before telling them that they would have to sample some cold roast beef (Bastian et al., 2012). Prior to the sampling, we asked half to write about the origins of meat and half to write about the origins of vegetables. Finally, we had them rate the cow a second time. Consistent with the earlier study, we found that anticipating meat consumption – in this case roast beef – led participants to see cows as less mindful and worthy of moral concern. This was especially pronounced when they had made an explicit link between meat and animal. Thus, these two studies show that when people sit to engage in meat consumption they activate a range of psychological defenses which help ease the process of consumption.

The separation of meat from animal is an important facet of the psychology of actually eating animals. In a series of studies, psychologists have examined how decreasing or increasing the meat/animal link changes people's willingness to eat animals (Kunst & Hohle, 2016; Kunst & Haugestad, 2018). For instance, presenting processed meat or carcasses without the head attached made people less empathetic towards meat animals killed for human consumption. By contrast, presenting meat (e.g., beef) alongside an image of the animal (e.g., a cow) led people to feel more empathy and reduced their willingness to eat meat. Finally, at a linguistic level, replacing terms like "beef" and "pork" with animal names like "cow" and "pig" also served to reduce willingness to eat animals. In short, any process that shrinks the gap between meat and animal – which changes eating meat into eating animals – works to undermine meat-eating behavior. So, could one solution be to show consumers more unprocessed meat, to force a meat→animal connect? Perhaps, but the gains may be short term. In a follow-up, researchers found that people with considerable experience seeing unprocessed animal products were less affected by these manipulations, indicating perhaps that we adapt to unprocessed meat to maintain our consumption practices (Kunst & Haugestad, 2018).

So, how does our psychology change when we actually eat meat? Both when we anticipate or have actually eaten meat we are more likely to see that meat animal as mindless and less worthy of moral concern. Culture works here to help us avoid feeling bad too, by separating animals from meat by presenting them without heads or recognizable features, independent of the animals they come from, and using language to create further separation.

Beyond the personal: Social barriers to the meat paradox

In the three sections above, we have outlined the processes people can engage in: how they think about themselves, animals, and eating to help them avoid the meat paradox. Although the list is extensive, it may also strike you as odd. If you are, or have been, a meat-eater you can probably recall engaging in some of these processes, however most of the time when you sit down to eat a beef burger you are probably not thinking "wow, aren't cows mindless?" or "geez, I can't wait to get the necessary iron in red meat to stay healthy." Rather, you are probably not thinking of animals at all, or perhaps unless meat is most of the meal (e.g., a steak), you are potentially not thinking about even eating meat, simply eating food! We argue that this is no accident, and that society is set up to protect us from even experiencing the meat paradox such that we do not need to engage psychological processes to reduce our discomfort (see Bastian & Loughnan, 2017 for a thorough review). Put simply, if society protects me from nasty self-realizations, I do not need to deal with them at all.

We argue that a range of protective mechanisms serve to prevent people from being made aware of the meat paradox at all. These mechanisms form an invisible, defensive shield around people's psyche which means they do not have to engage the psychological heavy lifting of justifying or defending their meat-eating behavior. Perhaps the most powerful of these is the invisibility of meat production. In Western countries, meat animals are raised and slaughtered at locations far from large population centers, often for good economic reasons. An additional effect of this process is that it renders the link between animal and meat invisible for most consumers. Indeed, work by Kunst and colleagues (2016, 2018) shows that when people with limited exposure to the meat-animal link are confronted with it, their desire to eat animals is reduced. For people with regular exposure to the meat-animal link, this effect is muted. Thus, by setting up a society that locates meat production away from most consumers, society has helped the population avoid confronting the meat paradox. When they do, they are more affected by it.

The notion that isolation fosters prejudice and mistreatment is an old one in psychology (Allport, 1954; Cook, 1971). Between human groups, a lack of contact has been found to generally support prejudice and dislike (Hewstone & Brown, 1986). Pleasingly, the reverse is also true; increasing the contact between previously isolated groups reduces prejudice and discrimination (Hewstone & Brown, 1986). Could it be that by isolating consumers and farm animals we have created psychological conditions for prejudice and discrimination against animals? If so, would bringing these group into contact reduce that enmity and exploitation? Recently, we examined whether contact helps reduce meat eating. We had participants come to a farm and engage in physical contact – touching, patting – meat animals such as cows, pigs, and lambs. We measured their beliefs and attitudes about animals before and after this experience. We found that contact with animals increased the amount of mind and moral concern they reported for the animal (Loughnan, Davies, Antal, & Zahareiva, in prep). Interestingly, we did *not* find that it robustly reduced their meat eating commitment. This failure to find any effect on meat eating itself is reminiscent of other work by Piazza and Loughnan (2016), where we found that convincing people that pigs were smart oddly did not increase their moral concern for pigs. Perhaps, in both cases (Piazza & Loughnan, 2016; Loughnan et al., in prep), these results indicate that when one psychological defense is knocked down others spring up. Regardless, it seems that by breaking the link between animal and meat, society can help us avoid the meat paradox.

In the same manner, signifiers of the animal base of meat – things like eyes, ears, skin, even bone – have been steadily disappearing from store-bought meat, at least in the West. Again, these could easily reflect shifts in how people eat, the time they have to prepare meals, their cooking skills, however, by rendering the animal and the meat into separate entities people are able to avoid thinking about animals when they prepare and eat them. Again, Kunst and Hohle (2016) found that people felt more empathy for chickens when presented with an intact chicken carcass than when presented with minced chicken. By presenting meat in ways that do not resemble the animal, societies help people avoid the psychological work of reducing dissonance at the personal level. Combined, these finding provides a good psychological basis for intervention. We know that society helps break the meat-animal link by locating meat production away from populations and breaking down animals before consumers see them. Thus, it is the people with the least contact with animals who will be most affected by having this link drawn to their attention. Targeting people with little contact and forcing this link could be a cost-efficient way to intervene.

In addition to the individual defenses that people possess to help them navigate the meat paradox, society also works to protect them from experiencing it at all. This presents both a challenge and an opportunity for the activist community. The challenge will be to pierce the willful ignorance around meat eating which protects many people living in major urban centers in Western nations, who are also eating the most animals. The opportunity is that these people may be lacking in robust personal defenses for their meat eating, relying instead on society to protect them from this meat-animal link, making them a ripe target for behavioral change.

Speculations beyond animals

So far we have outlined the psychology of eating meat, specifically the eaters, the eaten, and the eating. We have also just shown how people navigate the meat paradox, and those industries and institutions which navigate the meat paradox for us. In the final section of this chapter we wanted to think beyond the important but narrowly defined scope of humans eating animals. The fact that people seek justification for eating meat suggests killing animals for food is something people feel uneasy about. Indeed, by killing animals, we subtly or overtly tell ourselves and our children that it is okay to harm others if we benefit from it. We suggest that these broader consequences of meat eating shape the way we think about human-human interactions: in devaluing animals, we overvalue ourselves and desensitize ourselves to the plight of others (Hodson, MacInnis, & Costello, 2014). For the remainder of the chapter, we seek to make the following two points. We suggest that the product-production disconnect of animal consumption illustrates a broader phenomenon whereby people fail to think about the indirect consequences of their behavior. We then discuss whether animal exploitation may be a fundamental form of oppression, and speculate whether disrupting it may asymmetrically reduce other morally troublesome behavior.

For a behavior as commonplace and violent as killing animals, psychological theories which seek to explore how human-animal and human-human relations connect are surprisingly scarce. Carol Adams (1990; 2004) outlined the similarities between the portrayals of women and animals in the media; both groups are often reduced to objects to be consumed (as either food objects or sex objects). The link between the subordination of women and the subordination of animals in societies around the world also fits well theoretically with the association between meat and masculinity discussed earlier in this chapter. Keeping with the theme of linking perceptions of humans with perceptions of animals, Costello and Hodson (2014) developed the interspecies model of prejudice (see also Dhont et al., 2014). The interspecies model of prejudice shows both theoretically and empirically, that peoples' negative attitudes towards animals predict how much they dehumanize other groups (such as immigrants and ethnic others) (see also Hodson, Dhont, & Earle, this volume). In addition to the interspecies model of prejudice, Dhont, Hodson, and Leite (2016) developed the Social Dominance Human Animal Relations Model (SD-HARM) which illustrates that social dominance is the common ideological core underlying speciesism and ethnic prejudice (see also Dhont et al., this volume). We see our theorizing as complimentary to this line of thought.

However, while these theories do a good job of measuring and explaining the similarities between and common roots of prejudice towards humans and animals, they cannot explain why people who do not hold prejudicial attitudes or who do not value group dominance, still harm animals. Indeed, the vast majority of people who are low SDO, "left wing" or "liberal" are themselves meat eaters who participate in the slaughter of animals. This is because prejudice and harmful behavior do not necessarily equate: one can be prejudiced without harming others (e.g., simply holding prejudicial views), and one can harm others without being prejudiced (e.g., meat eaters who do not hold speciesist views). Because of this reasoning, we suggest that there will be limitations to the usefulness of lowering SDO as a strategy to reduce meat consumption. As we have shown in this chapter, people do not necessarily give prejudicial or dominance oriented views as a justification for harming animals; rather people simply tend not to talk or think about animals when asked about their meat consumption. The two (meat and killing animals) are psychologically distinct for most meat eaters in the developed world today. While we do not deny that speciesism is involved in the exploitation of animals, we believe meat consumption reflects but an example of a wider phenomenon of blind consumption. We spend the remainder of this chapter describing how this is so, before discussing the consequences of this widespread psychological disconnect. An example is useful to illustrate.

Suppose that you walk into a middle-priced electronics store on the local high street. After walking around you find the cell phone that you have been looking at online. It has got a good camera, decent battery life, and is even on sale in store. You walk up to the counter, and while making your purchase you donate some loose change to a charity box next to the cash register. This is a situation most of us have been in. Most people have bought a cell phone before, and many of us donate loose change to charity. This example is characteristic of the way people think about morality. If you asked any person walking out of that shop – having just paid for their new cell phone as mentioned, and even donated to charity – whether they had done anything immoral in the store, they would probably be quite offended. In fact, if they did not do anything immoral directly, then they probably believe that they did not do anything immoral. This effect is typified by the contact finding in moral dilemmas, which shows using a lever to drop someone to their death is experienced by the actor as less immoral than directly pushing someone to their death (Greene et al., 2009). Much like it would be immoral to push someone to their death, but deemed less immoral to pull a lever to drop someone to their death, it would be immoral to steal from a store, but it is perfectly fine to not ask questions about that business – questions which might raise some eyebrows among the morally conscious of us. If we did ask some questions, in any cell phone store in the world, we would find out that the cell phone we would like to buy may contain conflict resources – precious metals sourced during armed conflict (Eichstaedt, 2011). We would also find out that those conflict metals in our cell phone may have come from the Democratic Republic of Congo, a country torn apart by armed conflict which is in part fueled by the mining industry (Eichstaedt, 2011). Now, it would be pretty impressive for a retail assistant to possess that amount of information about the products they sell. However, far from being unique, this example represents a norm in industrialized nations of the psychological disconnection of products from their production – a connection people have almost complete ignorance of, and apathy towards knowing.

To be sure, the way people fail to think about animals when purchasing meat is conceptually similar to how people fail to think about the production of all the other products and services they buy. Like the meat paradox, the *human paradox* we have just outlined will have its own suite of defense mechanisms to justify our behavior in the face of evidence that anyone living in an industrialized nation is likely contributing to the ongoing wars in the Democratic Republic of Congo, irrespective of whether one thinks of themselves "pro human rights" or not. The simple fact is that almost all the electronics we use (cell phone, computer, car, etc.) have elements which can be sourced from areas in the midst of war. In addition to a human paradox, it is likely that there will also be an environment paradox whereby people claim to care for the environment, while behaving contradictory. Much like meat production and cell phone production, the consequences of burning fossil fuels, and the corruption oil revenue can fund (Ikelegbe, 2005), are often both psychologically as well as geographically removed from our consumption.

Therefore, finding a human paradox – where people report caring for people around the world yet engage in consumption which indirectly harms them (e.g., purchasing conflict minerals), or an *environment paradox* where they report caring for but yet help destroy the planet (e.g., burning fossil fuels) would be two extensions which our theorizing on human–animal relations can contribute to other disciplines. Whether the dynamics at the individual and social level are analogous to those we observe for meat consumption is an interesting question for future research. It is likely that the factors at the social level for the environment and the human paradox are like the meat paradox: industrialized nations are such that consumers seldom confront the unsavory backstory to their consumption – society does it for us. Just as factory farms and slaughterhouses are in rural areas, well hidden from any city centers, the mining of minerals and the drilling for oil often happens out of sight and consequently out of mind. However, when it comes to people navigating the environment and human paradoxes, we expect the dissonance reduction strategies to differ compared to the 4N's of meat eating justification. The justifications people give for contributing to the exploitation of humans and the environment may include benefits to the economy, and for the environment paradox in particular, denial that climate change is human made and happening. In sum, there is a psychological similarity between our participation in animal exploitation, and our participation in other exploitative practices (e.g., conflict minerals, environmental damage), which operates both at the societal and individual level. Having just outlined the paradoxical nature of everyday life, and the fact that it is not only meat eaters who experience cognitive dissonance about moral behavior, we now question how morally paradoxical behaviors might relate to each other. Specifically, whether the meat paradox is foundational to other paradoxical behavior, and what this might mean for animal rights activists. We begin by considering the implications human–animal relations and human–human relations have for each other.

Prior work has argued that recognizing meat eating is immoral can, via cognitive dissonance reduction, increase the frequency of this immoral behavior (Bastian & Loughnan, 2017). That is, after the guilt from harming animals is removed, people are free to not only continue the behavior, but to engage in it more often. In addition, effectively engaging in dissonance reduction in the human–animal domain may also lead to more immoral behavior in the human–human one. If the lesson learned from navigating the meat paradox is that it is justifiable and acceptable to exploit others to the point of death to fulfil our needs, then can this thinking buttress similar beliefs in the human sphere? It is possible that after reconciling with oneself that even though one gets other people to kill animals on their behalf, that they are still "good" people. More importantly, that because taking another animals life for food, fun, or fashion is justifiable, they will not think twice about buying that new pair of shoes, flying around the world, or wearing diamonds. Put simply, by justifying harming animals, we may be teaching each other, and indeed our children, that it is okay to degrade others (or the environment) so long as we benefit from that behavior.

The reverse could also be true; that dissonance reduction in the human sphere, could lead to further justification of animal exploitation. Put another way, if the psychological mechanisms at play in reducing cognitive dissonance from harming animals are activated when reducing the dissonance of other morally troublesome behavior, then justifying harmful behavior in the human sphere could be undermining efforts to reduce harm in the animal sphere (e.g., violations of human rights, environmental destruction). For example, if people can justify contributing to armed conflicts and the slavery of humans, then they could also think exploiting other animals is justifiable. In sum, we have just discussed how the psychology of justifying morally troublesome behavior may be conceptually the same, irrespective of the type of consumption (e.g., degrading animals, humans, or the environment). Whether or not these moral paradoxes (animal, human, or environment) buttress one another is an important question for future research. Let us now turn that thinking around. While this reasoning is concerning for the fact that these psychological processes may be closely connected, because of that very connection, there is also the potential for a united front for positive change.

If I learn that it is unacceptable to kill animals to meet my needs, can I then accept the exploitation of other humans? Moreover, if I learn that there is an unsavory back story to where meat comes from, and I spend time considering that connection, what does this do to the rest of my consumption (but see Piazza, this volume). In line with the SD-HARM theorizing (Dhont et al., 2016, this volume), it may be that by attacking the psychology of a *foundational* exploitation we may collapse an edifice of human exploitations. Whether this foundation is based on social dominance orientation (as per the SD-HARM prediction), or on the conceptual similarity in the psychology of consumer behavior (or a combination of both) is an additional avenue for future research. It is unclear at this stage whether challenging the exploitation of animals in isolation, or whether challenging a more general apathy towards consumption will improve our relationship with animals.

Following the theme of this book, bridging insights from academia and activism, there are strong practical implications of our theorizing for activists. However, we acknowledge that improving concern for other animals, and concern for other humans and the environment will be no small task. Indeed, millions of people including researchers, politicians, academics, artists, and activists already dedicate their lives to raising awareness of, and lessening the impact we have on the environment and on others. We think the difficulty they all face is a larger scale version of the meat paradox – people possess a suite of personal beliefs and society has structures in place, which serve to protect the everyday person from *encountering* the harm their daily behavior causes. Indeed, it is not only meat eaters who navigate moral paradoxes. To be sure, vegetarians, vegans, environmentalists, human rights advocates, and indeed us as authors of this volume, could also look inward and find paradoxical aspects of our own lives that we wrestle with. In line with the vision of this book, and just as the authors' research has been informed in part by animal rights activism, certain aspects of our theorizing on the psychology of the meat paradox may benefit animal rights activists.

Animal rights activists seem to be well equipped for drawing the connection between meat and animal exploitation. However, the psychology is increasingly clear – the challenge is getting the behavior to change, and the message to stick. As we have outlined in this chapter, people host a suite of interchangeable and often nonsensical beliefs to reduce dissonance when confronted by information by animal rights activists. For activists, there is a need to incorporate psychological research on the mechanism of cognitive dissonance such that they approach their activism with an understanding of what is likely to happen when they show people different information. As recent work has shown (Piazza, this volume), different strategies for different demographics of people are likely to be the most effectual. It may be that a different approach to the frequent highlighting of the cruel nature of exploiting animals is more effective for some. An example of such a strategy would be to make meat-eaters feel empowered by making more autonomous decisions about their dietary choices, or about human animal relations in general. As most people grow up eating meat, they never actually made a choice to do it, meat was given to them. Drawing attention to this would likely work more so for unconventional people (perhaps also libertarians) who might appreciate that aspect of animal rights activism which challenges the status quo; which highlights the arbitrary nature of the human-animal divide; or which promotes free thinking. For those leaning to the political right, it may be that providing positive portrayals of animals and influencing specific biases towards animal species may elicit less exploitative treatment of animals. For activists, using roundabout ways to reduce animal exploitation may feel like sacrificing integrity, but if the goal is to end animal exploitation then activists may need this nuance (see Leenaert, this volume). Consequently, animal rights activists may benefit from being critical of their activism and asking whether the forces that led them to behavioral change are actually the most effective form of activism for others. In addition, animal rights activists could benefit from looking at whether the activism that might feel good to do (e.g., storming into a restaurant with signs aimed at disrupting diners), is actually going to bring about change for those still eating animals.

While further research is needed to clarify such speculations, it may be that after people make the connection between meat and meat production, they begin to ask questions about the backstory to other components of their life and their consumption. One way to measure this in future research would be to look at whether people who abstain from eating meat for ethical reasons, are also more attuned to, or are at least more receptive to, the consequences of other parts of our consumer life. However, the reverse direction might also work. In this way, animal rights activists can work with academics, as this book is doing, and see where their efforts are best directed. It may be that drawing the connection between meat and harming animals is not the most efficient way to reduce animal exploitation for everyone.

Conclusions

We have shown how eating meat has a psychological impact on the eaters and the eaten. We have also discussed how eating meat represents a wider tendency to morally disengage from behavior we are not directly involved in. Consequently, understanding the psychology of eating meat may improve efforts to raise concern for the environment and for those on the fringes of our moral concern (such as drug addicts and the homeless, and the workers who produce what we consume). Indeed, a promising way of reducing slavery would be to get people to appreciate human-animal exploitation; if they accept that it is unacceptable to exploit animals, how could they endorse slavery? It is not unlikely, that in turn, greater concern for the environment, and greater concern for the production of the products we consume, will include greater concern for the animals that are exploited. In this way, the goals of environmentalists, human rights advocates, and animal rights activists share important common ground. To obtain advances in any of the domains, requires a connection between our own lives and that in the world which we contribute to, indeed between our daily lives and the societal order we perpetuate. To obtain advances in any of the domains will require the discomfort of shedding light on our everyday behavior – challenging both the order of society, and the order of our thinking.

Acknowledgment

This chapter was partially supported by a Philip Leverhulme Prize awarded to Steve Loughnan.

References

Adams, C. J. (2004). *The Pornography of Meat*. New York: Continuum.

Adams, C. J. (1990). *The Sexual Politics of Meat: A Feminist-Vegetarian Critical Theory*. New York: Continuum.

Allen, M., Wilson, M., Ng, S., & Dunn, M. (2010). Values and beliefs of vegetarians and omnivores. *The Journal of Social Psychology, 140*, 405–422.

Allport, G. W. (1954). *The Nature of Prejudice*. Reading, MA: Addison-Wesley.

Altemeyer, B. (1981). *Right-wing authoritarianism*. Winnipeg: University of Manitoba Press.

Bastian, B., Costello, K., Loughnan, S., & Hodson, G. (2012). When closing the human-animal divide expands moral concern: The importance of framing. *Social Psychological and Personality Science, 3*, 421–429.

Bastian, B., Loughnan, S., Haslam, N., & Radke, H. R. M. (2012). Don't mind meat? The denial of mind to animals used for human consumption. *Personality and Social Psychology Bulletin, 38*, 247–256.

Bastian, B., & Loughnan, S. (2017). Resolving the meat-paradox: A motivational account of morally troublesome behavior and its maintenance. *Personality and Social Psychology Review, 21*(3), 278–299.

Berndsen, M., & van der Pligt, J. (2004). Ambivalence towards meat. *Appetite, 42*, 71–78.

Bentham, J. (1789). An introduction to the principles of morals and legislation. UCL Bentham Project.

Bottomley, E., & Loughnan, S. (2017). Chickening out of change: Will knowing more about thinking chickens change public perceptions? *Animal Sentience, 17*.

Bratanova, B., Loughnan, S., & Bastian, B. (2011). The effect of categorization as food on the perceived moral standing of animals. *Appetite, 57*, 193–196.

Caviola, L., Everett, J. C., & Faber, N. (2019). The moral standing of animals: Towards a psychology of speciesism. *Journal of Personality and Social Psychology, 116*, 1011–1029. doi:10.1037/pspp0000182

Cook, S. W. (1971). *The Effect of Unintended Interracial Contact upon Racial Interaction and Attitude Change (Final Report, Project No, 5-1320).* Washington, DC: US Department of Health, Education, and Welfare.

Costello, K., & Hodson, G. (2014). Explaining dehumanization among children: The interspecies model of prejudice. *British Journal of Social Psychology, 53*(1), 175–197.

Dhont, K., & Hodson, G. (2014). Why do right-wing adherents engage in more animal exploitation and meat consumption? *Personality and Individual Differences, 64,* 12–17. doi: 10.1016/j.paid.2014.02.002

Dhont, K., Hodson, G., Costello, K., & MacInnis, C. C. (2014). Social dominance orientation connects prejudicial human–human and human–animal relations. *Personality and Individual Differences, 61,* 105–108.

Dhont, K., Hodson, G., & Leite, A. C. (2016). Common ideological roots of speciesism and generalized ethnic prejudice: The social dominance human-animal relations model (SD-HARM). *European Journal of Personality, 30,* 507–522. doi: 10.1002/per.2069

Dhont, K., Hodson, K., Leite, A. C., & Salmen, A. (this volume). The psychology of speciesism. In K. Dhont & G. Hodson (Eds.), *Why We Love and Exploit Animals: Bridging Insights from Academia and Advocacy.* Abingdon: Routledge.

Dowsett, E., Semmler, C., Bray, H., Ankeny, R., & Chur-Hansen, A. (2018). Neutralising the meat paradox: Cognitive dissonance, gender, and eating animals. *Appetite, 123,* 280–288.

Eichstaedt, P. (2011). *Consuming The Congo: War and Conflict Minerals in the World's Deadliest Place.* Chicago, IL: Chicago Review Press.

Graca, J., Calheiros, M., & Oliveira, A. (2014). Moral disengagement in harmful but cherished food practices? An exploration into the case of meat. *Journal of Agricultural and Environmental Ethics, 27,* 749–765.

Greene, J. D., Cushman, F. A., Stewart, L. E., Lowenberg, K., Nystrom, L. E., & Cohen, J. D. (2009). Pushing moral buttons: The interaction between personal force and intention in moral judgment. *Cognition, 111*(3), 364–371.

Hewstone, M., & Brown, R. (1986). Contact is not enough: An intergroup perspective on the 'contact hypothesis.' In M. Hewstone & R. Brown (Eds.), *Social Psychology and Society. Contact and Conflict in Intergroup Encounters* (pp. 1–44). Cambridge, MA: Basil Blackwell.

Hodson, G., Dhont, K, & Earle, M. (this volume). Devaluing animals, "animalistic" humans, and people who protect animals. In K. Dhont & G. Hodson (Eds.), *Why We Love and Exploit Animals: Bridging Insights from Academia and Advocacy.* Abingdon: Routledge.

Hodson, G., & Earle, M. (2018). Conservatism predicts lapses from vegetarian/vegan diets to meat consumption (through lower social justice concerns and social support). *Appetite, 120,* 75–81. doi: 10.1016/j.appet.2017.08.027

Hodson, G., MacInnis, C. C., & Costello, K. (2014). (Over)Valuing "humanness" as an aggravator of intergroup prejudices and discrimination. In P.G. Bain, J. Vaes, & J.-Ph. Leyens (Eds.), *Humanness and Dehumanization* (pp. 86–110). London: Psychology Press.

Ikelegbe, A. (2005). The economy of conflict in the oil rich Niger Delta region of Nigeria. *Nordic Journal of African Studies, 14*(2), 208–234.

Joy, M. (2010). *Why We Love Dogs, Eat Pigs, and Wear Cows: An Introduction to Carnism.* San Francisco, CA: Conari Press.

Kunst, J. R., & Haugestad, C. A. P. (2018). The effects of dissociation on willingness to eat meat are moderated by exposure to unprocessed meat: A cross-cultural demonstration. *Appetite, 120,* 356–366.

Kunst, J. R., & Hohle, S. M. (2016) Meat eaters by dissociation: How we present, prepare and talk about meat increases positivity to eating meat by reducing empathy and disgust. *Appetite, 105,* 758–774.

Leenaert, T. (this volume). Steakholders. How pragmatic strategies can make the animal rights movement more effective. In K. Dhont & G. Hodson (Eds.), *Why We Love and Exploit Animals: Bridging Insights from Academia and Advocacy*. Abingdon: Routledge.

Loughnan, S., Bastian, B., & Haslam, N. (2014). The psychology of eating animals. *Current Directions in Psychological Science, 23*, 104–108.

Loughnan, S., Haslam, N., & Bastian, B. (2010). The role of meat consumption in the denial of moral status and mind to meat animals. *Appetite, 55*, 156–159.

Loughnan, S., Davies, T., Antal, K., & Zahareiva, E. (in prep). The effect of contact on animals and meat. *Manuscript in preparation*.

Marino, L. (2018). Thinking chickens: A review of cognition, emotion, and behavior in the domestic chicken. *Animal Cognition, 20*, 127–147.

Niemyjska, A., Cantarero, K., Byrka, K., & Bilewicz, M. (2018). Too humanlike to increase my appetite: Disposition to anthropomorphize animals relates to decreased meat consumption through empathic concern. *Appetite, 127*, 21–27.

Piazza, J. (this volume). Why people love animals yet continue to eat them. In K. Dhont & G. Hodson (Eds.), *Why People Love and Exploit Animals: Bridging Insights from Academia and Advocacy*. Abingdon: Routledge.

Piazza, J., Ruby, M., Loughnan, S., Luong, M., Kulik, J., Watkins, H., & Seigerman, M. (2015). Rationalizing meat consumption: The 4Ns. *Appetite, 91*, 114–128.

Piazza, J., McLatchie, N., & Olesen, C. (2018). Are Baby Animals Less Appetizing? Tenderness toward Baby Animals and Appetite for Meat. *Anthrozoös, 31*(3), 319–335.

Piazza, J., & Loughnan, S. (2016). Intelligence matters, but only when it is not self-relevant. *Social and Personality Psychological Science, 7*, 867–874.

Plous, S. (2003). Is there such a thing as prejudice toward animals? In S. Plous (Ed.), *Understanding Prejudice and Discrimination* (pp. 509–528). New York: McGraw-Hill.

Pratto, F., Sidanius, J., Stallworth, L., & Malle, B. (1994). Social dominance orientation: A personality variable predicting social and political attitudes. *Journal of Personality and Social Psychology, 67*, 741–763.

Rothgerber, H. (2013). Real men don't eat (vegetable) quiche: Masculinity and the justification of meat consumption. *Appetite, 14*, 363–375.

Rozin, P., Hormes, J. M., Faith, M. S., & Wansink, B. (2012). Is meat male? A quantitative multimethod framework to establish metaphoric relationships. *Journal of Consumer Research, 39*, 629–643.

Ruby, M. B. (2012). Vegetarianism. A blossoming field of study. *Appetite, 58*, 141–150.

Ruby, M. B., & Heine, S. J. (2011). Meat, morals, and masculinity. *Appetite, 56*, 447–450.

Ruby, M. B., & Heine, S. J. (2012). Too close to home. Factors predicting meat avoidance. *Appetite, 59*, 47–52.

Ruby, M., Heine, S., Kamble, S., Cheng, T., & Waddar, M. (2013). Compassion and contamination: Cultural differences in vegetarianism. *Appetite, 71*, 340–348.

Stanford, C. B., & Bunn, H. T. (Eds.). (2001). *Meat-eating & Human Evolution* (Vol. 2). Oxford: Oxford University Press.

ASKHAM BRYAN
COLLEGE
LEARNING RESOURCES

12

HOW WE LOVE AND HURT ANIMALS

Considering cognitive dissonance in young meat eaters

Hank Rothgerber

Abstract

The son of Lawrence Kohlberg, a psychologist noted for his work on moral development, once spent six months while a four-year-old refusing to eat meat because he believed it was "bad to kill animals" (Singer, 2002). The father assessed this as a primitive failure to distinguish between justifiable and non-justifiable killing and eventually was able to convince his son to reclaim meat in his diet. How did Kohlberg accomplish this? What did his son come to believe about meat? About animals? Was he uncomfortable eating meat after his resistance phase? These questions inspired the present chapter. The basic issue is whether young meat eaters experience cognitive dissonance, and if so, what strategies do they adopt to reduce the tension? The chapter first reviews more established evidence on cognitive dissonance in adult meat eaters, considers more speculative information on whether young eaters experience cognitive dissonance, and if so, how they may alleviate such tension, and suggests future research to help clarify a number of unresolved questions. From the review, it seems likely that children experience varying degrees of cognitive dissonance from eating meat – even at fairly young ages – and activists would be wise to target this group for change before behavior becomes more deeply entrenched.

A brief overview of cognitive dissonance theory

Imagine that while holding this newly purchased book in your hands and glancing at your large library of animal rights and environmentalism volumes, you hear a podcast discussing how many trees are killed every year in the production of trad-itional print books. How would you feel? Anxious? Guilty? Embarrassed? Angry? Like hitting the mute button? You would be experiencing what Festinger (1957) termed *cognitive dissonance*, originally formulated as psychological tension felt by an individual who is aware of an inconsistency involving themselves. Most often the inconsistency results from discrepancies between what we think (i.e., attitudes or beliefs) and what we do (i.e., behaviors). In this particular instance, dissonance arises because although we believe that maintaining the health of the physical environment is important, our demand for paper products threatens to undermine that very well-being.

Dissonance is unpleasant, and individuals are highly motivated to avoid information that produces it. Thus, one may switch to a different podcast and actively avoid situations that discuss the ecological threat posed by paper production. But if avoidance is impossible, there are two general strategies to reduce dissonance once it is experienced. One might reduce the discomfort behaviorally (i.e., by not purchasing any more print books), but more often dissonance is dissipated psychologically by selectively thinking about information that reduces the inconsistency. Festinger (1957) labeled these motivated perceptions "consonant cognitions," that is, thoughts that are consistent with or help explain away the troubling behavior. The individual might imagine numerous good reasons for purchasing a library full of advocacy-oriented print books, such as rationalizing that their work promotes environmental well-being far more than their book reading compromises it, objecting that ebooks are too hard on their eyes, believing that most of the books they own are made from recycled paper or well-managed forests, expressing skepticism about the accuracy of the podcast, recalling tree-friendly actions they have recently undertaken, and even reminding themselves that tree deforestation is not as much a contributor to climate change as fossil fuel combustion. Alternatively, one may reduce cognitive dissonance by altering the importance of these various thoughts. Although highly unlikely in the case of advocates, the thoughts that give rise to dissonance in the first place can be made to seem less important (e.g., one's attitudes toward the environment or to trees are not that important to the individual), or more likely, the thoughts reducing dissonance can be made to seem more important (e.g., the work that one is doing is cutting-edge and particularly critical in curtailing climate change). Dissonance, therefore, revolves around negative emotions brought on by internal conflicts and working through those conflicts to feel better.

Dissonance in meat eaters

There is abundant evidence that individuals are deeply conflicted when it comes to eating meat. Noted researcher Paul Rozin once commented, "meat should be of special interest to psychologists, because it is a quintessential example of the interesting and important state of ambivalence" (2007, p. 404). This ambivalence is manifested in the frequent misalignment between expressed attitudes and behavior toward animals. For example, ninety percent of respondents disapprove of the use of inhumane slaughter methods, but only 41% disapprove of eating meat produced from such inhumane techniques (Braithwaite & Braithwaite, 1982). Of those who consider animal welfare to be highly important, only about half report thinking about farm animal welfare when purchasing meat (Mayfield, Bennett, Tranter, & Wooldridge, 2007).

That individuals love animals and wish them humane treatment yet simultaneously eat them has been termed the *meat paradox* (Bastian, Loughnan, Haslam, & Radke, 2012; Loughnan & Davies, this volume; Loughnan, Haslam, & Bastian, 2010). Because the above examples highlight discrepancies between behavior and various ideals, an increasing number of experts have focused on cognitive dissonance as a useful framework to explain the psychology of eating meat (e.g., Bastian & Loughnan, 2017; Rothgerber, 2014a). In this case, dissonance may arise from engaging in a behavior (eating meat) that conflicts with a belief ("I don't like to hurt animals"). Individuals confronted with dissonance resulting from eating meat have two ways of responding – (a) reducing or eliminating meat consumption to be in alignment with beliefs, or (b) changing beliefs ultimately to bolster meat eating. Before discussing these responses, this chapter first considers strategies designed to prevent meat-related cognitive dissonance from ever initially taking root. These tactics can be considered the psychological immune system's preliminary line of defense.

Dissonance reduction

In the case of the meat paradox, the dissonance prevention strategies are designed the help the individual forget that they eat meat or that eating meat entails animal suffering. These strategies have been termed "apologetic" (Rothgerber, 2013) and essentially seek to avoid recognizing and acknowledging that a behavior and belief are at tension. The individual acts ambivalently, without rationalizations, and merely attempts to proceed without confronting the issue. Incidentally, this approach characterizes women, who disdain more direct justifications for eating meat (Rothgerber, 2013) and who experience more negative emotions when reminded of animal-meat connections (Dowsett, Semmler, Bray, Ankeny, & Chur-Hansen, 2018).

The first line of psychological defense is to avoid dissonance by refusing to acknowledge or think about eating meat or animals. The sheer physical isolation of factory farms from the rest of society is one way that individuals are assisted in avoiding thoughts about the treatment of animals used for food. To add social pressure, the very topic of factory farms is considered a taboo conversation (Iaccobo & Iaccobo, 2006). Avoidance has also been culturally enabled by institutions in our society that make gaining more information about the welfare of farmed animals nearly impossible, including legal guidelines protecting the privacy and secrecy of factory farms (Foer, 2009). In essence, avoidance has moved from a personal strategy to a cultural norm protected by powerful institutions (see Bastian & Loughnan, 2017). It is unsurprising, therefore, that 67 percent of respondents in one study indicated that they do not think about animal suffering in factory farming when they purchase meat (Signicom, 1997).

Individuals can also avoid dissonance by pretending that no animal is involved during meat consumption. This disconnect is accomplished by *dissociating* the animal from the food product. Even though they consume animal flesh, a delusion suggesting otherwise is promoted. According to Adams (1990), one way that individuals render animals absent from their consciousness is to change language about them as food products. Words like bacon, hamburger, and sirloin become substitutes for the animal flesh people consume, allowing meat eaters to maintain the illusion that animals are not involved. Plous (1993) discusses a number of ways language depicts animals as inanimate (e.g., hunters do not kill animals, they "harvest" them), including usage by the military and legal system. The United States Department of Agriculture (USDA) has helped people feel more comfortable about eating meat by referring to chickens, cows, and pigs as "grain-consuming animal units." As Bandura (1999) notes, such euphemistic labeling is often used to disguise objectionable activities.

Supporting this dissociation strategy, many consumers do not like to think that meat comes from a live animal (Mayfield et al., 2007), and this explains why the more meat resembles the actual animal, in terms of being red, bloody, and fatty, the more individuals are disgusted by it (Kubberoed, Ueland, Tronstad, & Risvik, 2002a). Pieces of meat that clearly remind the consumer that they were from an animal (e.g., eyes, tongues, brains, etc.) are unwillingly handled by consumers (Kubberoed et al., 2002b). Explicitly reminding shoppers about the animal origins of meat lead them to purchase less meat or prefer free range and organic meat (Hoogland, de Boer, & Boersema, 2005). Kunst and Hohle (2016) manipulated dissociation experimentally in a series of five experiments. For example, in one study, participants saw either a relatively unprocessed whole chicken, a chicken cut into commonly consumed parts, or highly processed minced chicken meat. As predicted, those who viewed the highly processed meat expressed the least empathy toward the animal it once was because they had a harder time imagining the highly processed meat as a living animal. The other studies showed that greater dissociation led to less belief that animals could experience sensations (e.g., fear, pleasure, etc.) and intellectual states (e.g., thinking, imagining, etc.), less disgust in eating the animal, and more willingness to eat the animal. Ignoring the link between animals and meat, therefore, serves to spare individuals from experiencing uncomfortable dissonance by making it seem as though no victim was involved.

Dissonance may also be avoided by convincing oneself that even if meat involves animals, the individual does not consume very much meat. In one study, when threatened with the prospects of watching a PETA documentary about meat production, women (but not men) reported eating less meat than did a control condition not expecting to watch such a documentary (Rothgerber, 2019). As a substitute to actual change and in an effort to eliminate the belief "I eat meat" despite evidence to the contrary, individuals strategically attempted to convince themselves and others that they avoid meat consumption. This strategy may be common: At least a dozen studies have documented that some people claim they are vegetarian but then simultaneously acknowledge that they eat red meat, chicken, and/or fish (see Rothgerber, 2014b). Although some of this discrepancy may arise from genuine confusion about what defines a vegetarian, some of it likely reflects wishful thinking designed to make oneself feel less guilty about harming animals.

Another dissonance prevention strategy emphasizes that even if meat involves animals and that one consumes meat, the meat that one eats is not associated with harming animals because it is "humanely" produced. These so called "conscientious omnivores" (COs) have been on the rise since the merits of the position were argued in an essay by Michael Pollan (2002). Although evaluating the ethics of this movement is beyond the scope of this chapter, it is worth noting that compared to vegetarians, COs are less likely to perceive their diet as something that they absolutely need to follow and report violating their diet more (Rothgerber, 2015). Thus, in some ways the movement may be seen as providing a shield against dissonance for those who still consume meat derived from cruel treatment. To a lesser extent, individuals may convince themselves that traditionally raised food animals do not have it so bad ("the pigpens are nice and warm," "the chickens always have enough to eat"), reasoning identified by Te Velde, Aarts, and Van Woerkum (2002) in interviews with Dutch consumers and farmers. As further evidence of this wishful thinking, Worsley, Wang, and Ridley (2015) found that the majority of Australians believed that farmers take good care of livestock.

The reviewed strategies – diverse as they are – all help protect the individual from acknowledging they are eating animals that have been harmed, which constitutes the source of meat-related dissonance in the first place. Constructing this illusion that they play no part in the harming of consumed animals may be challenging. Several additional strategies acknowledge the reality of the behavior, but then attempt to reduce dissonance by focusing attention elsewhere on other entities. One of these strategies is to obscure personal responsibility by blaming others for the harm to animals. Consumers are cynical about their ability to improve farm animal welfare and in addition to blaming farmers, allocate responsibility to governments and to retailers, the former for failing to make protective laws and the latter for failing to offer humanely produced meat (Harper & Henson, 2001; Te Velde et al., 2002). Such moral outrage has been shown in part to be motivated by an effort to assuage personal guilt and cast dispersions on others (Rothschild & Keefer, 2017). Another diversionary strategy is to derogate a seemingly benign group – vegetarians – as a way to avoid feeling guilty for their own practices. MacInnis and Hodson (2017; see also Hodson, Dhont, & Earle, this volume) found that vegetarians and vegans were evaluated as or more negatively than other commonly studied targets of prejudice (e.g., homosexuals, Blacks) – only drug addicts were evaluated more negatively. Minson and Monin (2012) found that nearly half of meat eaters generated negative associations of vegetarians. Vegetarians undermine a number of direct justifications for eating meat, and their mere presence has been shown to activate dissonance in meat eaters (Rothgerber, 2014a).

If these prevention strategies and diversionary tactics are not effective in muting meat-related cognitive dissonance, individuals are forced to confront their dissonance by either changing their behavior or changing their perceptions. Despite their affection toward animals, most measures indicate that individuals are reluctant to reduce dissonance from eating animals by reducing their intake of meat. In fact, over the last 30–35 years, average meat consumption in the USA actually increased from 176 to 240 pounds a year (Herzog, 2010), increasing at the fastest rate in four decades during 2015. In animal terms, the number of creatures killed to satisfy a family of four ballooned from 56 animals a year to 132 animals during the same time period. Recent estimates by the USDA are that 2018 will witness the highest per capita meat consumption rate in US history. An upward trajectory is also predicted to continue in the rest of North America and globally, where meat consumption has likewise been on the rise (Organisation for Economic Co-Operation and Development, 2014); projections estimate that meat consumption will increase in North America by 8%, in Europe by 7% and in Asia by 56% from 2011–2020 (OECD). Conversely, rates of vegetarianism are low – the practice is estimated at 3% in the USA (Cunningham, 2009) and in Europe, between 9% in Germany to less than 2% in Denmark and France (European Vegetarian Union, 2007).

Psychological literature has focused not on behavioral change but almost exclusively on how the meat paradox is resolved psychologically through direct, unapologetic justifications, rationalizations, and other strategic perceptions. Researchers have developed several measures that identify the justifications that meat eaters endorse to convince themselves that eating meat is defensible, including but not limited to the Meat-Eating Justification Scale (MEJ; Rothgerber, 2013) and the 4N Scale (Piazza et al., 2015). The MEJ identifies seven direct strategies to justify eating meat including denial of animal pain ("Animals don't really suffer when being raised and killed for meat"), dichotomizing animals into different categories based on their utility to humans ("To me, there is a real difference between animals we keep as pets and animals we eat as food"), and justifications based on religion ("God intended for us to eat animals"), fate ("Our early ancestors ate meat, and we are supposed to also"), hierarchy ("Humans are at the top of the food chain and meant to eat animals"), health ("Meat is essential for strong muscles"), and pro-meat attitudes ("I enjoy eating meat too much to ever give it up"). The 4N scale measures common rationalizations individuals employ to defend meat eating, including beliefs that eating meat is natural ("It is only natural to eat meat"), normal ("It is normal to eat meat"), nice ("Meat is delicious"), and necessary ("It is necessary to eat meat in order to be healthy"). These two measures are highly correlated (Piazza et al., 2015), not surprising given that hierarchy, fate, health, and pro-meat justifications have strong overlap with beliefs that meat is natural, necessary, and nice. Endorsement of both scales was related to greater reported meat consumption (Piazza et al., 2015; Rothgerber, 2013), with the 4N scale also being correlated with lower drive to restrict animal products from diet, greater commitment to eating meat, more objectification of animals, and less guilt about animal-product decisions (Piazza et al., 2015).

From a dissonance perspective, endorsement of these justification scales helps individuals feel better about eating meat and reduces guilt over doing so, hence facilitating greater consumption (Piazza, this volume). These perceptions allow individuals to act on a moral imperative and maintain their view of themselves as moral actors who do not inflict harm on others (Bandura, 1999).

This argument has been more directly supported by several experiments demonstrating that when meat-eating dissonance is triggered by some manipulation, individuals will psychologically reduce dissonance by denigrating animals, which is frequently the option provided for them. Bastian, Loughnan, and colleagues have found that eating meat during an experiment (Loughnan et al., 2010), being reminded of a link between meat and animal suffering, being expected to consume an animal in the imminent future, and judging animals rated as more appropriate to eat (Bastian et al., 2012) all led participants to judge animals as less capable of experiencing sensations (e.g., hunger, pleasure, pain, etc.) and intellectual states (e.g., thinking, planning, imagining, etc.). These strategic perceptions help eliminate aversive consequences for behavior (disrupting a necessary condition for dissonance) and thus, help participants resolve the conflict between eating animals and concern for their welfare; in Bandura's (1999) terminology, they help individuals achieve moral disengagement through denigrating the victim. Rothgerber (2014a) found that participants perceived animals to experience emotions to a lesser degree than humans and believed animals had lowered mental abilities through a different stressor: reading a brief description of a vegetarian or reading about vegetarians who consistently followed their diet and were vegetarian on their own choosing, not because of allergies. These manipulations were thought to trigger meat scrutiny and discomfort, thus causing meat eaters to psychologically justify the practice even more strongly. These studies reinforce and advance earlier work demonstrating that meat eaters perceived less human–animal similarity than did vegetarians (Bilewicz, Imhoff, & Drogosz, 2011), a finding the authors attribute to dissonance reduction. They are also supported by recent work showing that the more meat eaters experience ambivalence toward meat, the more likely they are to downplay the ability of animals to possess emotional and intellectual states (Buttlar & Walther, 2018).

It is possible that individuals may view meat as involving animals that experience pain and have capacity and eschew moral justifications for meat consumption, yet still minimize dissonance in another fashion: perceiving themselves as having no choice in the matter. Early in dissonance theory it was discovered that lacking the freedom to avoid a dissonant act would serve as an important consonant cognition (Linder, Cooper, & Jones, 1967), and an update of dissonance theory also posits that individuals must feel personally responsible for bringing about the aversive event if dissonance is to occur (The New Look; Fazio & Cooper, 1984). This may explain why meat eaters strongly believe that it is unhealthy to forego meat and why the more this view is endorsed, the more individuals report eating meat (Piazza et al., 2015; Rothgerber, 2013). By convincing themselves that meat is necessary for survival, individuals do not feel responsible for harming animals.

In summary, dissonance has become a powerful tool for explaining the psychology of meat eating in the contemporary era when the invisibility of meat's origins and consequences have been lifted. It is broad in scope, helps situate meat as a moral dilemma bearing similarities to other moral phenomenon, generates testable predictions, and suggests interventions grounded in a robust theory. Importantly, the theory also accounts for why appeals to change may not always be successful, and in the case of meat, for why meat consumption is not declining despite increasing information about its hazards. Dissonance theory predicts that people will strategically avoid information related to meat consumption and the treatment of animals, and that they will be motivated to intensify justifications following exposure to information urging meat abstention or reduction. If not framed carefully, these appeals may in some cases backfire and ironically, increase meat consumption.

Part of the difficulty in persuading individuals to give up meat is that it has become such an ingrained habit (see Bastian & Loughnan, 2017). This suggests that there may be value in investigating meat-related dissonance from a developmental perspective. Theoretically, it is compelling to study how young children process moral dilemmas that activate dissonance in adults. Practically, studying the experience of young meat eaters may offer insights into interventions before meat eating and dissonance reduction become chronic and habitual.

To underscore the importance of these interventions, consider the following, non-exhaustive sampling of problems associated with meat consumption. Reports by the UN (2006) and the Pew Commission (2008) concluded that farmed animals contribute more to global warming than all transport combined – 40% more to be precise. The environmental impact of meat extends beyond global warming: The UN report characterized the livestock industry as one of the top sources of a broad range of environmental problems including the loss of biodiversity, land degradation, water shortage and pollution. There are a number of public health problems caused by meat production (and, by inference, consumption – see the Pew Commission for a detailed listing). And this does not even include the meat industry's treatment of animals, the veil of which has been lifted by popular works by Foer (2009) and others.

Dissonance in young meat eaters

While some scholars have examined the way that children think about meat and animals, few have explicitly couched their analysis in cognitive dissonance terminology. The purpose of the remainder of the chapter, therefore, is to review evidence suggesting that young meat eaters may experience dissonance from eating meat, consider how such dissonance may be reduced, and offer suggestions for future research and for activists concerned with curtailing meat consumption.

For young meat eaters to experience meat-related dissonance, four conditions must be met: (1) they must eat meat regularly; (2) they must love animals; (3) they must understand that meat is derived from animals; and (4) they must be capable of experiencing cognitive dissonance in a general sense. Once the above conditions are met, dissonance in this population is best thought of as occurring on a continuum, with all things being equal, older children more likely to have higher levels of meat-related dissonance than younger children because of their increased knowledge, awareness, and ability to make critical connections. I would suggest that a fifth condition is not necessary for dissonance to occur but will certainly intensify the magnitude of it: whether the young person recognizes or understands that animals are harmed and killed in order to produce meat. That is, knowing that you are eating something that you loved when it was alive creates some conflict, but the added knowledge that the living animal was harmed and suffered makes the conflict more intense. This point will be discussed further when the third condition is discussed and when dissonance reduction in young meat eaters is considered. I should also note that I will interchangeably use terms such as "children" and "young" meat eaters here to refer to individuals under the age of 18, in keeping with standard definitions of what constitutes a child. When reviewing existing studies, I will make reference to more specific age groups when that information has been provided by researchers.

Diet

Evidence from Australia and the USA suggests that many children regularly consume meat even at a very young age. An Australian study found that 14 year-old girls ate meat an average of slightly more than one meal per day for an average of 273 grams, and for boys meat was consumed nearly one and a half meals per day, for an average of 301 grams per day (Laing, 1999). These patterns were relative stable until ages 10–12 for boys and ages 14–16 for girls, although meat intake steadily increased with age. In a California sample, slightly more than half of the children ate hot dogs during their second year of life, and 64 percent were reported to eat beef/hamburger "occasionally" or "regularly" from birth to two years (Kwan, Block, Selvin, Month, & Buffler, 2004). A nationally representative American sample revealed that children and adolescents (n = 8,307) ate an average of 148.7 grams of animal flesh per day (Paeratakul, Ferdinand, Champagne, Ryan, & Bray, 2003). Although distributions were not broken down by specific age, the authors reported that the results from younger children showed a similar pattern to that of older children and adolescents. It seems reasonable to conclude that meat is an important part of diet for many children, even at a very young age.

Love for animals

At the same time society encourages children to eat animals, it also encourages them to love animals. Animals occupy a prominent role in the socialization of children. At young ages, they play central roles in both written and visual forms, as children's books and television programs feature animal characters, in many cases to the complete exclusion of humans. Children are socialized to have intense connections to these animals to the point that many are oblivious to conventional moral distinctions between humans and all other living creatures (Zerubavel, 1997). Stuffed animals are popular toys for young children and help intensity the emotional connection the child has for animal life. These bonds reach their full maturity when it turns to animals kept as pets – across different age groups, there is evidence that children are emotionally expressive and connected to their pets. When asked who they would turn to in emotional situations or when wanting to share a secret, nearly half of a sample of five-year olds without prompt, mentioned a pet (Melson & Schwarz, 1994). An even greater percent (75 percent) of 10–14 year-olds revealed that they turned to their pets when they were upset (Covert, Whirren, Keith, & Nelson, 1985).

Realizing the animal origin of meat

That children adore animals yet still regularly eat meat before they appreciate that they are eating the dead body of an animal suggests that this realization [what Pallotta (2008) termed a "meat epiphany"] may be an emotionally intense experience for children. An examiner of intellectual growth in young children commented nearly a century ago that "there is probably no moral field in which the child sees so many puzzling inconsistencies as here" (Isaacs, 1930, p.162). In contrast, the meat paradox likely takes on a different meaning for adults who likely feel less attached to animals in the first place, who have for some time eaten meat with full awareness of their actions, who are armed with more sophisticated strategies for dealing with dissonance, who are more fully socialized, and who are relatively desensitized following years of justifying meat consumption.

Despite its potentially traumatic impact on development, we know little about the experience of the meat epiphany and even about when and how it is communicated to children. Bray, Zambrano, Chur-Hansen, and Ankeny (2016) provided a number of reasons why parents may resist revealing the origins of meat to their children including that it may activate their own dissonance, may result in their children embracing vegetarianism, may confuse children that killing or hurting animals is morally acceptable, and may not be something that they even think about themselves. Nonetheless, they discovered that 93% of their Australian sample had discussed the animal origins of meat with their children. The majority considered it appropriate to have these discussions before their child was 5 years old, with 37% believing that from 0 to 2 years of age was most appropriate. Parents may expect that conversations at these young ages would be easier and less traumatic but still have enough impact to discourage the need for future conversations. Parents who believed conversations should be initiated when children were older than 5 years reported their children being more upset when told about the animal origin of meat. It is not clear whether this is because the children were older and understood more, or whether the parents waited because they knew that something about themselves or their children would make the conversation more upsetting in the first place.

There is reason, therefore, to believe that even very young meat eaters know something about the animal origin of food, but it is unclear when they first effectively process this information. It is also possible that there may be a difference between understanding that meat is derived from animals in general and knowing which forms of meat are derived from which particular animals. Plous (1993) reported that only 46% of 1st–5th graders properly identified hamburger as coming from a cow. This more specific understanding undoubtedly takes longer to acquire than a more global understanding that meat is derived from animals.

As alluded to earlier, dissonance will also be stronger if children not only connect meat to animals, but if they connect it to animals that have suffered and been harmed. It is difficult to assess the extent to which children understand that animals are harmed and that they suffer when made into meat. Harris (2012), a leading authority who has written about the ways in which children approach the morality of eating meat, assumes that most children have learned something about the way animals are farmed and slaughtered, even without direct experience. There is little evidence to support this conclusion one way or the other though.

Capable of experiencing dissonance

Several experiments have revealed that children as young as 4 years can experience dissonance for both self-driven behavior (Egan, Santos, & Bloom, 2007) and for induced behavior (Aronson & Carlsmith, 1963). However, these experiments involved dissonance from making relatively benign choices. For example, in the study by Egan et al. (2007), when first faced with choosing between several stickers, children later devalued a sticker they had previously not chosen, and in the study of Aronson and Carlsmith (1963) children rated a toy that they had not played with under a mild threat as less desirable than one they had avoided under a more severe threat. While most writers (e.g., Bray et al., 2016; Pallotta, 2008) on child-animal relations have assumed tension to occur following the meat epiphany, there is no experimental evidence directly addressing whether children can experience dissonance over morally relevant behavior in general, or in the specific case of the meat paradox.

Overall, although some questions remain, there is reason to believe that many children five years old or younger may experience some dissonance from eating meat. The magnitude of dissonance likely depends on a number of factors in addition to their age, including their values, intellect, affinity toward animals, knowledge of food production, their parents' beliefs about eating meat, etc. Some children may think about eating meat a great deal and others may scarcely give it a thought. The question becomes how do children reduce potential dissonance they experience? Do they seek to avoid it, favor behavioral changes, or psychologically seek changes in beliefs to reduce the discomfort? If psychological, what justifications and strategic perceptions are favored? How might these beliefs be different than the ones embraced by adults to reduce their own meat-related dissonance? What differentiates those choosing behavioral changes from those who continue eating meat?

Dissonance reduction in young meat eaters

Some children are presumably so upset by the meat epiphany that regardless of their family practices, they eschew meat and identify themselves as vegetarians. In one study, these so called "independent vegetarians" (i.e., young vegetarians from meat-eating families) all cited ethical motives for their meat abstention and rarely cited other concerns (Hussar & Harris, 2010), so it is not too difficult to imagine dissonance pressures at work. But this group may not be completely dissimilar to the meat-eating majority; that is, qualitative and anecdotal evidence suggests that it is not uncommon for children to turn against meat eating after having a meat epiphany as demonstrated in the abstract of this chapter with the example of Kohlberg's son. For example, Bray et al. (2016) reported that many children expressed empathy toward animals when learning of the origins of meat, and the response was so strong that some stopped eating meat, at least temporarily. Amato and Partridge (1989) claimed that meat resistance experiences are not uncommon in childhood, and Singer (2002) noted that children frequently stop eating meat following a meat epiphany, at least until parental pressure is levied. In a study of thirty adult vegan animal rights activists, the majority retrospectively reported that in childhood, they had a desire to abstain from meat following awareness that meat was derived from dead animals (Pallotta, 2008).

Despite behavioral proclivities of some children to abstain from meat, parental pressure, socialization, and cultural conditioning are generally successful in encouraging regular childhood meat eating, the high rates of which were examined earlier. Even among the activists studied by Pallotta (2008), who overwhelmingly recalled being sensitive to animal suffering during childhood and the majority of whom expressed a desire to abstain from meat, only half became vegetarian before age 18. Amato and Partridge (1989) observed that very few children become vegetarian when they have childhood meat resistance experiences. Dissonance experienced by the majority of children upon learning the animal origin of meat is likely, therefore, to be resolved through psychological techniques. Are the techniques used by children equivalent to the ones employed by adults?

Direct justifications would seem unlikely. Because of their strong affection toward animals, it seems doubtful that children would denigrate the capacity of animals. Even meat-eating children recognize that animals can suffer, and they perceive that an animal's ability for thought to be similar to that perceived by independent vegetarians (Harris, 2012), suggesting that this is an improbable outlet for their dissonance. Justifications based on human fate or hierarchies seem too sophisticated for many children, as does the strategy of underreporting their meat consumption. Rationalizations such as normal, nice, and necessary may create some pressure to continue eating meat but may be more effective for adults who likely find this conflict less traumatic. Religious justifications that God made animals for the purpose and pleasure of humans appear not to be too complex on the surface and may be a possible way to alleviate dissonance in some young meat eaters, but they still do not resolve the fact that animals are harmed and suffer in the conversion to meat. Thus, even those taught this religious doctrine may grapple with important questions.

The psychologist Scott Plous (1993), philosopher Peter Singer (2002), and sociologists Kate Stewart and Matthew Cole (2009) have written about how children are socialized to both love and eat animals. Their divergent analyses identify several strategies to resolve the conflict: (1) dichotomizing animals into different groups (e.g., pets versus farmed animals) and attaching more value to those categorized as non-food; (2) and avoiding the truth about the treatment of farmed animals served as meat.

One way to encourage children to eat meat and reduce dissonance is to subtly dichotomize animals into those that are consumed and those that are not. Stewart and Cole (2009) suggested that culturally appropriate relationships with non-human animals are socially transmitted in subtle, insidious, and powerful ways in childhood. They argued that all representations of nonhuman animals are defined according to their usefulness to humans and propose two independent dimensions by which humans define animals: in terms of their visibility and how objectified they are. The most privileged status is attached to pets who are both highly visible and treated as subjects with whom, as previously discussed, children form a close relationship. On the opposite extreme lie farmed animals, who are nearly invisible and who are objectified. To support the invisibility notion, Plous (1993) reported that only 3 percent of stuffed animal varieties from a major manufacturer were commonly eaten animals, i.e., cows, pigs, or chickens. Of the 34 types of stuffed animals offered by a large retail toy catalogue in the USA, none featured a cow, pig, or chicken.

Anderson and Henderson (2005) noted that in the majority of children's stories, there is a separation between consumed animals such as chickens, sheep, cows, and pigs, which are largely invisible and treated as replaceable commodities, and domesticated animals and pets, which become humanized. Stewart and Cole (2009) specify several story and communication techniques that blur the boundary between humans and these latter animals, thereby elevating our sympathy and fondness for them. Cruelty to domesticated animals is scorned, teaching children that while it is wrong to harm certain animals, it is simultaneously acceptable to eat others (Paul, 1996). In addition, factory farms are physically located away from public view and although many schoolchildren visit old-fashioned farms, few visit anything resembling a modern CAFO (Plous, 1993). Farmed animals are typically invisible and unmentioned in literature and film (Stewart & Cole, 2009), but when represented, it is often an inaccurate portrayal muting the harsh conditions faced by these animals (Singer, 2002). American children are socialized to believe that meat originates from happy farm animals living in peaceful settings (Plous, 1993). It is unsurprising, therefore, that only one-quarter of 1st–5th graders believe that farmed animals experienced unhappiness, about half the number that believe that pets or wild animals are sometimes unhappy (Plous, 1993).

The effect is to render farmed animals largely invisible, which discourages children from thinking about them and giving them moral consideration. As Singer (2002) expressed, "rather than having one unified attitude to animals, the child has two conflicting attitudes that co-exist carefully segregated so that the inherent contradiction between them rarely causes them trouble" (p. 214). Although this strategy does not help children directly justify meat consumption or prevent them from recognizing that they are consuming animals, it does make the consequences seem less adverse. By developing less affection toward farmed animals, children are less likely traumatized by the meat epiphany.

The dichotomization explanation is compelling, and well-illustrated by contextual analysis of media. It helps explain a number of paradoxes, including the fast food industry's use of promotional food tie-ins for children, where representations of popular animals are used to encourage children to consume dead animals (Stewart & Cole, 2009). The notion that animals are separated into those we love and those we eat has not been tested experimentally though, and may not lend itself to such tests easily. One key issue to resolve is whether the similar levels of empathy toward animals displayed by young independent vegetarians and meat eaters (Harris, 2012) generalizes to all animals. To the extent that young meat eaters use dichotomization to reduce dissonance, they would be expected to have similar levels of empathy toward unconsumed animals as do independent vegetarians but less empathy toward consumed animals.

I would suggest that another mechanism may play role a prominent role in helping young meat eaters reduce their dissonance: dissociating the living animal from the meat on the plate. Dissociation is a form of avoidance and represents an indirect strategy to facilitate meat consumption. These strategies are common in groups with more ambivalence toward eating meat (e.g., women; see Rothgerber, 2013; Rothgerber, 2018), and may be more appropriate for children who love animals dearly. Harris's (2012) analysis of what separates independent vegetarians from their counterparts focuses on differences in the inability to dissociate. Statements of independent vegetarians frequently linked eating meat to killing animals. Pallotta (2008) noted that his adult animal rights activists were emotionally disturbed by connecting meat to animals. Meat-eating children, in contrast, may be able to keep animals separate from meat, and internalize social standards framing meat as a pleasurable, social and healthy activity. Data reported by Plous (1993) showing that young children do not always know the exact origin of various meats offers the possibility that some of this ignorance is motivated, designed to block thoughts of specific animals during consumption. Dissociation has not been tested in young meat eaters. One possibility is to modify the procedure by Kunst and Hohle (2016) in which meat is presented in various degrees of processing. Researchers could then measure how such processing affects children's willingness to eat the meat as well as their perceptions of empathy and judgments of how emotionally and intellectually similar to humans they perceive the animals to be.

Another potential dissonance reduction mechanism is to dissociate not the animal from the meat, but to dissociate the animal from harm and suffering. That is, some young meat eaters may not connect the harm and suffering inflicted upon animals to consumed meat. As mentioned earlier, there is little evidence addressing this possibility. One study (Hussar & Harris, 2010) asked children to name a specific meat that they did not consume and reasons why they avoided eating it. Independent vegetarians (aged 6–10) almost unanimously mentioned the suffering and death of animals as their primary motive, responses shared by slightly less than half of young vegetarians whose families were also vegetarians, and motives completely ignored by young meat eaters. This suggests that the majority of children – who regularly consume meat – may reduce dissonance in part by not connecting their behavior to animal harm and suffering. This speculation is indirectly supported by a study of nine and eleven-year-olds outside of the specific context of meat, which found that across all children, there was a tendency to criticize the deliberate killing of animals the more the animal was deemed capable of suffering (Olthof et al., 2008). Because Harris (2012) notes that children judge an action as wrong if it causes distress, it would appear that young meat eaters do not extend concerns with harm and suffering to consumed animals. How could they not condemn their own behavior if they recognized the harm and suffering it caused?

But it may not simply be that vegetarian children are more knowledgeable and aware that animals are harmed and suffer when made into meat, but that they are more upset by this information and consider it more when they think about meat. Animal suffering may be more accessible to them and harder to forget. In support of this, when independent vegetarians were asked to consider animal suffering in the context of meat, they reported more distress than meat eaters and also were less likely to stop thinking about the distressing situation (Olthof, 2009). Harris (2012) cites a number of interviews with independent vegetarians in which they explicitly link meat with animal suffering. This idea that independent vegetarians are not more sensitive to animal suffering in general but toward the animal suffering that the practice of meat consumption entails has not been tested. It is still difficult to evaluate whether children intentionally disconnect meat from animal suffering as a potential way to reduce dissonance, or whether alternatively, those children who genuinely lack this awareness are more likely to eat meat in the first place in a blissful state of ignorance.

Speculation, future research, and implications

For many adult meat eaters, the behavior is so chronic and habitual that it is likely mindless, except when circumstances call meat eating into question. For the child meat eater, however, this experience is likely to be very different. Their admiration of animals is powerful and their awareness and understanding of what they are putting into their mouth is so minimal and new by adult comparisons. It is easy to imagine their bewilderment upon having a meat epiphany, deeply troubled by the inconsistency of their own actions yet likely noticing how little the adults around them seem to be bothered by it. Based on the example of independent vegetarians, it seems clear that some children are quite capable of experiencing dissonance from a moral dilemma, in this case involving the decision to eat an animal that they may hold in positive regard. As vegan activists and others concerned with rising rates of meat consumption attempt to curtail the behavior, they may be wise to consider that children likely have complex and ambivalent thoughts and feelings toward eating meat. This group lacks extensive experience eating meat and lacks a history of justifying the behavior – thus, a key point is that *children may be more amenable to behavioral change than adults.*

Our understanding of young meat eaters is rather limited though. When do children realize that meat comes from an animal? How does this understanding evolve over time? Gradually? All at once? Do children actively think about this, or only at selective occurrences? How do children react to this knowledge? What does their reaction depend upon? Do young meat eaters negatively change their views of animals upon learning of meat's origins? How do they respond to vegetarians? Further research on independent vegetarians would be extremely helpful because this group represents the decision activists would like to inspire in many others. A key question, therefore, is why do some young meat eaters become vegetarian even when their families eat meat?

Do independent vegetarians simply experience more cognitive dissonance than their peers? If so, perhaps activists should seek to heighten dissonance broadly speaking. But perhaps independent vegetarians do not feel greater dissonance pressures but are simply different in how they choose to reduce these feelings of inconsistency. If this is the case, researchers could examine the extent to which internal differences (e.g., values, beliefs, experiences, etc.) between independent vegetarians and their peers are responsible for behavioral differences as well as how much can be explained by differences in parental characteristics and other external factors. Harris (2012) outlines three possible differences between independent vegetarians and their peers that could explain the former's decision to forego eating meat: greater attachment to animals; greater belief that animals are similar to humans, especially in their ability to suffer; and greater sensitivity to the direct connection between eating meat and animal suffering. He opines support for the latter possibility, but there is insufficient evidence to form definitive conclusions. What is needed as a next step is a comprehensive, in-depth interview assessment of children and how they process eating meat, with analysis focused on dissonance and dissonance reduction techniques. Such a study should compare children of different ages in their understanding of and beliefs about eating meat as well as adopt a longitudinal approach following the same group over time to directly assess changes as individuals develop.

While there will always be ethical issues to carefully consider before exposing children to certain stimuli and messaging, the present review and results suggest that specific interventions to discourage (a) dichotomizing animals, (b) dissociating animals from meat, and (c) dissociating meat from harm and suffering are likely to have the largest impact. In broad terms, messages should be transmitted that attempt to blunt dichotomization, chiefly by emphasizing that all animals share certain capabilities and are worthy of respectful treatment. Examples of diverse pets around the world should be shared with children to emphasize that one can develop close relationships with many types of animals. While it may be uncomfortable, messages should remind young meat eaters that meat is derived from animals, and that these animals do not live in humane conditions. This would seem particularly important because "chicken tenders" and other heavily processed meat that bear no correspondence to the living animal seem particularly popular with children. These foods are likely to evoke less disgust and empathy for animals and to facilitate a pattern of meat consumption that may later be difficult to eradicate.

Activists may fail to realize or may underestimate the amount of ambivalence that young meat eaters experience from their behavior. Meat substitution products may benefit from tailoring specific foods to children, in terms of marketing, packaging, taste, presentation, etc., as the recent launch of the vegan McDonald's happy meal in the UK exemplifies. Seeking change at the institutional level may have merits as well. Many children consume a sizeable portion of their daily calories in school, and this makes the school system a prime candidate for lobbying. Although suggestions such as having a meatless day of the week are likely to meet resistance on several grounds, schools probably underestimate the percentage of their students who have some hesitation over consuming animals.

If one assumes from this chapter that children are actively struggling with justifying meat consumption and are open to change, it also highlights the need to effectively target parents and those raising children, as these individuals exert strong influence and control over dietary behavior. Specifically, adult caregivers need to understand that vegetarian and vegan diets can be nutritiously adequate (Melina, Craig, & Levin, 2016) and likely need assistance in successfully implementing such diets. Perhaps health concerns are not the only impediments to parents being willing to support a vegetarian diet in their children. Additional research is needed to better help understand how parents react to their children's resistance to eating meat, what predicts resistance from parents, and what specific concerns parents have for their children following vegetarian diets.

Conclusion

Despite growing recognition that meat eating is a social problem on numerous affronts, efforts to change individuals' behaviors have been met by limited success. A recent meta-analysis admittedly only based on six experiments by Animal Charity Evaluators (2017), for example, concluded that leaflets designed to spread information about the conditions of factory farming do little to change meat consumption (see Bockman, this volume). One large reason for the ineffectiveness of these appeals may be the cognitive dissonance they create. Dissonance pressures cause individuals to avoid the issue or when confronted with their own inconsistencies, cause them to defend meat eating by rationalizations and justifications often hostile to animal welfare. Dissonance theory perversely suggests that increasing the propensity of people to feel bad from eating meat may ironically cause them to consume more meat. The seeming ineffectiveness of individual-focused strategies such as leafletting and videos has led some to conclude that the movement needs to focus more on political change (see Bockman, this volume; Leenaert, this volume). As Che Green (2017) noted, "Changing the hearts and minds is not only a slow process, but it may have also have an upper limit," as he called for "focusing more on influencing both corporate and government policies…"While there is wisdom in this suggestion, it is important not to overlook the possibility that perhaps some interventions need to be targeted at children before dissonance has taken root and meat eating is deeply habitual and entrenched.

This chapter suggests that children are capable of experiencing dissonance from eating meat, and that they are mostly likely to reduce such negative tension from less direct and more apologetic approaches, such as dichotomizing animals into those we love and those we eat, dissociating the animal from the meat we eat, and possibly avoiding thoughts that the food on the plate came from an animal that was harmed and suffered. Trying to break these tactics, with opportunities for behavioral commitment to meat reduction or abstention, may be a promising approach for those wishing to curtail meat consumption.

References

Adams, C. (1990). *The Sexual Politics of Meat: A Feminist Vegetarian Critical Theory*. New York: Continuum.

Amato, P. R., & Partridge, S. A. (1989). *The New Vegetarians: Promoting Health and Protecting Life*. New York: Plenum Press.

Anderson, M.V., & Henderson, A. J. (2005). Pernicious Portrayals: The impact of children's attachment to animals of fiction on animals of fact. *Society & Animals, 13*, 297–314.

Animal Charity Evaluators (2017). Leafletting. Available at: https://animalcharityevaluators. org/advocacy-interventions/interventions/leafleting/#report Retrieved Nov. 21, 2018.

Aronson, E., & Carlsmith, J. M. (1963). Effect of the severity of threat on the devaluation of forbidden behavior. *The Journal of Abnormal and Social Psychology, 66*, 584.

Bandura, A. (1999). Moral disengagement in the perpetration of inhumanities. *Personality and Social Psychology Review, 3*, 193–209.

Bastian, B., & Loughnan, S. (2017). Resolving the meat-paradox: A motivational account of morally troublesome behavior and its maintenance. *Personality and Social Psychology Review, 21*, 278–299.

Bastian, B., Loughnan, S., Haslam, N., & Radke, H. (2012). Don't mind meat? The denial of mind to animals used for human consumption. *Personality and Social Psychology Bulletin, 38*, 247–256.

Bilewicz, M., Imhoff, R., & Drogosz, M. (2011). The humanity of what we eat: Conceptions of human uniqueness among vegetarians and omnivores. *European Journal of Social Psychology, 41*, 201–209.

Bockman, J. (this volume). Accomplishing the most good for animals. In K. Dhont & G. Hodson (Eds.). *Why We Love and Exploit Animals: Bridging Insights from Academia and Advocacy*. Abingdon: Routledge.

Braithwaite, J., & Braithwaite, V. (1982). Attitudes toward animal suffering: An exploratory study. *International Journal for the Study of Animal Problems, 3*, 42–49.

Bray, H. J., Zambrano, S. C., Chur-Hansen, A., & Ankeny, R. A. (2016). Not appropriate dinner table conversation? Talking to children about meat production. *Appetite, 100*, 1–9.

Buttlar, B., & Walther, E. (2018). Measuring the meat paradox: How ambivalence towards meat influences moral disengagement. *Appetite, 128*, 152–158.

Cooper, J., & Fazio, R. H. (1984). A new look at dissonance. *Advances in Experimental Social Psychology, 17*, 229–268.

Covert, A. M., Whirren, A. P., Keith, J., & Nelson, C. (1985). Pets, early adolescents, and families. *Marriage and Family Review, 8*, 95–108.

Cunningham, J. (2009). How many vegetarians are there? *Vegetarian Journal, 29*(4), Article 3. Retrieved from: www.vrg.org/journal/vj2009issue4/index.php

Dowsett, E., Semmler, C., Bray, H., Ankeny, R. A., & Chur-Hansen, A. (2018). Neutralising the meat paradox: Cognitive dissonance, gender, and eating animals. *Appetite, 123*, 280–288.

Egan, L. C., Santos, L. R., & Bloom, P. (2007). The origins of cognitive dissonance: Evidence from children and monkeys. *Psychological Science, 18*, 978–983.

Festinger, L. (1957). *A Theory of Cognitive Dissonance* (Vol. 2). Palo Alto, CA: Stanford University Press.

Foer, J. S. (2009). *Eating Animals*. New York: Little, Brown and Company.

Green, C. (2017). We are the low-hanging fruit. Available at: https://faunalytics.org/we-are-the-low-hanging-fruit/. Retrieved November 21, 2018.

Harper, G., & Henson, S. (2001). Consumer concerns about animal welfare and the impact on food choice. *EU FAIR CT98-3678, Centre for Food Economics Research, The University of Reading*.

Harris, P. L. (2012). *Trusting What You're Told: How Children Learn From Others*. Cambridge, MA: Harvard University Press.

Herzog, H. (2010). *Some We Love, Some We Hate, Some We Eat: Why It's So Hard to Think Straight About Animals*. New York: HarperCollins.

Hodson, G., Dhont, K, & Earle, M. (this volume). Devaluing animals, "animalistic" humans, and people who protect animals. In K. Dhont & G. Hodson (Eds.). *Why We Love and Exploit Animals: Bridging Insights from Academia and Advocacy*. Abingdon: Routledge.

Hoogland, C. T., de Boer, J., & Boersema, J. J. (2005). Transparency of the meat chain in the light of food culture and history. *Appetite, 45*, 15–23.

European Vegetarian Union (2007). How many veggies…? In *European Vegetarian Union*. Retrieved March 2009 from www.euroveg.eu /lang/en/info/howmany.php.

Hussar, K. M., & Harris, P. L. (2010). Children who choose not to eat meat: A study of early moral decision-making. *Social Development, 19*, 627–641.

Iacobbo, K., & Iacobbo, M. (2006). *Vegetarians and Vegans in America Today*. Westport, CT: Greenwood Publishing Group.

Isaacs, S. (1930). *Intellectual Growth in Young Children*. London: Routledge.

Kubberod, E., Ueland, O., Tronstad, A., & Risvik, E. (2002a). Attitudes toward meat and meat-eating among adolescents in Norway: A qualitative study. *Appetite, 38*, 53–62.

Kubberød, E., Ueland, Ø., Rødbotten, M., Westad, F., & Risvik, E. (2002b). Gender specific preferences and attitudes towards meat. *Food Quality and Preference, 13*(5), 285–294.

Kunst, J. R., & Hohle, S. M. (2016). Meat eaters by dissociation: How we present, prepare and talk about meat increases willingness to eat meat by reducing empathy and disgust. *Appetite, 105*, 758–774.

Kwan, M. L., Block, G., Selvin, S., Month, S., & Buffler, P. A. (2004). Food consumption by children and the risk of childhood acute leukemia. *American journal of epidemiology, 160*, 1098–1107.

Laing, D. G. (1999). The development of meat-eating habits during childhood in Australia. *International Journal of Food Sciences and Nutrition, 50*, 29–37.

Leenaert, T. (this volume). Steakholders: How pragmatic strategies can make the animal protection movement more effective. In K. Dhont & G. Hodson (Eds.), *Why We Love and Exploit Animals: Bridging Insights from Academia and Advocacy*. Abingdon: Routledge.

Linder, D. E., Cooper, J., & Jones, E. E. (1967). Decision freedom as a determinant of the role of incentive magnitude in attitude change. *Journal of Personality and Social Psychology, 6*, 245–254.

Loughnan S., & Davies, T. (this volume). The meat paradox. In K. Dhont & G. Hodson (Eds.), *Why We Love and Exploit Animals: Bridging Insights from Academia and Advocacy*. Abingdon: Routledge.

Loughnan, S., Haslam, N., & Bastian, B. (2010). The role of meat consumption in the denial of moral status and mind to meat animals. *Appetite, 55*, 156–159.

MacInnis, C. C., & Hodson, G. (2017). It ain't easy eating greens: Evidence of bias toward vegetarians and vegans from both source and target. *Group Processes & Intergroup Relations, 20*(6), 721–744.

Mayfield, L. E., Bennett, R. M., Tranter, R. B., & Wooldridge, M. J. (2007). Consumption of welfare-friendly food products in Great Britain, Italy and Sweden, and how it may be influenced by consumer attitudes to, and behaviour towards, animal welfare attributes. *International Journal of Sociology of Food and Agriculture, 15*, 59–73.

Melina, V., Craig, W., & Levin, S. (2016). Position of the Academy of Nutrition and Dietetics: vegetarian diets. *Journal of the Academy of Nutrition and Dietetics, 116*(12), 1970–1980.

Melson, G. F., & Schwarz, R. (1994, October). Pets as social supports for families with young children. Paper presented at annual meeting of the Delta Society, New York.

Minson, J. A., & Monin, B. (2012). Do-gooder derogation: Disparaging morally motivated minorities to defuse anticipated reproach. *Social Psychological and Personality Science, 3*(2), 200–207.

OECD. (2014). OECD/Food and Agriculture Organization of the United Nations (2014), *OECD-FAO Agricultural Outlook 2014*, OECD Publishing. Retrieved from http://dx.doi.org/10.1787/agr_outlook-2014-en

Olthof, T. (2009). How to explain vegetarian children's morally-grounded decision not to eat meat? Paper presented at the 35th Conference of the Association for Moral Education, Utrecht, The Netherlands.

Olthof, T, Rieffe, C. J., Meerum Terwogt, M., Lalay-Cederburg, C., Reijntjes, A. H. A., & Hagenaar, J. (2008). The assignment of moral status: Age-related differences in the use of three mental capacity criteria. *British Journal of Developmental Psychology, 26*, 233–247.

Paeratakul, S., Ferdinand, D. P., Champagne, C. M., Ryan, D. H., & Bray, G. A. (2003). Fast-food consumption among US adults and children: dietary and nutrient intake profile. *Journal of the American dietetic Association, 103*, 1332–1338.

Pallotta, N. R. (2008). Origin of adult animal rights lifestyle in childhood responsiveness to animal suffering. *Society & Animals, 16*, 149–170.

Paul, E. S. (1996). The representation of animals on children's television. *Anthrozoös, 9*, 169–181.

Pew Commission on Industrial Farm Animal Production (2008). *Putting meat on the table: Industrial farm animal production in America*. www.pewtrusts.org/~/media/assets/2008/pcifap_exec-summary.pdf

Piazza, J. (this volume). Why people love animals yet continue to eat them. In K. Dhont & G. Hodson (Eds.), *Why We Love and Exploit Animals: Bridging Insights from Academia and Advocacy*. Abingdon: Routledge.

Piazza, J., Ruby, M. B., Loughnan, S., Luong, M., Kulik, J., Watkins, H. M., & Seigerman, M. (2015). Rationalizing meat consumption. The 4Ns. *Appetite, 91*, 114–128.

Plous, S. (1993). Psychological mechanisms in the human use of animals. *Journal of Social Issues, 49*, 11–52.

Pollan, M. (2002). An animal's place. *The New York Times*. November 10, 2002.

Rothgerber, H. (2013). Real men don't eat (vegetable) quiche: Masculinity and the justification of meat consumption. *The Psychology of Men and Masculinity, 14*, 363–375.

Rothgerber, H. (2014a). Efforts to overcome vegetarian-induced dissonance among meat eaters. *Appetite, 79*, 32–41.

Rothgerber, H. (2014b). A comparison of attitudes toward meat and animals among strict and semi-vegetarians. *Appetite, 72,* 98–105.

Rothgerber, H. (2015). Can you have your meat and eat it too? Conscientious omnivores, vegetarians, and adherence to diet. *Appetite, 84,* 196–203.

Rothgerber, H. (2019). "But I don't eat that much meat": Situational underreporting of meat consumption by women. *Society and Animals, 27,* 150–173.

Rothschild, Z. K., & Keefer, L. A. (2017). A cleansing fire: Moral outrage alleviates guilt and buffers threats to one's moral identity. *Motivation and Emotion, 41*(2), 209–229.

Rozin, P. (2007). Food and eating. In S. Kitayama & D. Cohen (Eds.), *The Handbook of Cultural Psychology* (pp. 391–416). New York: The Guilford Press.

Signicom (1997). *Dierenwelzijn in de bio-industrie.* Amsterdam: Signicom.

Singer, P. (2002). *Animal Liberation.* New York: Harper Collins.

Stewart, K., & Cole, M. (2009). The conceptual separation of food and animals in childhood. *Food, Culture & Society, 12,* 457–476.

Te Velde, H., Aarts, N., & Van Woerkum, C. (2002). Dealing with ambivalence: farmers' and consumers' perceptions of animal welfare in livestock breeding. *Journal of agricultural and environmental ethics, 15*(2), 203–219.

United Nations (2006). *Livestock's long shadow: Environmental issues and options.* Rome: Food and Agricultural Organization of the United Nations.

Worsley, A., Wang, W., & Ridley, S. (2015). Australian adults' knowledge of Australian agriculture. *British Food Journal, 117*(1), 400–411.

Zerubavel, E. (1997). *Social Mindscape. An Invitation to Cognitive Sociology.* Cambridge, MA: Harvard University Press.

13

HUMANE HYPOCRISIES

Making killing acceptable

John Sorenson

Abstract

This chapter examines the contradiction between the values that many people claim to hold and their actual behavior, and how the system of producing animals as commodities under capitalism maintains their exploitation through construction of various myths, by the criminalization of animal activists, and by the pretense that those who oppress and exploit nonhuman animals are the victims of those activists. The chapter critiques the fetishization of animal body parts as commodities and advertising and industry propaganda that conceal actual relations of production and structural violence, and rejects the exculpatory narratives embraced by consumers, neo-welfarists, post-humanist academics, and do-it-yourself slaughter enthusiasts. It exposes the hypocrisy at the heart of the so-called "humane" use of non-human animals, showing that the humane myth is essentially a speciesist reaffirmation that the violent domination and murder of others is legitimate. Various rhetorical techniques (such as claims that nonhuman animals embrace their deaths, demonization of animal advocates, use of euphemisms, arrogation of victimhood) are examined. The chapter maintains that vegan animal rights activists present the rational alternative to this confused and hypocritical behavior: We must end the institutionalized exploitation of nonhuman animals and stop regarding them merely as property and resources for our use.

Hypocrisies of loving and killing

"Be kind to animals." As children, most of us have received instructions to be kind to nonhuman animals, to show them respect and not to harm them. Those instructions seem to reflect, and be aligned with, what is often termed the biophilia hypothesis, as popularized by Edward O. Wilson (1984) and Stephen R. Kellert (Kellert & Wilson, 1994), which suggests that humans are motivated by a love of living beings and are naturally drawn towards other forms of life. We generally consider these feelings to be positive; for example, the great nineteenth-century naturalist Charles Darwin (2006, p. 837) identified "disinterested love for all living creatures" as "the most noble attribute of man [sic]." Indeed, the love that many of us feel for other animals is powerful and profound, part of a deeply meaningful engagement with the world. Clearly, not all people are motivated to act on such feelings and, despite overwhelming evidence to the contrary, many believe in a sharp division between humans and other animals and think that these other animals exist to serve human purposes. A concomitant assumption is that they do not qualify for equal moral consideration, let alone feelings of love and compassion. This human-centred view of the world is the ideology of speciesism (See also Dhont, Hodson, Leite, & Salmen, this volume).

Furthermore, these instructions to care for other animals are delivered to us in the context of a global capitalist system of structural violence that, every year, kills billions upon billions of nonhuman animals for food and clothing, for experimental purposes in vivisection laboratories and for entertainment (Nibert, 2013; Sanbonmatsu, 2011; Sorenson, 2010). The suffering and death inflicted upon these animals within this system of institutionalized barbarism staggers the imagination, turning any notion of our supposed love for them into nothing but a grim and ghastly joke. Nevertheless, claims to love other animals are made consistently by many, even by those who work in those industries that are based on the exploitation and killing of nonhuman animals, those who consume their body parts or those, such as hunters, who deliberately set out to kill animals for entertainment or food (Kerley, 2017; CNN, 2017; Rudy 2011). Even those who do not actually feel compassion for other animals and who endorse and readily participate in these vast enterprises of exploitation feel social pressure to acknowledge that animals are not mere things and that they should be treated with respect, even if they are used and killed. In this chapter, I would like to consider both the contradiction between the values that so many people claim to hold and their actual behavior and how the system of producing animals as commodities under capitalism maintains their exploitation through the construction of various myths, by the criminalization of animal activists, and by the pretense that those who oppress and exploit nonhuman animals are the victims of those activists.

Faced with this striking contradiction between our professed values and our actual behavior, we must either reject what we have claimed to hold as our fundamental values, change our behavior, or find some means of characterizing our behavior as being consistent with our stated principles. Unfortunately, only a minority of people actually put their stated compassionate feelings and ethical principles into practice by becoming vegans, not only removing animal-derived products from their diet and making the effort not to use nonhuman animals in other ways but also taking active measures to oppose their exploitation. Vegans envision a society that is congruent with our highest principles of compassion and fairness, rejecting the instrumental use of others and calling for trans-species social justice that recognizes the inherent value of other beings (Matsuoka & Sorenson, 2014). (We may note that not all vegans are motivated by ethical concerns and that some adopt a plant-based diet for health reasons.) While ethical veganism presents the logical solution to the moral problem created by the instrumental use and killing of others, as well as the serious environmental harms and violence towards various human groups created by industrial animal exploitation (2014), it is not yet the choice that most humans make. Despite long and ongoing campaigns by animal advocacy groups and the growing availability of healthy and delicious alternative choices, the majority of humans continue to eat the corpses of a wide variety of nonhuman animals, most of whom suffer terribly during their short lives and are killed when it is judged that their bodies will yield the maximum profit. Even most of those who state their love for individual animals such as cats, dogs, or horses and identify them as companions or as part of the family continue to participate in the exploitation and killing of other animals by purchasing and consuming their corpses, devouring them at meal-times, wearing their skins, using their feathers in pillows and so on. These contradictions extend even to those who become activists for certain types of animals. For example, after animal rights activists denounced volunteers from the North Shore Animal League at a dog adoption event in Union Square, calling them hypocrites for wearing Canada Goose coats trimmed with coyote fur, events manager Kate Lombardi of the Bideawee no-kill animal shelter in New York asked volunteers not to wear fur or Canada Goose coats to events; however, Lombardi only made the request "so as to avoid harassment" and did not address the ethical contradiction itself (Fonrouge, 2018).

Yet, even while engaging in actions or practices that cause harm or death to nonhuman animals, few people are willing to explicitly reject the key ethical tenet of most religions, the "Golden Rule" of showing concern for others, and to expose themselves as selfishly prioritizing their most trivial desires over the most essential interest of others, such as that of staying alive. Instead, the preferred option is to find ways to make the violent exploitation of others seem acceptable and aligned with higher moral principles. Thus, we see the construction of elaborate and disingenuous narratives of justification and self-deception, designed to provide absolution for these harmful actions.

Industry propaganda

There are some obvious reasons for this choice, such as selfishness, inertia, force of habit, and the unthinking embrace of tradition. However, we should also pay attention to the discourse created by the various industries that profit from the exploitation and commodification of other animals. These industries represent unthinkable horrors, enacted on a colossal scale. Based on an institutionalized and savage cruelty without limit, one that constitutes a veritable cult of death, saturating every aspect of our culture, the consequences extend far beyond the direct and immediate victims to include the exploitation of human workers, environmental pollution, deforestation and habitat destruction, biodiversity loss and mass extinctions. These animal-exploitation industries face the task of making what is increasingly-recognized as violent, unethical and unnecessary seem acceptable, normal and desirable.

Knowing full well that for those who "love animals" veganism constitutes the only logical response to these destructive systems, industries have developed campaigns to undermine the compassionate message put forward by animal rights activists. Part of this involves persistent and well-funded efforts to represent those who care about other animals as criminals, extremists and terrorists (Driver, 2016; Farming UK, 2017; Lupica, 2018; Sorenson, 2016). The animal exploitation industries are extremely powerful, able to directly influence governments to enact policies and laws that function to serve their interests. The enormous profits they reap allow them to mobilize billions of dollars in lobbying efforts to buy political influence and advertising campaigns to convince consumers that their products are necessary and desirable, and to shape ideas about what sorts of relationships we should have with other animals and the environment. As part of their efforts to marginalize criticism, these industries have crafted extensive advertising and propaganda campaigns that mask the institutionalized violence carried out against non-human animals and deny the suffering they are forced to endure. These propaganda campaigns invert reality so that violence is normalized and presented as acceptable.

Corporate advertising campaigns are designed to construct elaborate alternative realities, allowing consumers to engage in fantasies about themselves so that they are able to enjoy the products of this violence while pretending that they are not paying for it to occur and to continue to think of themselves as animal lovers (Kunst & Hohle, 2016). On the one hand, advertising promises consumers that they can experience near-inconceivable delights and pleasure, epitomized in the elevation of strips of flesh cut from the corpses of pigs into "bacon," a commodity fetishized as being so unutterably delicious as to induce rapture in those who consume it while robbing them of any power to resist its temptations. Indeed, the thrust of many of these advertising efforts is to persuade consumers that they are freely building an attractive, independent and unique personal identity by expressing their fealty to the magical substance marketed to them. One task of these advertising campaigns is to conceal the negative health consequences for humans who consume these commodities. Keeping with the same example, bacon is high in calories, sodium, and saturated fat and treated with nitrates and other chemicals to preserve the shelf life of the product and enhance its flavor, and is linked with increased risk of heart attacks, strokes, kidney disease, asthma and several types of cancer (Macmillan, 2017).

The main task of these advertising campaigns, however, is to conceal the real relations of production, which are ones of domination, exploitation and horrific violence inflicted on pigs, who suffer terribly in the industry that reproduces and kills them on a massive scale, as well as on the human workers who are brutalized and degraded by the savagery they must inflict on others and who suffer high rates of physical injuries in an industry that demands the greatest extraction of surplus value. The persistent aim of these advertising fantasies is to present agony as joy and cruelty as compassion. Industries portray nonhuman animals as eagerly embracing their deaths, enraptured by the possibility of offering up their bodies for human consumption (ample evidence can be found, for example, in the thousands of advertising images portraying animals wishing to be consumed collected on the Suicide Food, 2011, blog, active from 2006–2011). In addition to this duplicity, advertising performs other magical operations. The process of consuming these animals is depicted as a guarantee not only of inexpressible sensory pleasure but also of health, friendship and harmonious family gatherings, masculine power, and general happiness. Again, one striking indication of the remarkable success of such campaigns is the frequency with which the most well-argued ethical critiques for animal rights and veganism are simply dismissed with the smug, reverential intonation of the single word, "bacon."

As animal rights activists have worked to expose the violence and suffering involved in the exploitation of nonhuman animals, those who profit from that exploitation have had to create new stories to legitimize their actions. The persistent, albeit patently nonsensical, claim made by all these industries is that they are deeply concerned about the welfare of nonhuman animals. For example, in a press release, the Western Australian Farmers Association (2017) stated: "Following the release of details surrounding the death of thousands of live sheep on route to the Middle East in July 2016, WAFarmers reminds the public that animal welfare is the top priority for the agricultural industry." Similarly, the National Pork Board, supported by the US Department of Agriculture's Marketing Service assures consumers that "animal welfare is a top priority" for those who raise pigs so that they can be killed to produce pork (Pork Checkoff, 2009). Organizers of the Methven rodeo in New Zealand identified animal welfare as their "utmost priority" (Harfield, 2015). Similar commitments to animal welfare as "top priority" are expressed by the American Mink Council (n.d.), AstraZeneca pharmaceutical corporation (2015), Calgary Stampede (n.d.), Canada's National Dairy Study (2015), the Foundation for Biomedical Research (n.d.) Finnish pharmaceutical company Orion (n.d.), Oakdell egg producers (Lillywhite, 2017), Smithfield Cattle Company (n.d.) and many others. Indeed, according to the propaganda produced by these animal-exploitation enterprises, profits are merely an afterthought for these industries, with their laser-like focus instead trained on caring for animals. As proof of their commitment to animal welfare, many slaughterhouses highlight the fact that they have adopted the ideas of Temple Grandin, a paid industry consultant who makes the dubious assertion that her own situation as an autistic person provides her with unique insight into the consciousness of other animals. Yet rather than advocating for the rights of these animals, Grandin designs systems to restrain, stun and kill them, systems that she claims are not only more "humane" but more efficient, and thus more profitable. While the effectiveness of her designs has been challenged (e.g., Eckhoff, 2011), Grandin nonetheless provides industries with a "humane" face that is useful for their marketing campaigns and reassures consumers that they are acting morally.

Perpetrators of violent acts can resort to a number of techniques of moral disengagement that assist them in excusing and justifying their behavior (Bandura,

1999; Lifton, 1986). One of the most common techniques is the use of euphemistic language. For example, one may speak of the processing, rather than the killing, of animals. Other terms frequently invoked to put a softer light on the exploitation of animals include cage-free, grass-fed, local, natural, sustainable and organic. In most cases, there are no regulatory controls to determine or verify the claims made. For example, in the USA, the Animal Welfare Institute pointed out that the United States Department of Agriculture (USDA) Process Verified Program (PVP) allows businesses to present their products as USDA Certified but the PVP is only a voluntary marketing program that identifies that a business has met the standards it sets for itself, effectively allowing businesses to use the USDA logo to misrepresent themselves:

> Conventional producers are using the program to claim that their products come from animals who enjoy exceptional treatment. In fact, the animal husbandry guidelines currently certified under the PVP are equivalent to conventional production methods – methods which no reasonable consumer would consider 'humane'.
>
> *Mathews, 2012, p. 2*

Indeed, as the report notes, the "humane" treatment of nonhuman animals includes the intensive confinement of animals in conditions where high ammonia levels cause respiratory problems, routine mutilations such as debeaking and other painful procedures, forced insemination, the mass killing of male chicks in the egg industry, the separation of mothers and infants, and the exertion of total control over every aspect of animals' lives, including their reproduction and socialization, the power to kill them and the decision of when to do so. Investigations of so-called humane operations reveal them to be just as abusive as conventional systems (e.g., Animals Australia, 2014; Compassion Over Killing, 2015; Hsiung, Goldberg, & Goldberg, 2015).

Happy meat

Animal rights advocates point out the deceptive character of the terms deployed in this "humane" discourse by identifying their basic purpose: to make people feel more comfortable about consuming other animals. While proponents of various "humane" practices say that what they do is based on respect and compassion for other animals and that these practices are better than what is inflicted upon animals in factory farms, they still subject animals to unnecessary harm by killing them (since it is not necessary for humans to consume animal products). In some cases, there may be some differences between so-called humane and conventional practices but while it is certainly preferable to inflict less, rather than more, pain on another individual, it is the use, rather than simply the treatment, of animals that is the essential problem and the enslavement of other animals is ethically unacceptable regardless of how "humane" it is purported to be; if we agree that nonhuman animals matter morally then we cannot simply use them for our own purposes and end their lives when we consider that it is advantageous for us to do so (Francione, 2009). It is the use itself that is the abuse, as we make them into resources for ourselves, thwarting their desire to live as they wish. Davis (2017) points out the deception involved in the promotion of these humane forms of exploitation: people are offered permission to hurt one group of animals (the "humanely raised" ones) in exchange for refraining from the consumption of others (the factory farmed ones). In fact, as noted above, in many cases there are few differences between "cage-free" or "free-range" systems and factory farms: terminology is vague and misleading, access to the outdoors may be extremely limited, conditions may be filthy, overcrowded and stressful, mutilations are still standard practice, treatment is typically just as brutal, and, in the end, all the animals are killed, often in the same facility regardless of whether they came from a so-called humane farm or a factory farm (Bohanec & Bohanec, 2013; United Poultry Concerns, n.d.). At the heart of the humane myth is the essential speciesist reaffirmation of violent domination and murder, the conviction that we have the unquestionable right to kill other animals for our pleasure. For example, food writer Tamara Haspell (2017) describes the pigs she raises for food:

> These pigs will have short lives. In about four months they'll reach slaughter weight … and we'll take them to a nearby USDA slaughterhouse. Meantime, we want to make their lives as happy and comfortable as we can…We do this because we like both pigs and pork, and we want to make sure that any pig we eat was raised well.

While claiming to "like both pigs and pork," Haspell clearly prioritizes the latter, in reality reducing living individuals to a commodity and killing them when it suits her.

Essentially, what is promised in the discourse of happy meat is nothing new at all but simply a reassertion of the animal welfare position that it is ethically acceptable to kill other animals as long as they have been treated "well" up to that point. Despite abundant evidence to the contrary, animal exploitation industries have always claimed that they have been concerned about the well-being of the animals they control, a clear absurdity given that the animals are killed for profit. The happy meat narrative is simply a reassertion of the same claims that the industry has always used, but modified to incorporate and negate the ethical criticisms raised by those who truly do care about nonhuman animals. The humane myth is designed to allow speciesist exploitation to continue in a way that allows consumers not to recognize themselves as exploiters but rather to see themselves as people who "love animals."

While the animal exploitation industries have long used the discourse of animal welfare to legitimize their operations and make them seem less horrific by claiming that their concerns for the animals they kill are their top priority, a more recent innovation has been the outright embrace of violence towards other animals as a demonstration of one's alleged love of them. Those who deploy this rhetorical turn use terms such as *humane slaughter* or *ethical butchering* to make the killing of other individuals seem less horrific and, indeed, praiseworthy. The discourse of hunting, both for food and for sport, is replete with claims of love for murdered animals (Luke, 2007). The humane myth is an assertion that eating the body parts of non-human animals is natural, normal and acceptable. The object is to portray veganism as extreme and utopian. Purveyors of the humane myth dismiss veganism on the grounds that it is utopian and that people will not change their behavior in order to stop harming animals but what is clear is that it is the happy meat consumers themselves who refuse to change their behavior even though they acknowledge that it is deeply disturbing and violent.

Do-it-yourself killing

In some cases, the act of killing nonhuman animals is itself commodified and marketed as a spectacle, as in the case of Wild Abundance (see below). In these more recent elaborations of the humane myth, some individuals have taken up the practice of do-it-yourself killing and butchering of nonhuman animals. They present this as an alternative to conventional systems of exploitation, claiming to have disentangled themselves from industrial food systems; however, the real alternative is veganism, not some minor modification to the practices of exploitation. While the proponents of "humane" methods claim to demonstrate more compassion for nonhuman animals, their actions instead demonstrate just how limited their compassion truly is. While offering little to nonhuman animals, the humane myth offers much to those humans who subscribe to it, providing a means of attaining personal empowerment through the claim of having accepted responsibility for how other animals are used. The discourse of these "ethical" killers asserts that by removing animals from the clutches of huge corporations and slaughtering the animals themselves, they have reduced the animals' suffering and gained deep insights about their food.

For example, Beth Krause, formerly national media relations coordinator for Whole Foods Corporation, and Becky Faudree, formerly Whole Foods' global meat buyer, accepted an invitation to visit Tamar Haspell's farm in order to kill some turkeys themselves. Krause explained in an interview that "I love animals (to look at, to hang out with and yes, to eat)" (Gabriel 2012); provided with the opportunity to kill animals directly, she "wondered if the experience would make me vegetarian (and if so, what would that mean for my job representing our Meat Department?)" but instead "felt a progression of fear to sadness, compassion and gratitude" while Faudree (quoted in Krause, 2013) described a "profound experience":

> my turkey slaughter experience ... sharing how close it made me feel to that animal in that moment – and the tremendous amount of appreciation I had for the turkey as I held the knife and cut the artery with tears rolling down my cheeks[.]

On slaughter day, I had an attachment to my turkey, once killed. From plucking to eviscerating, taking part in the entire process was truly important and an experience I will not forget. We watched it change from bird to meat.

Obviously, the "attachment" and "compassion" these women felt for the turkeys did not lead them to spare the lives of those animals. Instead, killing animals is transformed into a self-indulgent piece of theatre in which the killers can see themselves not as murderers but as animal lovers as they convert living individuals into dead objects.

The narrative of happy meat and humane slaughter is marketed by businesses such as Wild Abundance in North Carolina. Run by Natalie Bogwalker, the organization offers classes in neo-primitivist skills, providing workshops on topics such as homesteading, carpentry, so-called humane slaughtering and butchering which instructs participants on how to kill a sheep "in a sacred way" (Wild Abundance, 2018) as well as "non-violent communication" (Brunton, 2015), a type of interaction evidently not extended to those individuals of other species who are killed in the workshops. It is difficult to determine what it is about the killing of animals by do-it-yourselfers such as Bogwalker that makes it "sacred" and an act of "honoring" the victims, as claimed on the Wild Abundance website. A photograph posted on the North American Animal Liberation Press Office website that shows Bogwalker straddling a sheep, pinning the frightened animal to the ground in order to cut her throat, indicates that the process is less spiritual and reverent than suggested by the killers.

Appalled by such obvious cruelty, animal protection groups One Protest and the Let Live Coalition campaigned against the killing classes held at Wild Abundance in 2016. Adam Sugalski of One Protest stated: "We're deeply disturbed by the DIY animal slaughter and butchering class at Wild Abundance and appalled by Wild Abundance's effort to put an ethical and 'sacred' spin on for-profit, gruesome backyard slaughter for inexperienced hobbyists" (Ammons, 2016).

Attacking vegans

On the online petition site he created, Bogwalker's partner Frank Salzano (n.d.) denounced the criticism as "a fierce campaign of harassment" from "an extremist vegan animal rights organization." As other animal users also objected to accurate characterizations of their activities, corporate media seized on the extremist vegans vs noble farmers trope. For example, Northumberland "trainee farmer" Alison Waugh complained to British media that vegans had described farmers who forcibly breed nonhuman animals in order to slaughter them for profit as "murderers and rapists" (Cashmore, 2018). "One of Ireland's top farmers," Peter Hynes, protested that "vegan activists" had called him a rapist because of the use of artificial insemination on his dairy farm (Donnelly, 2018). Dairy farmers Jonny and Dulcie Crickmore complained that they were being abused online by vegans who used "words like rape, murder and slavery" to describe the forced insemination, killing and confinement of nonhuman animals (Tingle, 2018). Rather than having their treatment of animals described accurately, farmers prefer to present themselves as caring protectors of animals. For example, Alison Waugh claimed that for those practicing "good stocksmanship; to learn compassion for another being was the first lesson for any of us" (Utton, 2018).

Along with misrepresenting their own violent activities as ones of compassion, farmers portrayed themselves as victims rather than perpetrators of violence. For example, Alison Waugh said she had received death threats from vegans while the *Telegraph* reported that "Farmers live in fear with death threats from 'violent activists'" (Rudgard & Pearlman, 2018), prompting the Association of Independent Meat Suppliers [AIMS] to meet "with the National Counter Terrorism Police Operations Centre team to discuss how best to respond to the actions" (Utton, 2018). Norman Bagley, of the AIMS, said: "The activists over the past 18 months have become much more violent" although the only examples he cited were not violent actions at all but instead peaceful "situations of activists lying down in front of lorries, holding their children up to pig wagons" while dairy farmer Paul Tompkins said that farmers were "distressed" because animal advocates had filmed their operations (Rudgard & Pearlman, 2018).

These complaints invert actual relations of power and violence. For example, Bagley's words make the real victims, nonhuman animals, disappear and in their place he substitutes the farmers who exploit them. Even in terms of the humans involved, it is the activists who place themselves at risk by "lying down in front of lorries" rather than endangering farmers. Such claims misrepresent the animal rights movement, for example by ignoring the fact that the Animal Liberation Front (ALF), supposedly the most radical part of that movement, is pledged to nonviolence (although the ALF accepts the destruction of property used to harm and exploit animals and does not categorize that as violence). Virtually every animal advocacy organization adopts a position of nonviolence and most reject the destruction of property. A further indication of deliberate misrepresentation was provided by Waugh herself, when she later admitted that she had lied about receiving death threats ("I've not had people making specific death threats towards me but I certainly know people who have received really horrible threats" [Utton, 2018]). Just like their industrial-scale counterparts, those who profit from small-scale farming of nonhuman animals are ready to characterize vegans as violent "extremists" in order to deflect and discredit their criticisms. Seeking to provide further evidence of this supposed extremism on the part of animal activists, producers and consumers of animal body parts complain that while they demonstrate their even-handed tolerance by respecting the choices of vegans not to consume animals, vegans do not reciprocate by respecting their own choice to consume those animals and instead seek to force their personal views on others. This argument is an attempt to move the issue out of the field of social justice and present it as merely a matter of personal choice. In doing so it strips nonhuman animals of their inherent value as sentient beings with agency, interests and voice and reduces them to the status of objects, to be argued over by humans with differing opinions.

Some of these same themes are promoted by another propagandist for the humane myth, Meredith Leigh. Formerly a vegan but now promoting her work as a butcher and the author of *The Ethical Meat Handbook,* Leigh acts as one of the instructors in the "humane slaughter" activities at Wild Abundance. Leigh clarified in an interview that in reality she has no interest in ethical issues and that she began killing animals in order to profit from it:

> I'm not interested in having that moral conversation [about killing animals]… We started Foothills Meats and I almost feel that it was a business decision to add meat.
>
> *Sandford, 2016*

Asserting her "compassion" for the animals she kills and butchers, Leigh (2016) presents herself as the target of "an international coalition of militant vegan organizations" and as the victim of "bullying," "fundamentalism" and "ideological terrorism." Substituting herself as the victim once again works to render invisible those nonhuman animals who are the actual victims of real violence, and presents those seeking to rescue animals from people like Leigh who harm them as dangerous and violent. In other bizarre musings Leigh uses this appropriation as a device to further legitimize that violence:

> If, when my time comes, someone lays me down in green grass and soothes my body, and someone I love then lovingly cuts my throat, and my blood runs into the grass and fertilizes the soil, and my body is used for food and necessity, I could handle that.
>
> *Leigh, quoted in Capps, 2017*

Clearly, this is a fantasy scenario that will never occur. Cannibalism is considered improper and it is very doubtful that "someone… [she] love[s]" will use Leigh's body for food. At the time of writing, even euthanasia for merciful reasons remains illegal in most states in the USA and even in cases of where it is legal for humans to be assisted in the process of dying by physicians, the techniques typically involved include passive measures such as "do not resuscitate" orders or the withholding of food and water, or the administration of lethal drugs. I am aware of no case in which cutting a person's throat, whether done "lovingly" or not, has been employed for these reasons and the prospect of this being either accepted or becoming standard practice seems highly unlikely. Depending upon how one's throat is cut, a person might die quickly but if only a single artery is cut, death could take up to ten minutes (Vidanapathirana & Samaraweera, 2016) while massive spurting of blood and convulsing spasms of the body would be extremely disturbing to any onlookers who could be persuaded to participate in such a grisly event. It is true that public executions were popular spectacles in the past, drawing thousands of onlookers, and that they still continue in places such as Iran, North Korea, Saudi Arabia and Somalia but these are considered as punishments for heinous crimes and the death penalty in general is depicted as "cruel, inhuman, and degrading" (Amnesty International, 2016). Thus, Leigh proposes a scenario that is not only bizarre but, for all intents and purposes, impossible, as a means of creating a false equivalence. Her own assent to an imagined act of violence towards her in the future, one that will never happen, is presented as a legitimization of real acts of violence that are, in fact, inflicted on others who have expressed no such assent and who struggle against their killers. Of course, the most significant point here is that Leigh has a choice in these matters and her nonhuman victims do not.

Leigh's narcissistic projection conforms to a well-known strategy of abusers, that of portraying themselves as victims in order to justify their own victimization of others. Those who promote humane slaughter frequently describe their own spiritual experiences and the difficulties they face in actually killing other animals. However, it is Leigh who retains the power in this situation, choosing to dominate the animals she kills, "co-opt[ing] the language of vulnerability [and] … presenting herself as an icon of compassion while committing murder" (Eure & Van Kleeck, 2016). This strategy of posing as a victim is also a means to appeal for public support. For example, in an October 10, 2017 post on her *Ethical Meat Handbook* Facebook page, Leigh castigates vegans as "fundamentalists" who spread "fake news" about Bogwalker and are "screwing up her business ratings." Bogwalker and Leigh both present themselves as innocent victims of vegan bullies, a serviceable narrative widely-reported in various food-related blogs and mainstream media. Readers' comments appended to these online sites consisted of a torrent of angry, hostile remarks and violent threats directed at vegans and animal activists, denouncing their efforts to prevent the killing of nonhuman animals as foolish sentimentalism and extremism.

Adopting one of the standard rhetorical devices used by perpetrators of violence to excuse their behavior, Leigh claims to be motivated by a noble purpose, that of reducing suffering of animals who are killed for food. She maintains that she is working to create ethical changes in society by speaking to those who consume animals: "At this point in our culture, the people who will have the greatest influence on the lives and deaths of animals are the people who eat meat, and not the people who don't" (quoted in Weingarten, 2016). This is the same justification given by other former vegetarians and vegans who have resumed eating the flesh of nonhuman animals (Sorenson & Matsuoka, 2016, pp. 437–438). Their assertion is that their financial support for the happy meat business means that animals will experience better lives and suffer less. However, in many cases, any improvements in the treatment are negligible and this does not address the fact that, regardless of how they are raised, all the animals are killed. Even leaving aside this fundamental issue, even the most optimistic assessments of the "grass-fed beef" business acknowledge that it constitutes less than one per cent of the total US beef market (Held, 2017). Even if this were to expand, the impact would be disastrous. As Monbiot (2017) points out, "0.01m^2 of land [is needed] to produce a gram of protein from beans or peas, but 1m² to produce it from beef cattle or sheep: a 100-fold difference" and where grazing land cannot be used to produce crops for human consumption it constitutes ecosystems that are home to various types of wildlife who would be wiped out to make room for cows. Furthermore, the term "grass-fed beef" is an ideological construction, part of what Adams (2015) identifies as the absent referent, which obscures living animals and speaks only of the commodities derived from their bodies. Such substitutions, intended to make the individual victims seem invisible, are essential to the industry, part of the effort to conceal the cruelty, killing and violence upon which it is based.

The statements made by farmers and butchers and the sympathetic coverage given to them by industry-friendly media follow a definite pattern, suggesting a coordinated strategy to marginalize and vilify animal activists, depicting them as angry extremists forcing their opinions on others and making outrageous accusations against hard-working farmers. This discourse rejects terms such as murder and rape, insisting that they are inapplicable to the treatment of nonhuman animals, but this is simply a reassertion of speciesism and human exceptionalism. While legally it is the case today that only humans can be murdered there is growing support for applying this term to the premeditated killing of other animals as well (Bekoff, 2017). Of course, it may be argued that the rape of humans is based on the urge to dominate, degrade, and humiliate and to demonstrate the rapist's power and that this differs from what is done to nonhuman animals. While there is certainly no shortage of undercover video from factory farms and other institutions showing humans taking pleasure in abusing other animals, it is probably true that the main reason for the artificial insemination of nonhuman animals is to impregnate them in the most efficient and profitable way rather than to show domination or to derive sexual pleasure. In this sense some might see the analogy being better made to sport and trophy hunters who kill animals for pleasure; as Kalof, Fitzgerald, and Baralt (2004) have shown, the discourse of sport hunting is replete with sexualized imagery. However, even if one refuses to acknowledge farmers' forced artificial insemination of captive nonhuman animals as rape, it is clear that the production and consumption of meat and other commodities derived from animals' bodies is thoroughly saturated with the most vulgar and violent sexual images (Adams, 2015).

Problems within the "animal movement"

Even in "the animal movement" and mainstream academic animal studies, strategies of moral disengagement are found. A striking example is the assertion made by philosopher Peter Singer, author of *Animal Liberation* and widely regarded as a key figure in the contemporary animal rights movement. Singer (1975, pp. 228–230) claims that, at least in theory, nonhuman animals can legitimately be used for our purposes if they live good lives, are killed painlessly and replaced by other animals, although he acknowledges that such conditions do not exist in practice. Nevertheless, Singer has expressed support for various large corporations such as Whole Foods that claim to have instituted humane reforms while continuing to profit from the killing of nonhuman animals for sale as food. Radical abolitionists argue that Singer's utilitarian emphasis on reducing suffering, rather than upholding the rights of animals not to be used as property and to be killed actually perpetuates the exploitation of animals by suggesting that this can be done humanely (Frost, 2016). Davis (2011) and Francione (2016) argue that Singer's position is not one of rights but of welfare and they maintain that Singer perpetuates the speciesism he supposedly challenges, while Stanescue (2016) points out how Singer is mistaken in his idea that so-called humane meat is more compassionate and creates less harm to animals. One means by which perpetrators of violence attempt to legitimize their actions is to lower the status of their victims. Davis and Francione note that Singer does just this, creating a moral hierarchy by characterizing nonhuman animals as not being self-aware over time and not having an interest in continuing to live. They argue that the effects of this welfarist view have been disastrous for the animal movement and have led to a focus on "humane" reforms rather than to the establishment of animal rights.

Other former vegetarians and vegans have gone further and present the happy meat movement as a means to help animals. For example, embracing posthumanism's claims, Jody Emel, Connie L. Johnston, and Elisabeth Stoddard (2015, p. 164) suggest that it is time to "rethink veganism as the only just position with respect to domesticated farmed animals" and describe their own moves away from veganism to include the consumption of animal body parts. Like Meredith Leigh, Emel and colleagues employ the device of claiming a noble purpose to excuse their support for violent exploitation. Purporting to be concerned about these animals they are now willing to consume, they advocate for what they say is a "flatter hierarchy" in which those who wish to exploit, kill, and consume nonhuman animals will be able to do so by creating "livelier livelihoods" of mutual dependency (the term is an inversion of reality, since the livelihoods remain deadly ones for the animals). In these making these dubious claims they use the ideas of American posthumanist philosopher Donna Haraway. Haraway is famous for her critique of the anthropocentrism of humanist thought, and her view of humans and certain other animals, such as dogs, as companion species that have co-evolved and are co-constitutive.

Emel et al. (2015) apply this notion to the situation of farmed pigs, making the preposterous suggestion that they can be regarded as "coworkers" in industries where their whole lives are controlled by the humans who own them and determine when they will be killed and in which their bodies constitute the commodities upon which the entire system of exploitation is based. Seeming to recognize that there is some problem with their formulation, Emel et al. (p. 165) say that they "are still debating this amongst ourselves." Essentially, their debate, to whatever extent it exists, is like a discussion about replacing one type of capitalist with another who may be a bit nicer while leaving unexamined the basic question of the exploitation of labor.

Weisberg (2009; 2014) has identified some of the serious shortcomings in Haraway's approach, with its acceptance of the instrumental use of other animals. By presenting what are in reality relations of power and domination, such as those between laboratory scientists and the nonhuman animals they experiment on, as ones of partnership and collaboration, Haraway engages in what Weisberg (2009, p. 23) characterizes as "collusion with the very structures of domination that she purports to oppose" and provides "ideological cover" for various forms of violent exploitation of other animals.

Haraway's work is often linked with that of French philosopher Jacques Derrida, whose deconstructionist approach examined how philosophical discourse constructed an unbridgeable gap between humans and other animals. While he acknowledged that human relations with other animals could be described as genocidal, Derrida (2008) never suggested any practical or political measures that should be undertaken in order to intervene against these genocidal practices. Steiner (2013, p. 8) points out that Derrida is contemptuous and mocking in his dismissal of any efforts to actually change the situation of nonhuman animals:

> Derrida scrupulously avoids making any definitive claims about the right of animals not to be eaten by human beings. The closest Derrida comes to making any such announcement is a rather tepid statement at a conference to the effect that "I am a vegetarian in my soul." In his soul. To my knowledge, Derrida was neither a vegan nor a vegetarian. Moreover, even though Derrida purports to believe that "we must reconsider in its totality the metaphysico-anthropocentric axiomatic that dominates, in the West, the thought of just and unjust," he punctuates his remarks about the wrongness of killing animals with the qualification that he is not recalling this in order to start a support group for vegetarianism, ecologies, or for the societies for the protection of animals.

The most that Haraway (2008) can muster in defence of animals is to modify Derrida's advice on "eating well" (Nancy 1991) into a suggestion of "killing well." The gist of their thinking is because killing cannot be avoided completely, efforts to reduce violence by becoming vegan and actively working for the rights of animals are somehow illegitimate. Thus, like other versions of the happy meat narrative, the posthumanist vision presented by Haraway and Derrida offers a way for their theoretical followers to continue to exploit and consume other animals while viewing themselves as humane and radical and even as animal lovers.

Implications and conclusions

Butchers, chefs, consumers, farmers, neo-primitivists and posthumanist academics are all eager to embrace the humane myth and adopt the narrative of happy meat. While noting that in terms of the overall suffering and death of nonhuman animals, the so-called humane movement is small potatoes, accounting for less than one per cent, Stanescue (2017, p. 211) points out that the humane myth is dangerous nonetheless because it draws attention away from the reality of the lives of farmed animals and substitutes a false and idealized romantic picture of benevolence and care. This is the same false picture offered up by the discourse of animal welfare in conventional farming, as well as other industries based on the exploitation of animals. Furthermore, while the humane movement can only exist as a token because it utilizes much more resources than conventional farming, it nevertheless operates to naturalize and normalize the killing of nonhuman animals and make it seem inevitable. Stanescue maintains that humane meat and other animal products function as a kind of atonement for affluent consumers who continue to purchase most of the animal products they consume from conventional sources. In this sense, the (neo)welfarist promoters of cage-free eggs, happy meat, and organic milk operate in a similar fashion as did the medieval Catholic church with its sale of indulgences (Francione, 2009).

All of these and further mental and rhetorical gyrations stem from and reflect a peculiar paradox. Convinced of our superiority, humans have come to dominate the world, decimating other species through forms of violent exploitation that are now changing the planet in ways that threaten the future of our own civilization (Ripple et al., 2017). We are driven by the ideology of speciesism, which proposes a sharp distinction between ourselves and all other beings and operates as "one of the few truly fundamental *existential* structures of human life ... [shaping] every human culture, economy, and society" with murder at its core (Sanbonmatsu, 2014, p. 31). Yet at the same time, our biophiliac tendencies give us the nagging sense that what we are doing is wrong, that we are committing terrible crimes. Thus, we must engage in utterly surreal efforts to disguise what we do, desperately grasping at inane arguments in an effort to elevate the most sordid, violent behavior into something that appears not only necessary but noble. We try to fool ourselves, to portray these violent atrocities as something other than what they are, to characterize the exploitation and killing of other animals as something respectful, ecological, or even as compassionate efforts to help those animals. Vegan animal rights activists present the rational alternative to this confused and hypocritical behavior: we must end the institutionalized exploitation of nonhuman animals and stop regarding them merely as property and resources for our use. Doing so will allow us to uphold the Golden Rule that we all claim to respect, to live up to our better impulses and to become our best selves.

References

Adams, C. J. (2015). *The Sexual Politics of Meat: A Feminist-vegetarian Critical Theory* (25th anniversary edition). New York: Bloomsbury Academic.

American Mink Council (n.d.). Animal welfare. Retrieved from https://amcouncil.com/animal-welfare/

Ammons, J., (2016, November 23). Let Live Coalition vs. Wild Abundance: Vegan activists rally against a local permaculture school. *Mountain Xpress*. Retrieved from https://mountainx.com/news/let-live-coalition-vs-wild-abundance-vegan-activists-rally-against-a-local-permaculture-school/

Amnesty International (2016). Death penalty. Retrieved from www.amnesty.org/en/what-we-do/death-penalty/

Animals Australia (2014). If this is the "best," what is the worst? Retrieved from http:// animalsaustralia.org/features/not-so-humane-slaughter/

AstraZeneca, (2015). Animal research. Retrieved from www.astrazeneca.com/content/dam/ az/our-company/Documents/Research-with-animals---May-2015.pdf

Bandura, A. (1999). Moral disengagement in the perpetration of inhumanities. *Personality and social psychology review, 3*(3), 193–209. doi: 10.1207/s15327957pspr0303_3

Bekoff, M. (2017, February 12). Murder, she didn't write: Why can only humans be murdered? *Psychology Today.* Retrieved from www.psychologytoday.com/blog/animal- emotions/201702/murder-she-didnt-write-why-can-only-humans-be-murdered

Bohanec, H., & Bohanec, C. (2013). *The Ultimate Betrayal: Is There Happy Meat?* Bloomington, IN: iUniverse.

Brunton, M. H. (2015, September 30). Natalie Bogwalker front woman for Wild Abundance. *Sophie Online.* Retrieved from http://sophiemagazine.com/uncategorized/ natalie-bogwalker-front-woman-for-wild-abundance/

Calgary Stampede (n.d.). Animal care. Retrieved from https://corporate.calgarystampede. com/animal-care

Capps, A. (2017, November 3). Dear Asheville: Your 'humane slaughter' instructors are not your victims. *Free From Harm.* Retrieved from https://freefromharm.org/animal- rights/humane-slaughter-instructors-are-not-victims/

Cashmore, A. (2018, January 29). "When they call you murders and rapists, it's not OK": Farmers tell of their fears after being targeted by militant vegans who compare them to slave owners. *Daily Mail.* Retrieved from www.dailymail.co.uk/news/article- 5324905/Farmers-talk-targeted-militant-VEGANS.html

CNN (2017, January 12). Ranch owner: Hunters love animals. Retrieved from www.cnn. com/videos/us/2018/01/12/lavandera-rhino-trophy-pkg.cnn

Compassion Over Killing (2015). Foster Farms exposed. Retrieved from http://cok.net/ inv/foster-farms/

Darwin, C. (2006). The descent of man, and selection in relation to sex. In E. O. Wilson (Ed.), *From So Simple a Beginning* (pp. 767–1254). New York: W. W. Norton.

Davis, K. (2011, April 14). What happened to Peter Singer? *United Poultry Concerns.* Retrieved from www.upc-online.org/thinking/peter_singer.html

Davis, K. (2017, October 27). Fantasy farming vs. factory farming: A Faustian bargain. *United Poultry Concerns.* Retrieved from www.upc-online.org/alerts/171027_fantasy_ farming_vs_factory_farming.html

Derrida, J. (2008). *The Animal that Therefore I Am.* (trans. D. Willis). New York: Fordham University Press.

Dhont, K., Hodson, G., Leite, A., & Salmen, A. (this volume). The psychology of speciesism. In K. Dhont & G. Hodson (Eds.), *Why We Love And Exploit Animals: Bridging Insights from Academia and Advocacy.* Abingdon: Routledge.

Donnelly, M. (2018, February 6). Farmer subjected to "frightening" online abuse by vegan campaigners. *Farm Ireland.* Retrieved from www.independent.ie/business/farming/ rural-life/farmer-subjected-to-frightening-online-abuse-by-vegan-campaigners- 36569548.html

Driver, A. (2016, November 28). NPA raises animal rights concerns in Sunday Times. *National Pig Association.* Retrieved from www.npa-uk.org.uk/NPA_raises_animal_ rights_concerns_in_Sunday_Times_article.html

Eckhoff, V. (2011, December 6). Horse slaughterhouse investigation sounds food safety and cruelty alarms. *Forbes.* Retrieved from www.forbes.com/sites/vickeryeckhoff/2011/ 12/06/horse-slaughterhouse-investigation-sounds-food-safety-and-cruelty-alarms/ #77e1238456bf

Emel, J., Johnson, C., & Stoddard, E. (2015). Livelier livelihoods: Animal and human collaboration on the farm. In K. Gillespie & R. Collard (Eds.), *Critical Animal Geographies* (pp.164–183). New York: Routledge.

Eure, C., & Van Kleeck, J. (2016). The hypocrisy of a butcher's "vulnerability." *Striving With Systems.* Retrieved from https://strivingwithsystems.com/2016/10/14/the-hypocrisy-of-a-butchers-vulnerability/

Farming UK (2017, August 18). Farmers to promote dairy benefits amid vegan week of anti-dairy activism. Retrieved from www.farminguk.com/News/Farmers-to-promote-dairy-benefits-amid-vegan-week-of-anti-dairy-activism_47231.html

Fonrouge, G. (2018, February 9). No-kill shelter has to remind volunteers not to wear fur, Canada Goose items. *New York Post.* https://nypost.com/2018/02/09/no-kill-shelter-has-to-remind-volunteers-not-to-wear-fur-canada-goose-items/

Foundation for Biomedical Research (n.d.). Animal care is a top priority to the scientific community. Retrieved from https://fbresearch.org/animal-care/

Francione, G. (2009, July 14). A revolution of the heart. *Animal Rights: The Abolitionist Approach.* Retrieved from www.abolitionistapproach.com/a-revolution-of-the-heart/

Francione, G. (2016, February 18). Challenging Peter Singer's paternity claim. *Animal rights: The abolitionist approach.* Retrieved from www.abolitionistapproach.com/challenging-peter-singers-paternity-claim/

Frost, B. (2016, December 9). How the mainstream "animal movement" has destroyed veganism. *Ecorazzi.* Retrieved from www.ecorazzi.com/2016/12/09/how-the-mainstream-animal-movement-has-destroyed-veganism/

Gabriel, A. (2012, April 20). Whole Foods' Beth Krause enjoys the sweet life. *American Statesman.* Retrieved from www.statesman.com/lifestyles/society/whole-foods-beth-krauss-enjoys-the-sweet-life/CmLzbehbmVA7DZeKTlSOXP/

Haraway, D. (2008). *When Species Meet.* Minneapolis, MN: University of Minnesota.

Harfield, R. (2015, October 23). Animal welfare top priority. *Guardian Online.* Retrieved from www.guardianonline.co.nz/news/animal-welfare-top-priority/

Haspell, T. (2017, July 29). Pigs! Starving Off the Land. Retrieved from https://starvingofftheland.com/2017/07/pigs/

Held, L. (2017, May 7). Despite many challenges, grass-fed beef could go mainstream.. *Civil Eats.* Retrieved from https://civileats.com/2017/06/07/despite-many-challenges-grassfed-beef-could-go-mainstream/

Hsiung, W., Goldberg, L., & Goldberg, M. (2015). A deadly feast. *Direct Action Everywhere.* Retrieved from https://static1.squarespace.com/static/515cca87e4b0bca14d767b61/t/5656a274e4b01a74ead529aa/1448518260139/A+Deadly+Feast.pdf

Kalof, L., Fitzgerald, A., & Baralt, L. (2004). Animals, women and weapons: Blurred boundaries in the discourse of sport hunting. *Society and Animals, 12*(3), 237–251. doi: 10.1163/1568530042880695

Kellert, S. R., & Wilson, E.O. (1994). *The Biophilia Hypothesis.* Washington, DC: Island Press.

Kerley, P. (2017, November 24). I love animals but I kill them too. November 24. *BBC News.* Retrieved from www.bbc.co.uk/news/resources/idt-sh/I_love_animals_but_I_kill_them_too_hunting_alaskan_style

Krause, B. (2013, January 20). Being grateful for the meat we eat. *Whole Story.* Retrieved from www.wholefoodsmarket.com/blog/being-grateful-meat-we-eat

Kunst, J.R., & Hohle, S.M. (2016). Meat eaters by dissociation: How we present, prepare and talk about meat increases willingness to eat meat by reducing empathy and disgust. *Appetite, 105,* 758–774.

Leigh, M. (2016, November 14). Vegan bullying and the new world. Retrieved from www.mereleighfood.com/blog/2016/11/14/vegan-bullying-and-the-new-world

Lifton, R. J. (1986). *The Nazi Doctors: Medical Killing and the Psychology of Genocide.* New York: Basic Books.

Lillywhite, C. (2017, September 14). Animal welfare. Retrieved from www.oakdell.com/faq-items/animal-welfare-policy/

Luke, B. (2007). *Brutal: Manhood and the Exploitation of Animals.* Chicago, IL: University of Chicago Press.

Lupica, D. (2018, January 3). "Battling vegan activism" among "challenges" for pig industry in 2018. *Plant-Based News.* Retrieved from www.plantbasednews.org/post/battling-vegan-activism-challenges-pig-industry-2018

Macmillan, A. (2017, February 2). Yes, you need to cool it on the bacon. *Time.* Retrieved from http://time.com/4657475/bacon-processed-meat-nutrition/

Mathews, R. (2012). Humanewashed: USDA process verified program misleads consumers about animal welfare marketing claims. Washington, DC: *Animal Welfare Institute.* Retrieved from https://awionline.org/sites/default/files/publication/digital_download/fa-humanewashedreportonusdapvp.pdf

Matsuoka, A., & Sorenson, J. (2014). Social justice beyond human beings: Trans-species social justice. In T. Ryan (Ed.), *Animals in Social Work* (pp. 64–72). London: Palgrave Macmillan.

Monbiot, G. (2017, December 11). Mass starvation is humanity's fate if we keep flogging the land to death. *Guardian.* Retrieved from www.theguardian.com/commentisfree/2017/dec/11/mass-starvation-humanity-flogging-land-death-earth-food. Retrieved July 24, 2018.

Nancy, J-L. (1991). Eating well or the calculation of the subject: An interview with Jacques Derrida. In E. Cadava, P. Connor, & J-L. Nancy (Eds.), *Who Comes After the Subject?* (pp. 96–111). New York: Routledge.

National Dairy Study (2015). Animal care. Retrieved from www.nationaldairystudy.ca/animal-care/

Nibert, D. A. (2013). *Animal Oppression and Human Violence.* New York: Columbia University Press.

Orion (n.d.) Bioethics and animal welfare. Retrieved from www.orion.fi/en/Orion-group/Sustainability/policies/pharmaceutical-rd-ethics-policy/bioethics-and-animal-welfare/

Pork Checkoff (2009). Ethical treatment of pigs. The Pig Site. Retrieved from www.thepigsite.com/video/vars/a/68/

Ripple, W. J., Wolf, C., Newsome, T. M., Galetti, M. Alamgir, M. Crist, E., Mahmoud, M. I., & Laurance, W. F. (2017). World scientists' warning to humanity: A second notice. *BioScience,* 67(12), 1026–1028. doi: 10.1093/biosci/bix125

Rudgard, O. & Pearlman, J. (2018, January 30). Farmers live in fear with death threats from 'violent activists'. *Telegraph.* Retrieved from www.telegraph.co.uk/news/2018/01/29/vegan-campaigner-says-farmers-complain-death-threats-abuse-playing/

Rudy, K. (2011). *Loving Animals.* Minneapolis, MN: University of Minnesota Press.

Salzano, F. (n,d.). Stop misguided vegan activists from harassing new mom who runs sustainable living school. Change.org. Retrieved from www.change.org/p/one-protest-let-live-stop-misguided-vegans-activists-from-attacking-primitive-skills-permaculture-school

Sanbonmatsu, J. (2011). *Critical Theory and Animal Liberation.* Lanham, MD: Rowman & Littlefield.

Sanbonmatsu, J. (2014). The animal of bad faith: Speciesim as an existential project. In J. Sorenson (Ed.) *Critical Animal Studies: Thinking the Unthinkable* (pp. 29–45). Toronto: Canadian Scholars Press.

Sandford, J. (2016, February 5). With the "ethical meat handbook" Asheville author cuts her own path toward sustainability. *Ashvegas*. Retrieved from http://ashvegas.com/with-the-ethical-meat-handbook-asheville-author-cuts-her-own-path-toward-sustainability

Singer, P. (1975). *Animal Liberation: A New Ethics for Our Treatment of Animals*. New York: HarperCollins.

Smithfield Cattle Company (n,d.). Animal welfare. http://smithfieldcattleco.com/feedlot/animal-welfare/

Sorenson, J. (2010). *Animal Rights*. Black Point: Fernwood.

Sorenson, J. (2016). *Constructing Ecoterrorism: Capitalism, Speciesism and Animal Rights*. Black Point: Fernwood.

Sorenson, J., & Matsuoka, A. (2016). The New carnivores. In J. Castricano & L. Corman (Eds.) *Animal Subjects 2.0* (pp. 429–455). Waterloo, ON: Wilfrid Laurier University Press.

Stanescue, V. (2016, November 3). An open letter to Peter Singer. *Animal Liberation Currents*. Retrieved from www.animalliberationcurrents.com/an-open-letter-to-peter-singer/

Stanescue, V. (2017). New weapons: Humane farming, biopolitics and the postcommodity fetish. In D. Nibert (Ed.) *Animal Oppression and Capitalism, V. 1 The Oppression of Nonhuman Animals as Food* (pp. 209–228). Santa Barbara, CA: Praeger.

Steiner, G. (2013). *Animals and the Limits of Postmodernism*. New York: Columbia University Press.

Suicide Food (2011, December 27). Five years: An announcement.. http://suicidefood.blogspot.com

Tingle, R. (2018, February 5). Furious viewers blast vegan campaigner who claims the dairy industry rapes and "sexually violates" cows on *This Morning* as he sits next to farmers who have been hit by death threats from his militant followers. *Daily Mail*. Retrieved from www.dailymail.co.uk/news/article-5353603/Furious-viewers-blast-vegan-campaigner-Morning.html#ixzz56LMkjh00

United Poultry Concerns (n.d.). "Free range" poultry and eggs: Not all they're cracked up to be. Retrieved from www.upc-online.org/freerange.html

Utton, D. (2018, February 7). "I'll always speak my mind": Meet the young farmer taking the fight to militant vegans. *Express*. Retrieved from www.express.co.uk/life-style/life/915596/young-farmer-militant-vegans-alison-waugh-northumberland-agriculture-livestock

Vidanapathirana, M., & Samaraweera, J. C. (2016). Homicidal cut throat: The forensic perspective. *Journal of Clinical and Diagnostic Research: JCDR, 10*(3), GD01–GD02. doi: 10.7860/JCDR/2016/17997.7411

Weisberg, Z. (2009). The broken promises of monsters: Haraway, animals and the humanist legacy. *Journal for Critical Animal Studies*, 7(2), 21–61.

Weisberg, Z. (2014). The trouble with posthumanism: Bacteria are people too. In J. Sorenson (Ed.) *Critical Animal Studies: Thinking the Unthinkable* (pp. 93–116). Toronto: Canadian Scholars Press

Wild Abundance (2018). The cycles of life: Humane slaughtering and butchery class. Retrieved from http://wildabundance.net/classes/the-cycles-of-life-humane-slaughtering-and-butchery-class/

Weingarten, D. (2016, May 27). How a former vegetarian became a butcher and ethical meats advocate. *Civil Eats*. Retrieved from http://civileats.com/2016/05/27/how-a-former-vegetarian-became-a-butcher-and-ethical-meat-advocate/

Western Australian Farmers Association (2017, August 16). Animal welfare is top priority. Retrieved from www.wafarmers.org.au/animal-welfare-top-priority/

Wilson, E. O. (1984). *Biophilia*. Cambridge, MA: Harvard University Press.

14

THE END OF FACTORY FARMING

Changing hearts, minds, and the system

Gene Baur

Abstract

Challenging the abuses of factory farming will require diverse approaches, including efforts to raise awareness as well as to transform social and economic infrastructures. Large quantities of meat, dairy and eggs are commonly consumed in affluent societies, which has led to an industrialized agricultural system that causes enormous animal suffering, environmental destruction, and human health risks. For decades, efforts to mitigate the harm of factory farming have sought to appeal to people's conscience, but these have had limited success. Intensive animal agriculture is entrenched within social, economic and political institutions, which must be reformed. Entrepreneurs are now a growing influence in reshaping our food industry, toward plant-based agriculture. Vegan alternatives are gaining in popularity and becoming widely available, and mission-driven companies have been founded with the goal of replacing animal products with plant-based alternatives. Some of these ventures are attracting significant capital, including from agribusiness giants. Investments in these businesses coupled with technological advances, infrastructure reforms, and redirected government programs can make plant foods more widely accessible and ubiquitous in the marketplace, which is critically important to ending factory farming.

The first time I was unsettled by the fact that I was eating animals was one evening during high school when my mother had cooked a chicken for the family's dinner. I saw the body of a dead bird on his or her back, with the legs and wings attached, and I consciously realized that this had been a living, feeling individual. I did not eat chicken that night, and I avoided meat for a while. But the memory of that dead chicken faded, and I went back to eating meat regularly. Then, during my last year of college, I travelled around the USA and became involved with environmental and human rights organizations. I met and worked with activists who educated me about factory farming and how it harms people, animals, and the earth. I went vegan in 1985 and cofounded Farm Sanctuary in 1986.

Farm Sanctuary was founded to raise awareness and combat the abuses of factory farming. We started by investigating farms, stockyards and slaughterhouses to document the horrific conditions, and hoped that educating consumers about the intolerable cruelty of factory farming would inspire them not to support it. At Farm Sanctuary's shelters, with the largest one being over 360 acres located in upstate New York, we rehabilitate and care for hundreds of animals rescued from slaughter, including pigs and turkeys. These are intelligent and emotional, sentient beings. They are often frightened of people when they first come to the sanctuary because they have only experienced cruelty at human hands, but as they learn to trust us and feel safe, they heal both physically and emotionally. Eventually, they are able to express themselves, and their personalities blossom. Many come to enjoy positive interactions with people. Turkeys at Farm Sanctuary follow you around like puppy dogs. Pigs love belly rubs, and it is an amazing experience to see a 500-pound pig flop over on his or her side when you touch their tummy for a belly rub, during which they grunt and communicate their enjoyment as if to say "keep going." Interactions like these are mutually beneficial, good for both human and nonhuman animals (see also Baur & Stone, 2015).

I still believe it is important to inform people about the abuses of animal agriculture, and I believe people can make conscientious choices about how they eat. But, I have also come to believe that social norms and belief systems, and economic infrastructures, play critically important roles in influencing how people eat.

Meat eating as part of a carnist system

For decades, social scientists have contemplated the interplay between how attitude influences behavior, and how behavior influences attitude. When behavioral norms are aligned with established beliefs and attitudes, efforts to create change face myriad obstacles (Festinger, 1957; see also Rothgerber, this volume). Attitudes reinforce behaviors, and vice versa in a mutually calcifying process that maintains the status quo. Understanding the interplay between attitudes and behaviors is relevant when it comes to advancing animal rights and veganism in a world where animals are seen as commodities and consuming animal products is a societal norm, bolstered by entrenched economic and political interests. At the same time, preventing cruelty to animals is also a widely held societal norm, and numerous opinion polls have found that citizens oppose factory farming practices (e.g., Harvey 2018; NRG Research Group, 2017; Reese, 2017). There is a stark and obvious conflict between aspiring to be kind to animals, while also eating them and supporting factory farming. The dissonance is embodied in the inelegant concept of "humane slaughter," which represents a widely held Orwellian approach to the inconsistency. But, can the words "humane" and "slaughter" really fit together?

Psychologist Dr. Melanie Joy (2010) has coined the term "carnism," which she defines as "the invisible belief system, or ideology, that conditions people to eat certain animals." She explains:

> Because carnism is invisible, people rarely realize that eating animals is a choice, rather than a given. In meat-eating cultures around the world, people typically don't think about why they eat certain animals but not others, or why they eat any animals at all. But when eating animals is not a necessity, which is the case for many people in the world today, then it is a choice – and choices always stem from beliefs… As long as we remain unaware of how carnism impacts us, we will be unable to make our food choices freely – because without awareness, there is no free choice.
>
> *www.carnism.org/carnism*

Dr. Joy's important work helps elevate the discussion. She frames our animal eating habits more broadly as part of a socioeconomic system, underpinned by unconsidered beliefs and prejudices. Joy's approach here can be compared to the women's rights movement being framed as challenging a sexist social system. Individuals who have spoken against sexism have been derided, dismissed, and labelled, but when the focus turns to examining a "sexist" system, a more objective and reasoned discussion is possible. The invisible prejudices and beliefs that lead people to label and deride "feminists" with various unflattering adjectives becomes more visible. When change agents are marginalized, as vegan activists often are today, the important issues they present commonly go unquestioned. Rather than labeling advocates with terms like "feminist" or "vegan" and dismissing concerns that counter mainstream thinking, Joy's work helps us see meat eating as part of a larger social system. Veganism is a minority point of view trying to break through mainstream prejudices.

The role of wealth and power

Human beings have a long history of meat eating, which has traditionally been associated with wealth and status. In the past, elite members of society who had power over others and a disproportionate ability to control and appropriate resources ate lots of meat, while most other citizens ate mainly plants. Nourishing ourselves with plants requires fewer resources and is far more ecologically efficient than consuming meat (e.g., Godfray et al., 2018; Pimentel & Pimentel, 2003). Even today, across nations, the correlation between wealth and animal consumption is obvious (Bruinsma, 2003; Tilman & Clark, 2014; World Health Organization, n.d.). The United States is a wealthy nation and the population consumes a lot of meat and other animal products, and the country also uses a disproportionately large percentage of the earth's natural resources. As nations' economies grow, it is common for the population to eat more meat, as we are seeing today in countries like China and India. Wealth and meat eating typically appear together, and they are reflections of power.

It is often said that "power corrupts," and I have observed this on numerous occasions. When humans have power over others, whether they are human or non-human, there is a tendency to mistreat and disrespect the less powerful (e.g., Gwinn, Judd, & Park, 2013; van Kleef et al., 2008). This has physical, psychological and sociological manifestations. In the case of farm animals, we confine them in factory farms and subject them to egregious physical and emotional abuse, denying them their most basic needs. We then rationalize this cruel and violent behavior, which is validated and supported by mainstream societal beliefs. Humans are social animals, and we learn from and influence those around us. If everybody is behaving a certain way, we tend to pick up those habits as well (e.g., Cialdini & Goldstein, 2004; Crandall, Eshleman, & O'Brien, 2002; Terry & Hogg, 2001).

I recently came across the work of Dacher Keltner, a psychology professor at the University of California, Berkeley who has studied power dynamics (e.g., Keltner, 2016). His work illustrates the corrupting influence of power, and he even says that "having power makes people more likely to act like sociopaths" (Keltner, 2007). His research found that "people with power tend to behave like patients who have damaged their brain's orbitofrontal lobes (the region of the frontal lobes right behind the eye sockets), a condition that seems to cause overly impulsive and insensitive behavior. Thus the experience of power might be thought of as having someone open up your skull and take out that part of your brain so critical to empathy and socially appropriate behavior." He continues, "Power may induce more harmful forms of aggression as well. In the famed Stanford Prison Experiment, psychologist Philip Zimbardo randomly assigned Stanford undergraduates to act as prison guards or prisoners – an extreme kind of power relation. The prison guards quickly descended into the purest forms of power abuse, psychologically torturing their peers, the prisoners. Similarly, anthropologists have found that cultures where rape is prevalent and accepted tend to be cultures with deeply entrenched beliefs in the supremacy of men over women."

Our relationship with the animals who we exploit for food is disrespectful and abusive and marked by a lack of human empathy. Farm animals are among the most exploited animals on earth, both in terms of scale and intensity (Bockman, this volume). We control every aspect of their lives from conception to consumption. We genetically manipulate, artificially inseminate, mutilate, confine and mass produce these living, feeling animals, and then slaughter and consume billions of them every year (trillions when you include fish).

Our power and cruelty over the animals we eat is unmitigated and unchecked, and like in other abusive relationships, there is a tendency for the abusers to denigrate their victims in explicit ways as noted above, and in less conscious ways as well. Calling someone a "pig" or a "turkey" is usually aimed at disparaging a person, but it also implicitly maligns pigs and turkeys (see also Haslam at al., this volume; Hodson, Dhont, & Earle, this volume). People who make such statements have probably not interacted much, or at all, with pigs and turkeys, except by eating them. Statements like this illustrate how we unwittingly promote prejudice and discrimination against victims of our abusive food system.

Animal agriculture is an exploitive and extractive industry that embodies our sense of human entitlement and superiority, where those with power (i.e., us) hurt and take from those without (i.e., farm animals). In the case of dairy cows, for example, we take semen from bulls and artificially inseminate cows, because in order for cows to produce milk, they must give birth. Cows, like other mammals, lactate to feed their young, and on modern dairies, they live a constant cycle of forced impregnation, birth, and reimpregnation. The baby calves are taken from their mothers shortly after birth, and the mothers are hooked up to milk machines two or three times a day, which extract milk from their udders so it can be sold to humans. Dairy cows are forced, sometimes with hormone injections, to produce ten times more milk than they would in nature, and thus are pushed to their biological limit. In a healthy environment, cows can live more than 20 years, but in modern dairy production, their bodies wear out after just a few years, at which time they are taken to slaughter.

Exploiting and killing animals for food requires callousness and undermines our empathy, which is a very important part of our humanity. Can you imagine working in a slaughterhouse, cutting the throats of animals, one after the other, all day long? Doing this kind of work has negative effects on people, which can impact their wellbeing as well as that of their families and the community (see Fitzgerald, Kalof, & Dietz, 2009).

Coping and rationalization strategies

When the topic of factory farming and slaughter is raised, people often respond by saying "don't tell me, I don't want to know." Most consumers are uneasy about the violence and cruelty inherent in animal agriculture, and a common way to manage their discomfort is to look the other way, literally and figuratively. Agribusiness for its part is happy for consumers to remain uninformed about the gruesome realities behind the production of meat, milk and eggs. In fact, the farming industry has lobbied to pass "ag gag" (agriculture gag) laws to prevent undercover investigations and keep its inhumane and irresponsible practices hidden from the public.

Besides looking the other way, consumers also employ various rationalizations to deal with the conflict between acting with compassion as opposed to supporting animal exploitation and slaughter (see also Loughnan & Davies, this volume; Piazza, this volume). Eating animals is sometimes excused as a natural behavior that homo sapiens have participated in for thousands of years, a position often validated by pointing out that other animals eat animals (Joy, 2010; Piazza et al., 2015). There may have been times historically when people needed to eat meat to survive, but that is not the case anymore, especially for the vast majority of citizens in developed countries. And, just because we have participated in a particular activity for thousands of years does not mean it should continue. If we applied that reasoning, reviled practices and institutions like slavery would still be accepted as normal. It is true that other animals kill and eat other animals, but they do not have a choice. Carnivores need to eat meat, but human beings can live well eating only plants (see Melina, Craig, & Levin, 2016).

In natural ecosystems, there are carnivorous predators who hunt and eat prey animals. Historically, humans have been both predator and prey, as well as scavengers, but with our advancing technology, expanding population, and increased power on earth, we rarely find ourselves in the position of prey anymore. Homo sapiens have become the planet's most dangerous and pervasive predator. We kill wild animals, and we destroy natural habitats and ecosystems in order to create grazing and cropland to raise domesticated animals for slaughter. We mass produce and subject over nine billion farm animals to atrocious cruelty every year in the United States alone (see e.g., https://animalclock.org; Animal Charity Evaluators, 2016), and we control every aspect of their lives from conception to consumption.

Some people believe that consuming meat and other animal products is necessary for our health, so we have no choice but to eat it (Joy, 2010; Piazza et al., 2015). This helps relieve consumers' feelings of guilt. Of course, it is in the interest of agribusiness for people to believe that they need to consume meat for protein and cows' milk for calcium. It is easy to rationalize and defend certain behaviors, even unsavory ones, when there is no choice. But the idea that we need to eat animal foods is a myth. We can live well without eating meat, milk or eggs, so eating these products is a choice (Melina et al., 2016). And, if we knowingly and willfully eat food that causes unnecessary harm, we cannot evade taking responsibility. In fact, killing and eating animals also hurts us. It has been estimated that we could save billions of dollars on health care costs in the US by shifting to a whole foods plant-based diet (Bansal, 2018; Springmann, Godfray, Rayner, & Scarborough, 2016; see Schepers & Annemans, 2018, for estimates in the UK and Belgium).

Unfortunately, consumers in affluent countries like the U.S. support immense and routine animal cruelty by eating meat, milk and eggs. Looking away or rationalizing such behavior does not make it any less cruel. I ask people to consider, "if we can live well without causing unnecessary harm, why wouldn't we?"

Indeed, human beings do not need to slaughter or otherwise exploit other animals for food. We can fulfill our nutritional needs entirely with plants, so the violence and misery inflicted by animal agricultures is completely unnecessary. Yet, it is common and persistent. Meanwhile, factory farm profiteers wrap themselves in a cloak of honor, claiming that intensive animal agriculture is necessary to feed the world's growing human population. A mantra commonly spouted by Sonny Perdue, the Secretary of Agriculture in the Trump Administration is, "Do the right thing, and feed everybody." Of course, providing wholesome, nutritious food for everybody is a respectable goal, but doing so through animal agriculture is not the "right thing," as Sonny Perdue would have us believe.

Moreover, raising animals for food is inherently inefficient. We can feed more people with fewer resources through plant-based agriculture (e.g., Tilman & Clark, 2014). It makes sense to eat plants directly, instead of growing corn, soy and other crops to feed farm animals. Our animal-based food systems are squandering and depleting topsoil, fossil fuel, water and other precious resources (e.g., Eshel, Shepon, Makov, & Milo, 2014). The quest by agribusinesses for more land to feed more farm animals to feed our growing human population is destroying vast swaths of our planet (Gerber et al., 2013; Godfray et al., 2018). Rainforests and other diverse ecosystems are vanishing to create cropland or grazing land, and native species are going extinct (Godfray et al., 2018).

The United Nations has reported that animal agriculture is a leading cause of our planet's most significant environmental threats, locally and globally (Gerber et al., 2013; Steinfeld et al., 2006). It is at the center of the loss of biodiversity, and contributes more to climate change than the entire transportation industry (Godfray et al., 2018). Factory farms contaminate the land, water and air, and people living nearby experience myriad health problems (see also Hodson & Costello, 2018). Their quality of life is diminished along with property values because of the foul odors and toxic emissions.

Indeed, the fecal pathogens in factory farm waste have been made more hazardous by the widespread use of antibiotics and other drugs, which are used to keep animals alive and growing in filthy stressful conditions. The majority of antibiotics used in the United States are fed to farm animals, not used to treat human disease. The irresponsible use of drugs in animal agriculture has resulted in the development of anti-biotic resistant pathogens, which present risks to consumers who eat contaminated food. They also pose a risk in the environment, where antibiotic resistant bacteria have even been found in groundwater. When people are sickened with these virulent pathogens, formerly live saving drugs can be rendered useless.

The environment, local communities and the common good are undermined by industrial animal agriculture, which is bolstered by legal, social, and economic institutions. Agribusiness is very influential in Washington, DC and in state capitols, and it has worked closely with its government allies to enact laws and policies to serve its short sighted interests. It has passed "right to farm" laws across the USA, for example, which allow pollution and other factory farm offenses to continue without proper oversight or regulation. This industry, which exploits animals who have no power or legal protection, is also quick to take advantage of citizens with limited power, locating factory farms and slaughter houses in economically disadvantaged communities where citizens lack the resources to challenge them.

Social norms and social contagion

When speaking about the inhumane treatment of animals in modern agriculture, the renowned animal welfare expert, Dr. Temple Grandin says "bad has become normal." This phrase could apply to our entire food system, which also causes enormous preventable human suffering, as well as ecological destruction. And those with the least power usually suffer the most. Low paid workers and residents living near factory farms face elevated health risks, and so do people living in urban settings where there is a lack of wholesome food.

Human beings are greatly influenced by our social environment and by the people around us (Cialdini & Goldstein, 2004; Crandall et al., 2002). Just as a bacteria and other biological agents can pass from one person to another, so too can behaviors and beliefs. We rub off on those around us. If everybody we know acts and believes certain things, we tend to do so as well. And, if everybody around us is suffering from heart disease, diabetes, or other health problems, we assume these problems are normal. Even doctors and health professionals start accepting bad as normal, without attributing diet related ailments to our dysfunctional food system. Instead, they are quick to prescribe medications for preventable illnesses, which furthers the belief that these illnesses are a normal part of life, as are medications to treat them. If everybody around us is behaving and believing similarly, it is hard to do differently, unless we are exposed to others acting and thinking differently.

Social scientists talk about "social contagion" or "behavioral contagion," which has been defined as "The spread of ideas, attitudes, or behaviour patterns in a group through imitation and conformity" (Colman, 2015; see e.g., Ugander, Backstrom, Marlow, & Kleinberg, 2012). It is often used in reference to behavioral waves and trends within populations, but could also apply to widespread social norms like meat eating. The "monkey see, monkey do" phrase could just as accurately be "people see, people do." Everybody around me was eating meat as I grew up, and I adopted the habit without thinking very much about it. Most of us have a similar experience and adopt the dietary habits of our families and communities.

However, it is possible for people to act with conscience, and to choose not to support a bad system. In Farm Sanctuary's early days, we conducted undercover investigations to expose and educate people about factory farming cruelty. We hoped that people would decide to stop supporting this industry once they learned about it. But change is not so simple.

In recent years I have been influenced by the work of Dan Buettner, a National Geographic explorer and author of books about the Blue Zones. Dan has travelled the world to identify and study places where people live the longest and are the happiest, and he believes that environmental conditions, which include physical infrastructures as well as social components, are more important than personal choices in bringing about longevity and happiness. Although it is possible for humans to act as conscientious agents, the evidence seems to suggest that we mainly behave the way we do because it is easy and because our environment encourages it.

Unfortunately, the US food system and supporting infrastructure makes bad food ubiquitous and accessible. Government policies have encouraged the excessive production and consumption of processed foods and animal products, which are marketed through a robust network of fast food restaurants and other retailers. Subsidy programs encourage industrial agriculture and spur the overproduction of crops, which are fed to farm animals, and allow meat, milk and eggs to be mass produced.

Towards a plant-based food system

Preventing factory farm cruelty, and protecting the environment and our own health and well-being, ultimately requires that we change our food system, and reform the infrastructures and socioeconomic conditions that sustain it. We need to shift to eating plants instead of animals, and our government should stop subsidizing and enabling our broken food system. Instead of mass production operations that deplete natural resources and employ vast quantities of petrochemical inputs to produce calorie-rich and nutrient-poor commodities, we should support more diversified community oriented farms, including in urban and suburban locations. Some cities are even establishing food forests. I am encouraged by this and the growth of farmers markets and community gardens in cities across the USA, as well as by the "food not lawns" movement, wherein homeowners are turning their lawns into gardens and growing veggies, greens and other food. Instead of mowing grass, they are picking produce. Growing one's own food and connecting more closely with nature is healthy and empowering.

The role of education

I am inspired by the work of Stephen Ritz, a teacher in the South Bronx and founder of Green Bronx Machine. He teaches at a school plagued by crime, poor attendance and dismal graduation rates. His students face many challenges, but he was able to turn things around when (according to a jacket note in his book *The Power of a Plant*): "He flipped his curriculum to integrate gardening as an entry point for all learning and inadvertently created an international phenomenon." I attended Mr. Ritz's TEDx talk in New York City where he described how attendance rates improved from less than 50% to over 90% as students now had a purpose in coming to school. Growing plants and food also helped to grow hope and opportunity. In addition to improving academic performance, this program has spurred entrepreneurism and jobs. One of my favorite parts of his talk was when he described how his inner-city students had been hired to build gardens in the Hamptons.

For decades, Dr. Antonia Demas, founder of the Food Studies Institute, has worked to educate children and instill healthy eating habits, and like Steve Ritz, she has seen students empowered physically and emotionally through food-based education (www.foodstudies.org). She taps into children's natural curiosity, and introduces them to healthful foods through the traditions and arts of various cultures. And she works with students on practical life skills, which include gardening and preparing food. Dr. Demas is a leader in the movement for food literacy, which has been defined as "Understanding the impact of your food choices on your health, the environment, and our economy." (www.foodliteracycenter.org). I think it is critically important for citizens to become more food literate and to understand the impacts of their food choices, so they can ultimately make informed choices that are aligned with their values and interests. Indeed, there is a burgeoning opportunity in the vegan food market.

I also believe that government programs, such as those guided by the United States Farm Bill, should stop supporting and enabling factory farming and the status quo, and should instead promote plant-based agriculture, which can improve the lives of children and adults alike. Farming and gardening and eating good food can play a huge role in empowering and healing people, including veterans suffering from PTSD. It can also help people who have struggled with addiction and other challenges.

People incarcerated in prisons can also benefit from programs that connect people with good food. In addition to improving inmates physical and emotional health, such programs save money, help teach life skills, and give hope. A Louisiana prison featured in a National Public Radio Marketplace story reported benefits after establishing a farm and food service program operated by inmates. The warden quoted in the report said, "since the prison started feeding farm vegetables to inmates, the jail has spent less on medicine for hypertension and diabetes." And the warden said "more inmates have signed up for things like GED and anger manage-ment classes" (Abrams, 2018).

The explosive growth of the vegan food market

In recent years, we have seen the development and marketing of numerous plant based alternatives to animal products. Many different types of nondairy milks made from soy, almonds, coconut, cashews, and other plants are replacing cows' milk in supermarkets across the United States and Europe, and even traditional dairy businesses are investing in these alternatives. Meanwhile, start-up companies like Beyond Meat are producing "meat" from plants, which is inherently more effi-cient and more humane than slaughtering animals for food. Plant-based meats are becoming available in restaurants and grocery stores throughout the USA and most European countries, and traditional meat industry giants like Tyson are beginning to invest in these new companies and products.

A mission driven vegan start-up, which was called "Hampton Creek" and recently changed their name to "Just," made headlines when their product, Just Mayo, attracted a lawsuit from an established market leader, Hellman's mayonnaise (Kaufman, 2014; Strom 2014). Shortly afterwards, they received notice from the Food and Drug Administration (FDA) who sided with Hellman's in asserting that the vegan product, Just Mayo, could not be called "mayo." The lawsuit and govern-ment action garnered significant unfavorable media attention for Hellman's and the FDA, especially since agribusiness and the government appeared to be working together to undermine the small vegan company. Ultimately, the matter was settled after Hellman's dropped their lawsuit and the FDA agreed to allow Just Mayo to keep the name "mayo." Then, ironically, Hellman's launched their own vegan may-onnaise product (Associated Press, 2016).

I think that entrepreneurs, as well as established businesses and investors, are going to play an increasingly important role in building an infrastructure and making vegan food more widely accessible. I am grateful for the many inspired efforts of people and companies working to create a food system that serves the common good, and I believe this entrepreneurial activity will play an important role in changing agriculture. Metropolis Farms in Philadelphia (PA) is growing healthy food and generating employment opportunities in the inner city. Jack Griffin, the founder, envisions creating "green collar" jobs, especially for disenfranchised members of society, including people who have been imprisoned. His vertical indoor farms can grow large quantities of fresh healthy produce efficiently, and they can be located in urban communities where healthy food is most needed. His company started by growing greens in an abandoned warehouse in downtown Philadelphia and is now developing urban farming technology intended to create a food revolution and an infrastructure with "hyper local farms." The goal is to create vibrant businesses and produce fresh local food efficiently without the chemical inputs or vast resources, including water and energy, used on conventional farms. According to the Metropolis Farms website: "We use less than 2% of the water and 12% of the energy required by conventional outdoor farming." The company believes that: "A strategic network of local farms is the answer to many of the world's biggest problems." These problems go beyond food, and also address areas of citizen empowerment. Metropolis wants to put "power back into the hands of local communities by allowing forward-thinking individuals to operate a profitable farm anywhere, regardless of location or climate."

Towards a kinder and just food system

Along with various efforts to create a better food system, there is also a growing awareness about inequities and power dynamics around food and agriculture. The food sovereignty movement was created to address food injustice and to support and empower those who have been disenfranchised. It has been defined as "the right of peoples to healthy and culturally appropriate food produced through ecologically sound and sustainable methods, and their right to define their own food and agriculture systems" (The Declaration of Nyéléni, 2007, p. 9). The Food Empowerment Project, based in California, is a vegan organization that incorporates food justice, eco-feminist, and animal rights principles. It works to prevent injustices to animals, people and environment, and "seeks specifically to empower those with the fewest resources" (www.foodispower.org). I support this work, which includes making healthy food available in low-income areas as well as advocating for farm workers who pick our food. I believe it is very important for the vegan movement to work in concert with various aligned interests.

To me, being vegan is an aspiration to live as kindly as possible, which includes being kind to people and other animals, as well as to the earth and ourselves. Vegan advocates, like other activists, can become overwhelmed in addressing the immense harm caused by human callousness, which can trigger anger and take a toll on one's physical and emotional well-being. In such cases, it helps to take heart from any step in a positive direction, no matter how small, and to realize that change usually happens incrementally. Humans are fallible and make mistakes, and good people do bad things, so I try to remember the axiom to "love the sinner, hate the sin."

Our factory farm food system is the opposite of kindness. It abuses animals, destroys natural resources, and harms people in myriad ways. The good news is that citizens are becoming aware and consciously seeking to make food choices that are better aligned with their values and interests. People would rather not cause unnecessary suffering to other animals. They would rather avoid food that makes them sick, and they would prefer to support a food system that is just and does not destroy the planet. People want to live and eat better, and that inclination needs community and infrastructure support, and well as practical "how to" advice, like I included along with 100 vegan recipes, in my 2015 book, *Living the Farm Sanctuary Life* (Baur & Stone, 2015).

Over the years, I have come to believe that easy access to vegan food, and an infrastructure that supports it, is the key to preventing farm animal suffering. We are creatures of habit, and while our conscience plays a role in our food choices, accessibility and convenience probably have a greater influence. When people stop eating animals, whether it is because of compassion towards animals or any other reason, the result is that animals are not made to suffer for our appetites. Businesses, nonprofits, government officials and individuals may have different motivations, but it is encouraging to see many moving in a similar direction.

With a growing amount of readily available plant-based meats, milks, and eggs coming to market, along with familiar fruits, vegetables, grains, beans and other plant foods, eating plants instead of animals is easier than ever. New advocacy organizations, like the Good Food Institute and the Plant Based Foods Association, have recently formed to accelerate the replacement of animal products in the marketplace. Plant-based companies are also reaching consumers with recipes and other practical advice in promoting their food, and vegan chefs and food service professionals are working with businesses, schools, and institutions to provide more plant-based foods. I am heartened by these developments, and the critically important shifts underway in infrastructure, which are making vegans food more widely available.

Government food policies that have propped up our inefficient animal based food system need to be reformed, and agribusiness should be made liable for the vast harms it causes. Programs that encourage the mass production of cheap crops to feed farm animals and those that pay for the industry's externalities should be discontinued. The Supplemental Nutrition Assistance Program (SNAP), which "offers nutrition assistance to millions of eligible, low-income individuals and families and provides economic benefits to communities,"[1] should promote healthy eating habits instead of being used as a dumping ground for illness-promoting commodities, like government issue cheese. SNAP could also support community gardens, farmer's markets, and other endeavors that provide meaningful employment and fresh healthy food in economically disadvantaged areas. Tax rules and financial structures should be changed to discourage the concentration of wealth and power into fewer, larger farms. Land, water and other resources should be protected from the offenses of factory farming, and made more accessible to sustainable community oriented farming operations. There should be opportunities for entrepreneurial agricultural workers to own their own businesses.

Structural impediments should be removed to incentivize plant-based agriculture in urban, suburban and rural areas, and economic development programs should encourage a holistic approach that recognizes the many value-added opportunities in agricultural goods and services. In addition to growing, distributing and preparing food, there is a need to educate citizens. Opportunities exist in both the sciences and the humanities in fields that include agronomy, engineering, design, landscape architecture and eco-tourism, as well as hospitality, urban planning, wellness, and the culinary arts. The opportunities are vast and expanding.

The average age of farmers in the USA has been increasing for decades, and is now approaching 60 (Kurtzleben, 2014). We need to create pathways that attract younger generations to the profession, and preserve farmland, which we are losing to development and other non-farming purposes. We can do a better job of networking young agricultural entrepreneurs with aging farmers nearing retirement, and consider creative succession and retirement plans. For example, I would love to see dairy farms transitioned into plant-based agriculture, rather than acquired by a larger dairy. Older dairy farmers and their cows could retire and be allowed to live out their natural lives on the farm, while the facility becomes a sanctuary and plant-based agriculture operation.

The farm sanctuary movement has grown and spread significantly since 1986 when Farm Sanctuary was founded. Besides caring for animals, farm sanctuaries have visitor and educational programs, and some have overnight accommodations for guests. I would love to see farm sanctuaries continue to evolve beyond animal rescue and to expand into growing food, and doing more to attract and encourage plant-based farming.

Conclusion

Agriculture sustains us, and it reflects our attitude and relationship to the earth and other animals. The dominant factory-farming paradigm is abusive and extractive. It is bad for animals and the earth, and it is bad for people too. The experience of working in or visiting a factory farm or slaughterhouse is very different than visiting Farm Sanctuary. Interacting with animals in cruel and violent ways affects us negatively, while treating them with kindness helps improve our lives and theirs, reducing stress, while building empathy and connection. Gardening and growing our own food has similar benefits, plus it is good exercise.

We have nascent opportunities to transform our agriculture system, and there are glimpses of light emerging. Shifting to a more holistic plant-based food system will nurture healthier relationships with other animals, the earth and ourselves. Industrial animal farming developed over time, and ultimately became the dominant paradigm. But people are now recognizing the profound harm it causes, and they are looking for alternatives. Forward-thinking citizens and businesses are taking the initiative, and creating new opportunities. Investments are coming into plant-based agriculture, along with passion and a contagious enthusiasm. Incrementally, cracks are appearing in a factory farm system and a more healthful, compassionate, and ecologically sound food system is beginning to emerge.

Note

1 www.fns.usda.gov/snap/supplemental-nutrition-assistance-program-snap

References

Abrams, E. (2018, February 8). Farm-to-table comes to a Louisiana jail. *Marketplace*. Retrieved from www.marketplace.org/2018/02/08/life/farm-table-comes-louisiana-jail

Animal Charity Evaluators (2016). Why farmed animals? Retrieved from https://animalcharityevaluators.org/donation-advice/why-farmed-animals/

Associated Press (2016, February 2). Hellmann's gives vegan a go after suing eggless spread maker. *Mail Online*. Retrieved from www.dailymail.co.uk/wires/ap/article-3428302/Hellmanns-gives-vegan-suing-eggless-spread-maker.html

Bansal, A. (2018, May 30) A medical case for a whole food, plant-based diet. *Op-Med*. Retrieved from https://opmed.doximity.com/articles/a-medical-case-for-a-whole-food-plant-based-diet

Baur, G., & Stone, G. (2015). *Living the Farm Sanctuary Life. The Ultimate Guide to Eating Mindfully, Living Longer, and Feeling Better Every Day*. New York: Rodale Inc.

Bockman, J. (this volume). Accomplishing the most good for animals. In K. Dhont & G. Hodson (Eds.). *Why We Love and Exploit Animals: Bridging Insights from Academia and Advocacy*. Abingdon: Routledge.

Bruinsma, J., ed. (2003). *World Agriculture: Towards 2015/2030. An FAO Perspective*. London: Earthscan Publications Ltd. Retrieved from www.fao.org/3/a-y4252e.pdf

Cialdini, R. B., & Godlstein, N. J. (2004). Social influence: Compliance and conformity. *Annual Review of Psychology, 55*(1), 591–621.

Colman, A. M. (2015). *A Dictionary of Psychology*. Fourth Edition. Oxford: Oxford University Press.

Crandall C. S., Eshleman, A., & O'Brien, L. (2002). Social norms and the expression and suppression of prejudice: the struggle for internalization. *Journal of Personality and Social Psychology, 82*, 359–378.

Eshel, G., Shepon, A., Makov, T., & Milo, R. (2014). Land, irrigation water, greenhouse gas, and reactive nitrogen burdens of meat, eggs, and dairy production in the United States. *Proceedings of the National Academy of Sciences of the United States of America, 111*, 11996–12001. http://dx.doi.org/10.1073/pnas.1402183111.

Festinger, L. (1957). A theory of cognitive dissonance. Stanford, CA: Stanford University Press.

Fitzgerald, A. J., Kalof, L., & Dietz, T. (2009). Slaughterhouses and increased crime rates: An empirical analysis of the spillover from "the Jungle" into the surrounding community. Organization and Environment, 22, 158–184. doi: 10.1177/1086026609338164

Gerber, P. J., Steinfeld, H., Henderson, B., Mottet, A., Opio, C., Dijkman, J.,…Tempio, G. (2013). *Tackling Climate Change Through Livestock – A Global Assessment of Emissions and Mitigation Opportunities*. Rome: Food and Agriculture Organization of the United Nations (FAO).

Godfray, H. C. J. et al. (2018) Meat consumption, health, and the environment. *Science, 361*, eaam5324.

Gwinn, J. D., Judd, C. M., & Park, B. (2013). Less power = less human? Effects of power differentials on dehumanization. *Journal of Experimental Social Psychology, 49*(3), 464–470.

Harvey, F. (2018, April 17). The way some pigs are reared is 'upsetting and wrong', say shoppers. *The Guardian*. Retrieved from www.theguardian.com/environment/2018/apr/17/the-way-some-pigs-are-reared-is-upsetting-and-wrong-say-shoppers

Hodson, G., & Costello, K. (2018). Psychological implications of undervaluing animals: Dominance-based ideologies and systems of oppression. In A. Matsuoka & J. Sorenson (Eds.), *Critical Animal Studies: Towards Trans-Species Social Justice* (pp. 184–206). London: Rowman and Littlefield International.

Hodson, G., Dhont, K., & Earle, M. (this volume). Devaluing animals, "animalistic" humans, and people who protect animals. In K. Dhont & G. Hodson (Eds.), *Why We Love and Exploit Animals: Bridging Insights from Academia and Advocacy*. Abingdon: Routledge.

Joy, M. (2010). *Why We Love Dogs, Eat Pigs, and Wear Cows: An Introduction to Carnism*. San Francisco, CA: Red Wheel/Weiser.

Kaufman, A. C. (2014, December 19). Hellmann's mayo drops lawsuit against eggless 'Just Mayo'. *HuffPost*. Retrieved from www.huffingtonpost.co.uk/entry/just-mayo-lawsuit_n_6354824

Keltner, D. (2007, December 1). The Power Paradox. *Great Good Magazine*. Retrieved from https://greatergood.berkeley.edu/article/item/power_paradox

Keltner, D. (2016). *The Power Paradox: How We Gain and Lose Influence*. New York: Penguin Press.

Kurtzleben, D. (2014, February 24). The rapidly aging U.S. farmer. *U.S. News*. Retrieved from www.usnews.com/news/blogs/data-mine/2014/02/24/us-farmers-are-old-and-getting-much-older

Leenaert, T. (this volume). Steakholders: How pragmatic strategies can make the animal protection movement more effective. In K. Dhont & G. Hodson (Eds.), *Why We Love and Exploit Animals: Bridging Insights from Academia and Advocacy*. Abingdon: Routledge.

Loughnan S., & Davies, T. (this volume). The meat paradox. In K. Dhont & G. Hodson (Eds.), *Why We Love and Exploit Animals: Bridging Insights from Academia and Advocacy*. Abingdon: Routledge.

Melina, V., Craig, W., & Levin, S. (2016). Position of the Academy of Nutrition and Dietetics: vegetarian diets. *Journal of the Academy of Nutrition and Dietetics, 116,* 1970–1980. doi: https://doi.org/10.1016/j.jand.2016.09.025

NRG Research Group (2017). *Broiler Chicken Welfare Survey*. Retrieved from https://mercyforanimals.org/files/MFA_2017_Survey_US.pdf

The Nyéléni 2007 International Steering Committee (2007). *Nyéléni 2007: Forum for Food Sovereignty*. Retrieved from https://nyeleni.org/DOWNLOADS/Nyelni_EN.pdf

Piazza, J. (this volume). Why people love animals yet continue to eat them. In K. Dhont & G. Hodson (Eds.), *Why We Love and Exploit Animals: Bridging Insights from Academia and Advocacy*. Abingdon: Routledge.

Piazza, J., Ruby, M. B., Loughnan, S., Luong, M., Kulik, J., Watkins, H.M., & Seigerman, M. (2015). Rationalizing meat consumption: The 4Ns. *Appetite, 91,* 114–128. http://dx.doi.org/10.1016/j.appet.2015.04.011

Pimental, D., & Pimentel, M. (2003). Sustainability of meat-based and plant-based diets and the environment. *The American Journal of Clinical Nutrition, 78,* 660S–663S doi: https://doi.org/10.1093/ajcn/78.3.660S

Reese, J. (2017, November 20). *Survey of US Attitudes Towards Animal Farming and Animal-Free Food*. www.sentienceinstitute.org/animal-farming-attitudes-survey-2017

Rothgerber, H. (the volume). How we love and hurt animals: Considering cognitive dissonance in young meat eaters. *Why We Love and Exploit Animals: Bridging Insights from Academia and Advocacy*. Abingdon: Routledge.

Schepers, J., & Annemans, L. (2018). The potential health and economic effects of plant-based food patterns in Belgium and the United Kingdom. *Nutrition, 48,* 24–32.

Springmann, M. H., Godfray, H. C., Rayner, M., & Scarborough, P. (2016). Analysis and valuation of the health and climate change cobenefits of dietary change. *Proceedings of the National Academy of Sciences of the United States of America, 113*(5), 4146–4151. https://doi.org/10.1073/pnas.1523119113

Steinfeld, H., Gerber, P., Wassenaar, T., Castel, V., Rosales, M., & de Haan, C. (2006). *Livestock's long shadow –Environmental issues and options.* Rome: Food and Agriculture Organization of the United Nations (FAO).

Strom, S. (2014, November 10). Hellmann's maker sues company over its Just Mayo substitute mayonnaise. *The New York Times.* Retrieved from www.nytimes.com/2014/11/11/business/unilever-sues-a-start-up-over-mayonnaise-like-product.html

Terry, D. J., & Hogg, M. A. (2001). Attitudes, behavior, and social context: The role of norms and group membership in social influence processes. In J. P. Forgas & K. D. Williams (Eds.), *The Sydney Symposium of Social Psychology: Social Influence: Direct and Indirect Processes* (pp. 253–270). New York: Psychology Press.

Tilman, D., & Clark, M. (2014). Global diets link environmental sustainability and human health. *Nature, 515*, 518–522.

Ugander, J., Backstrom, L., Marlow, C., & Kleinberg, J. (2012). Structural Diversity in Social Contagion. *Proceedings of the National Academy of Sciences, 109*, 5962–5966.

van Kleef, G. A., Oveis, C., van der Löwe, I., LuoKogan, A., Goetz, J., & Keltner, D. (2008). Power, distress, and compassion: Turning a blind eye to the suffering of others. *Psychological Sciences, 19*, 1315–1322. doi: 10.1111/j.1467-9280.2008.02241.x

World Health Organization (n.d.). Global and regional food consumption patterns and trends. www.who.int/nutrition/topics/3_foodconsumption/en/

ASKHAM BRYAN
COLLEGE
LEARNING RESOURCES

15

STEAKHOLDERS

How pragmatic strategies can make the animal protection movement more effective

Tobias Leenaert

Abstract

As many as 70 billion farmed animals (including birds, pigs, cows, goats, and sheep, but excluding marine animals) are raised and killed each year for food. Especially when "factory farmed," most of these animals lead short and miserable lives and are subjected to pain, stress, anxiety, and boredom on a daily basis. While many people may deem it important to improve the living (and dying) conditions of these animals, the animal rights or animal protection movement wants to abolish the use of animals for human consumption altogether. If this movement wants to achieve its objective of abolition, a highly idealistic "go vegan for the animals" approach will not be sufficient. Given the extreme dependency on the use of animal products today, I suggest that the animal protection movement requires a lot more pragmatism. A pragmatic approach includes asking for reduction of animal products consumption, using non-moral arguments to motivate people, investing in creating an environment that facilitates change, and creating a larger tent.

Introduction

Most people, at least in Western countries, eat parts (meat or fish) or products (mainly dairy and eggs) of animals every day, often during three meals.[1] In these countries, there is barely any tradition of cooking meals without them. When, during public talks, I ask people if they can think of a plant-based recipe for a main dish which they received from their parents or grandparents, they almost invariably come up with a blank. Hence they are often quite unable to imagine how some people – vegans – can live without using animal products. In spite of the growing popularity of plant-based eating, many still believe that avoiding animal products is a way of living for ascetics only. Most people are what we may call *steakholders*: they are heavily invested in eating animal products and as such have a *stake* in not truly grasping the situation that farmed animals are currently in.

In the same way, at a more structural level, our economy, or a big part of it, is a steakholder too. I am talking about the industries of raising animals, feeding them, transporting them, slaughtering them, processing them or producing equipment to do all of these things. I am talking about companies that sell meat products or foods where animal products are an important ingredient (from yoghurts to pizzas), as well as about the whole distribution sector: stores and supermarkets, but also all kinds of restaurants, cafeterias, and caterers. And then there are chefs, teachers, food photographers, cookbook writers, TV personalities and so on, who may all be very invested in animal products because they too derive part of their livelihood from working with them.

In a situation like this, with a very high investment at both individual and societal levels, a plea to take animal interests and suffering into account will often fall on deaf ears. Moreover, even if it did not, and even if everyone accepted the idea that it is morally wrong for us to raise and kill animals for food – at least in the way that we do structurally – it would still not be obvious how we would move in a new direction. We have – or at least think we have – become very dependent on the use of animals for the continuation of our society. If an intelligent extraterrestrial race would study us and ask the question what *fuels our civilization*, one possible answer would be: non-human animals, which are raised and exploited for the benefit of *Homo sapiens*, the apex predator of this planet.

Idealism versus pragmatism

The movement that tries to rid the world of factory farming and the use of animals for human purposes – I will call this the *animal rights, the animal protection,* or the *vegan* movement interchangeably – is a relatively recent phenomenon[2] and has, as a moral movement, been relying mostly on moral arguments.[3] The movement's message is that it is wrong to use animals for food or other human purposes, and that, basically, we should cease doing that, today rather than tomorrow. Thus, the animal rights movement actually has a double demand: it asks that people change their *attitude* towards animals (from not sufficiently caring about them to believing that they are morally relevant and have interests) as well as their *behavior* regarding animals (moving from meat eater to vegan).

I call the focus on this double demand (basically "go vegan because animals have interests") an *idealistic* approach. It is focused on what the movement would ideally want, and it emphasizes that people should do the right thing for the right reasons.[4] I contrast this idealistic approach with a *pragmatic* approach. I believe that the animal rights movement needs a high dose of pragmatism today, because it faces an enormous challenge. I have already stated that we are extremely dependent on the use of animals today, but there are several other reasons why the challenge animal advocates face is so huge. One is that, unlike with other movements, the animal rights movement does not have the support of the victims it is trying to help.

The struggle would be easier if chickens, cows, and pigs joined animal advocates, but that will never happen. Furthermore, the challenge is all the bigger in that this movement is about animals, not about people. While vegans may not see much of a morally significant difference there, most people do, and will accept more confrontational tactics for human rights issues than for animal issues. Finally, eating meat (and animal products) is something very ancient, something that, to our knowledge, our species has done for a very long time. This obviously does not justify the practice, but it helps to explain that it may not be easy to get rid of it. Indeed behavior change in the field of food is notoriously tricky. In her book *Meathooked: The History and Science of Our 2.5-Million-Year Obsession with Meat*, Marta Zaraska provides an excellent overview of all the reasons – from genes to culture – why people are so hooked on meat.

For the above reasons, pragmatism, or an approach that focuses on reality and what is possible right now instead of on what is ideal, should get a more prominent role in the animal rights movement's outreach. Note that I am *not* saying, by any means, that pragmatism is the *only* approach that will work, but merely that it should be necessary part of the global efforts of the animal protection movement.

I will, in what follows, suggest four ways in which this movement can be more pragmatic. These are:

1.   a focus on reduction rather than elimination (the behavior)
2.   a focus on non-moral arguments (the attitude)
3.   a focus on creating a facilitating environment
4.   a focus on being more inclusive

Four ways to be pragmatic

The ask: Incremental steps can lead to abolition

The most obvious examples of incremental steps to improve the plight of farmed animals are (1) improving their living conditions (so-called "welfare reforms") and (2) reducing consumption and thus production (lowering supply by creating a lower demand), so that eventually fewer animals will come into the world only to lead miserable lives.

Many vegans and animal rights activists feel – idealistically – that it is morally wrong to suggest or imply anything else than the complete abolition of animal products, on both a personal and a societal level. Their argument is that if we do not want to be speciesist, we cannot voice demands that we would find unethical were we talking about humans. In the case of humans, so the argument goes, we would be abhorred at anyone asking for regulating, rather than abolishing, child abuse or slavery. If we allow these kinds of demands in the case of animals, we are implying – still according to this reasoning – that animals are less than humans, and that exploiting them is basically okay if we do less of it, or if we do it in a less bad way.

Although such comparisons may make some kind of sense in a theoretical way, in practice they are quite unhelpful. Eating animal products is not just condoned but actively celebrated by the great majority of the population, whereas child abuse is universally condemned (and is illegal). A rigid-idealistic approach that puts these different issues on the same level might be appropriate some decades in the future, but it is not now – or at the very least it cannot be considered the only right way to advocate for farmed animals – for the reasons stated above.

I will not talk about the advantages or disadvantages of campaigning for welfare reforms here, but I will focus on the other example of an incremental strategy that asks for the reduction of the consumption of animal products (rather than their elimination). Asking for reduction is strategic (for the aim of abolition) for several reasons. First of all, for most people, reducing the consumption of animal products (on both an individual and societal level) is a much easier thing to do, at this point, than completely eliminating them. There will hence be more people following up on an ask to reduce than an ask to go vegan.[5] Once people reduce, it is easier, from that position, to shift further towards eating plant-based (see the next paragraph). Moreover, because there are many more reducers than vegans (even including vegetarians), the reducers altogether have a bigger impact on reducing animal suffering than the small number of vegans.[6]

The most important reason why reducers are crucial, however, is that a critical mass of them may be the fastest way to tip the system towards a plant-based norm. To see this more easily, let us look at another phenomenon: the gluten-free "revolution." In the last five or ten years, most supermarkets and restaurants, not just in Western but also in other countries, have developed sizeable offerings of gluten-free products and dishes (usually at least as large as or larger than vegetarian or vegan products). These products cater not just to the one percent of the population that is gluten-intolerant or gluten-sensitive and requires a gluten-free diet, but also to the much bigger segment (close to ten percent of the population in some countries) which chooses to largely avoid gluten (McCarthy, 2017). This latter group consists of people who do not have a medical condition and do not *need* to be gluten-free, so they are less strict about it. We can call them gluten-reducers. It is not difficult to imagine that it was this bigger group who made it interesting for producers to really invest in the development of decent gluten-free products. It is thanks to them that gluten-free took off and is now everywhere.

Most importantly, the result is that for the really gluten-intolerant, their strict diet has become much easier to follow, and much tastier. The same is happening in the vegan domain. Demand creates supply, and with an adequate supply, shifting towards a plant-based diet will get easier and easier. If it has become a lot easier to be vegan today compared to a decade ago, vegans have in the first place reducers to thank, because it is mainly reducers who helped create demand. The many brands of dairy milk alternatives one can find in the supermarket, for instance, are not there because of ethical vegans, but because health conscious reducers and lactose intolerant people, or customers who just wanted variation, created a demand.[7] This is not to say vegans will not play a more important part in all of this in the future, when they have grown in numbers, but for now, it is the reducers who have the biggest impact on the market. Later in this article, we will discuss the importance of creating an environment that facilitates change.

The arguments: We do not necessarily need to lead with moral arguments

Above, I talked about the double demand of animal advocates: we want behavior change – which the previous paragraph was about – *and* attitude change. The desired behavior (i.e., reducing or ideally eliminating animal products) can be demonstrated for a variety of reasons. I roughly distinguish moral arguments from non-moral arguments. The main moral argument is that animals have interests (or rights) and can suffer, and therefore should not be killed, harmed, or even used. Non-moral arguments for reducing or avoiding animal products include especially health, but also financial reasons, trendiness or peer pressure (doing it for your significant other), etc.

The arguments that many vegan/animal rights advocates want to use, and the motivation they want people to have, is exactly this moral one, the concern for animals. This, for many of them, is the "right" reason. These advocates want people to do the right thing for the right reasons, so much so that sometimes people who demonstrate the right behavior for the wrong reasons are not considered part of the club. According to some, there is no such thing as "a health vegan," because a vegan is *by definition* motivated for ethical reasons. Other people, the argument goes, may label themselves as being on a plant-based diet, but not as vegans (Lampert, 2016).

As I stated before, this whole approach is highly idealistic and probably unrealistic. More importantly, it is not *necessary* that people do the right thing for the right reason. My view is that animal advocates – or advocates of any kind, really – should at least "allow" people to start out with doing the *right* thing for the *wrong* reasons. Doing the right thing for the wrong reasons is acceptable not just because doing the right thing is good for other beings irrespective of the motivation, but also and especially because reasons can evolve, and can develop out of new behaviors. I will come back to this point below.[8]

As indicated above, the animal rights movement – being a moral movement – has traditionally invested most of its advocacy resources in moral arguments. The cliché that vegans are preachy and make people feel guilty (e.g., Minson & Monin, 2012) obviously does contain some truth (though it should be noted that meat eaters, for a variety of reasons, may feel accused rather easily). Advocates and organizations spend a lot of time on ethical arguments, in their pamphlets, websites, documentaries, demonstrations, petitions, online and offline discussions, etc. There is also a lot of confidence in the idea that these moral arguments *work*, that they make a difference and can shift people's attitudes. And surely they may. But certainly not in every case. Part of the faith in moral arguments may come from the fact that many or most people who are vegans or animal rights advocates at this point were swayed themselves by exactly these type of moral arguments. Hence, they expect that these arguments will have the same effect on other people. Advocates who became vegan after watching the documentary *Earthlings* (basically a horror show of all the awful things people do to animals) presumably want others to watch that film, expecting or at least hoping it will have the same effect on them.

People, however, differ greatly in terms of their interests, degree of empathy, discipline, health situation, tastes, financial situation, etc. It may be argued that a big part of the population in many cultures cares about animals – witness cases of public outrage like the shooting of Cecil the lion in 2015 by the American dentist Walter Palmer – but does not care enough where caring would imply significant and inconvenient behavior change. In this sense, people who are vegan (or even vegetarian) at this point in time, can be considered low-hanging fruit. They were relatively easily swayed because they cared enough and were able to put this caring into practice. This clearly does not apply to the majority of the population – or else we would have a totally different situation by now and at least factory farming would no longer exist.

Advocates also derive confidence about the effectiveness of moral arguments from historical examples. We love to believe that when a society (or humanity) achieved something *good,* we got to that point primarily because we believed it was the right thing to do. We like to believe that slavery[9] was abolished first and foremost because people believed it was wrong, and a group of us stood up and was outraged at the practice.

I do not claim here that moral arguments and outrage were unimportant, but rather that people in idealistic movements may often rely on them too heavily and be partly blind to other, more mundane factors that played a part in the outphasing of certain practices. In the fight against slavery, for instance, we should consider the impact of the industrial revolution. Mechanized labor was, in some cases, cheaper than slavery (slaves have to be housed and fed). It is not hard to see how this factor helped push forward the movement for abolition.

Take, as another example, whale hunting. In 1986, whale hunting was banned in all but a few countries worldwide. Especially in the nineteenth century, whale oil, made from whale blubber, was popular as a fuel and as a food condiment. In 1849 Canadian physician and geologist Abraham Gesner developed kerosene. As a fuel, kerosene was cheaper, less dirty and longer lasting than whale oil. As a consequence of his invention, and of the later development of vegetable oils, the demand for whale oil declined and whaling was less and less profitable. The point here is that, had we not had these alternatives, abolishing whaling in 1986 would have been a lot more difficult, if not impossible. Gesner and other inventors obviously did not set out to abolish whaling, but their inventions paved the way for people to be touched by moral arguments in a later phase.

Moral revolutions are not made by moral arguments alone. Although moral arguments are an important and necessary part of vegan and animal rights advocacy, it is important to understand their limitations. Moral arguments in themselves also may create a lot of resistance. When they hear someone say that eating meat is wrong or bad, many meat eaters may feel that this implies a judgment or condemnation of their own (i.e., the listener's) choices. Indeed, the mere presence of a vegan, independent of their arguments, may be enough to evoke negative feelings. People want to think of themselves as morally good, and "because of this concern with retaining a moral identity, morally-motivated minorities may be particularly troubling to the mainstream, and trigger resentment" (Minson & Morin, 2012, p. 201). The problem is that the feeling of being judged often goes together with a defensive reaction: when our moral identity is threatened, we may try to put down the source of the threat (Minson & Morin, 2012, p. 201). This phenomenon has been called "do-gooder derogation": the putting down of people we consider doing something good. Research by Zane, Irwin, and Walker Reczek (2016) has shown that when meat eaters as much as *anticipate* moral reproach by vegetarians – that is, when meat eaters *think* that vegetarians would morally condemn them – they will tend to increase their derogation. What should concern animal advocates, however, is not in the first place the fact that they might be ridiculed or treated unfairly, but that the denigrators themselves will be less committed to changing their behavior in the future. Being derogated does not just offend the vegans, but prevents the meat eaters – out of some kind of self-protection – from taking steps towards veganism themselves (Zane et al., 2016).

But would we not, in the end, prefer that people do the right thing for the *right* reason? Fortunately, there is more than one way to help reach people develop the "right" attitudes. Usually, most people – and especially ethics-focused advocates, think of change as transitioning from attitude change to behavior change. We inform people with all kinds of statistics and stories, hope that they will discover that animals indeed matter (attitude change) and from there change their consumption (behavior change). Cognitive dissonance theory (Festinger, 1957), however, teaches us that in many cases, people will adapt their beliefs so that they are in line with their behavior. Meat eaters, who experience the dissonance between eating meat and actually caring about animals and realizing that they feel pain, may resolve the dissonance by denying animal pain or animal mind (Rothgerber, 2010; Rothgerber, this volume). Fortunately, we can also go about change the other way round: a change in behavior can lead to a change in attitude. In many cases, this might actually be an easier way to create change. This brings us back to how I started this article. Our incredible investment in and dependence on animals – we are steakholders, remember – makes it difficult for us to think about them through a different or new lens. In other words: where we stand depends on where we sit.

I offer several examples of behavior change preceding attitude change in my book *How to Create a Vegan World* (Leenaert, 2017), but let's briefly look at a few cases here. One famous example of behavior change preceding attitude change is what happened with seat belts. While many people opposed having to wear a seat belt initially, their attitudes around it seemed to have changed after having had to wear the belt for some time (Fhaner & Hane, 1979). One can imagine the same shift in attitudes regarding the laws on smoking in public spaces. Although the ban, where it was introduced, initially caused a lot of outrage, nowadays many of us cannot even imagine that smoking once was allowed in those very places where it was only recently banned, and may actually think the ban is a good idea, just like they can now appreciate the protection a seat belt gives them.

Back to our topic of interest: one paper examining the relationship between behavior and attitude carries the very interesting title *The role of meat consumption in the denial of moral status and mind to meat animals* (Loughnan, Haslam, & Bastian, 2010). In one of the studies, participants had to fill out a survey with questions on the moral status and mental states of cows. One half of the participants were given nuts to nibble on while they answered the questions, whereas the other half were given beef jerky. Judging from their answers, the second group attributed the cow with a lower moral status and a less developed mental life than the first group. The researchers concluded that the study "provides direct evidence that eating meat leads people to withdraw moral concern from both animals in general and the animals they ate." Their behavior (eating meat) influenced their attitude towards animals (see also Loughnan & Davies, this volume). If behavior change can precede and influence attitude change, we have to make it as easy as possible for people to change their behavior. And that is why we need an environment that facilitates that change.

The environment: Facilitating compassion

In his book *The Happiness Hypothesis*, psychologist Jonathan Haidt (2006) offers a useful model of behavior change:[10] a rider tries to direct an elephant through the forest. The rider stands for our rational, decision-making self. The much bigger and stronger elephant is the emotional part of us. Often, rider and elephant are in conflict: our inner rider decides we want to be healthier, while our inner elephant is unable to say no to one more piece of pie. While this conflict is exhausting, the efforts that both rider and elephant are required to make can be reduced when there is an easy path through the forest. This path is our environment. By adapting the environment, we can make it easier for people to change, even without them having the necessary motivation.

Crucial to creating a facilitating environment are commercial companies creating the alternatives for animal products. They can be supported (or sabotaged) by civil society (including vegan and animal protection organizations) and government. Companies creating vegetarian and vegan products, like tofu, sausages, and burgers, have been around for decades now.[11] In the past five years or so, however, we have seen the arrival of a new breed of company in this space. They are startups with a number of characteristics: they are well funded – often by venture capital; they are highly technological (many of them originating in Silicon Valley); and they are very ambitious, wanting to "disrupt" the conventional way of doing things.

Let us consider a couple of examples, all from California. The company Beyond Meat, backed by Bill Gates and Leonardo DiCaprio among others, produces the Beyond Burger that is now available all over the United States and already in many other countries as well (Going Global, 2018). Another company, Impossible Foods, was developed by world renowned former Stanford professor of chemistry Patrick Brown. His burger, so real that it actually bleeds, has received rave reviews and has become somewhat of a hype. By putting their products on the market, these companies obviously can make an enormous difference. The startup Just (previously called Hampton Creek), for instance, has managed to convince chain 7/11, as well as Compass Group, the biggest caterer in the world, to only use Just's egg-free mayonnaise all over the USA. That is a lot of chicken suffering avoided.

Potentially the biggest game changer could be the so-called *clean meat*: it is based on cultured cells and as such is real meat, but – after overcoming some remaining hurdles – it should involve no animal suffering whatsoever. Dutch researcher Mark Post, at the University of Maastricht, presented the first edible prototype at an internationally covered event in 2014. He has since formed his own company, called Mosa Meats, to bring the product to market in what he hopes will be just a couple of years. Back in California, the team of Memphis Meats, another start-up, has developed the first clean meatball and expects to be able to sell it or a similar product within five years.

Although for most of the general population, the potential impact and the value of these products will be clear, the same cannot be said for those within the animal rights movement. More generally, commercial endeavors cannot always count on the support of idealists. Here, however, is another chance to be pragmatic. Once more, advocates can either insist that these companies be in it for the right reasons, or can accept that the right thing is being done for less than ideal reasons.

The startups mentioned were founded by people who were at least partly motivated by the idea to make our food system more compassionate and sustainable. This probably does not apply – at least in most cases – where the meat industry itself is getting in on the plant-based wagon. Take the case of Tyson Foods, a big meat company which not so long ago bought a stake in Beyond Meat (Pellman-Rowland, 2017). As a vegan, it is easy for me to sympathize with the sentiment of many other vegans: that it is extremely frustrating to see that a company that has made billions out of the suffering and killing of chickens will now start to profit from plant-based foods. The same can be thought of McDonald's, which seems to be cautiously wetting its feet in the plant-based pool, having recently launched a real vegan burger in a few countries (Hosie, 2017). But we can obviously look at all this also through a pragmatic lens. First of all, companies like Tyson or McDonald's will probably not go bankrupt overnight because of lack of demand. Neither will they become a vegan company in the very near future. Obviously everyone is free to boycott or protest these companies, but realistically we can only expect them to change gradually, by developing and focusing on plant-based alternatives step by step. Many people, especially those with anti-capitalist ideas, will deplore the mere size of multinationals like them, but in these cases, size might be helpful. McDonald's, Tyson, and others investing into more plant-based options, have massive research and development departments through which they can make these products better. And they have huge advertising budgets, to bring them to a much bigger audience than smaller idealistic companies can. They can also use their huge network of contractors: suppliers, distributors, etc. Lastly, we may expect that as soon as a company is buttering its bread on two sides and is profiting from the growth of the plant-based pie, it will be less resistant to change, and less out to sabotage the growth of the vegan category or the animal rights movement.

The fact that today profit is the newest driver – after health, animal welfare, and sustainability – to develop and bring to market plant-based products may be deplored by the very idealistic, but is in fact a very good thing. The animal rights movement may have to allow the enemy to join the club and turn them into allies, rather than fighting them.

The commercial sector is crucial because it provides the alternatives. If there is nothing palatable to eat, we will not be very successful in making people shift their diets. But of course it is not the only sector that the vegan movement can support and work together with. An environment that is conducive to change consists of all kinds of institutions that offer ways for people to demonstrate the desired behavior (again, without necessary having the right attitude). I am talking here about all kinds of places where people eat, from schools to company cafeterias to restaurants, and the places where people shop, from small stores to supermarkets. In addition to that, there are governmental institutions that, given the right push by civil society, may help create additional incentives for businesses to develop and offer alternatives, and for customers or citizens to try them.

Choice editing is one more promising area to bring about behavior change. In public cafeterias, but also in restaurants, supermarkets, and other places where people buy food, we can make the desired choices easier to make and more damaging choices harder to make. We can make it easier, through the positioning of items on the menu, of products on shelves, of prices and discounts, for people to choose a plant-based option. An example of this is what EVA (Ethical Vegetarian Alternative), an organization I worked for, implemented in the city school system of Ghent, Belgium. Within the context of a campaign called *Thursday Veggieday*, the default dish in all city funded schools in the city is vegetarian or vegan. Students could still get a meat dish on that day, but they would have to go through some extra trouble. The result is that around 95% of the students eats vegetarian on Thursday (Traynor, 2009). It is a pragmatic, behavior-first option, that does not take away people's choices entirely.

The movement: Creating a bigger tent

The fourth and last way to be pragmatic is to be more inclusive and create a bigger tent. The vegan movement, almost by nature of the very concept of veganism, often takes a quite binary approach to things. For many vegans, either one is part of the solution, or one is not. Avoiding animal products is often seen as the single most important thing one can do to bring about change, and the one thing through which one declares allegiance to the cause. This obviously makes for a very small club. Only a small percentage of the population in most countries is (intentionally) vegetarian, and the number of vegans is smaller than the margin of error with which the polls are executed (Simcikas, 2018). If the abolition of the use of animals for food and other ends is to be achieved, this small group of people will need to expand and it will need allies.

The movement can choose to only include vegans and see the rest as the enemy, or can look at things on a spectrum rather than as divided in two different camps (binary). It can be inclusive rather than exclusive. Some polls seem to show that the amount of people that significantly agree with the ideas and direction of the animal rights movement is a lot higher than the number of people who demonstrate behavior change (i.e., being vegetarian or vegan). A recent survey conducted by researchers at Oklahoma State University, for instance, shows an incredible 47% of Americans to be in favor of a ban on slaughterhouses (Norwood, 2018). Results from a poll by the Sentience Institute tell us that 49% of US adults support a ban on factory farming, 47% support a ban on slaughterhouses, and 33% support a ban on animal farming (Reese, 2017). Obviously, these results need to be interpreted with caution. Nevertheless, they may very well indicate that the support for the goals of the animal movement might be higher than assumed, and further progress may largely be a matter of eliminating barriers. Many of these respondents may be the kind of people who know that something is wrong but for some reason cannot or will not stop doing it, but who might be entirely fine with government interventions, requiring *everyone* to alter their behavior. To idealists, this might seem weak, but it should in fact inspire them and give them hope.

Although being vegan is a very useful and praiseworthy thing to do or be (as I mentioned, I have been vegan myself for 20 years, at the moment of writing this), overfocusing on being vegan – and certainly on being perfectly vegan – may be misguided. The impact of one's own consumption on animal suffering and killing may actually be quite small compared to the impact one can have on other people's consumption. This impact can be achieved in different ways. One can be, for instance, a really good ambassador and communicator for animals, motivating a lot of people to decrease their consumption of animal products. Being vegan for this purpose is not a necessity. Arguably, it could, in some cases, actually constitute a hindrance, as it may increase the distance with the audience, and puts messenger and listeners in different categories. One can also achieve a big impact by spending time for all kinds of animal causes while not being vegan. One can have a restaurant or other business with great vegetarian and vegan options while not being vegan. One can make small or large donations to animal groups, with which these groups can fund actions and campaigns. It is clear that non-vegans – people who technically, in the eyes of many vegans – are still part of the problem, can contribute tremendously to the vegan cause, and presently probably have more of an impact than the vegans themselves, given that there are so many more of them.

Another way to make the tent bigger is to allow slightly more people to stick the term vegan on themselves (should they wish to do so). Also in thinking about its own ideology, rules and definitions, the vegan movement can use a lot more pragmatism. This is a message that is not always accepted with enthusiasm. The movement is quite invested in labels, and in making sure that people carrying a label can rightfully carry it. To suggest less rigidity seems to engender all kinds of fears, of watering down the ideology, of confusion, of inconsistency. Still, it is important to realize that there is no use in excluding a 99% vegan from the club of vegans. To insist on 100% purity and consistency is unnecessary, unproductive, impossible and insufficient.

Such insistence is *unnecessary* because when even a 95% vegan world is achieved, the final 5% will take care of themselves. At that point, the remaining animal ingredients or services will, if they have not become too expensive, be banned, as by then the public support for that will be huge. It is futile to focus on the final percentages, or the last few steps, *first*.

Second, goals of purity and 100% consistency can be *counter-productive*; although consistency on the whole is praiseworthy and non-vegans may sometimes admire a vegan's principledness and consistency, going too far in that direction may alienate people and be met with incredulity. To insist on avoiding any animal product in any circumstance, even in cases where consuming that product has no real-world impact on demand at all (think, for instance of "dumpster diving," where food is consumed that is going to be thrown away), is not just useless but can actually be damaging. Helping to create an image of veganism as something feasible and attractive is much more important than being pure to the final 0.01 percent.

Third, being 100% consistent is *impossible*. Animal ingredients are everywhere and cannot be avoided entirely. In this sense, it could be more truthful for vegans to say that they are 99% vegan. Avoiding the pretense of 100% consistency might help avoid the accusations of inconsistency and the gotchas that meat eaters like to throw at vegans. Finally, being 100% vegan is *insufficient* to reduce suffering. Although the damage one does with one's consumption will be greatly reduced, one will still cause suffering to both humans, animals, and the environment even eating only plant-based products. We should not let the perfect be the enemy of the good, as Voltaire said.

Conclusion

For the first time in the history of the known universe, one animal species is radically and intentionally starting to change its own diet. *Homo sapiens* is doing what no other mammal or other animal has done before: at least a small group of us is choosing to eat lower in the food chain, for ethical, environmental and health reasons. This is an incredible place to have arrived at, and it is something to be celebrated. There is, however, a very long way to go if we want to even make a dent in the suffering of farmed animals. I have argued that the movement that wants to help animals – and especially farmed animals – is highly idealistic in its approach and requires more pragmatism. We need a *Homo emphaticus*, but empathy and the right attitudes towards animals may follow behavior (and institutional) change rather than precede it, for the majority of the people. We already love some animals, like cats and dogs, and "charismatic megafauna" like tigers and elephants. Other animals we exploit, and it is exactly our exploitation of them that makes it harder for us to care about them. We may yet learn to love all animals once exploiting them has become redundant.

Whether one ideologically agrees with the project of the vegan/animal rights movement or not, to create a better situation for the farmed animals in this world, the movement needs to reach a lot more people. In this chapter, I suggested a pragmatic approach to do exactly that. Here are the implications of the above, in summarized form:

* Asking for reduction is an essential part of the road to abolition of animal use;
* all arguments that can help lead to change should be incorporated;
* the development of alternatives to animal products is crucial to make change easier, and the movement needs to support the commercial sector;
* and the movement needs to be more inclusive and count as its members not just people that are already vegan, but everyone looking in the same direction.

The animal rights movement is, fortunately, no longer the only player creating change in this space. Health and environmental organizations are telling the public that a high consumption of animal products is neither healthy nor sustainable (Health and Environmental Implications, n.d.) and the commercial sector is creating the products that will make meat, eggs and dairy redundant. The animal rights movement should work in tandem with other domains in society, which may have different motivations and put out different messages to the public, but which nevertheless will contribute, though often unintentionally, towards the same goal of reducing animal suffering and eliminating the use of animals for human purposes.

There is a lot we do not know about how best to change people and society in the domain of animal products consumption. Fortunately, more and more resources are going to research, and meta-charities like Animal Charity Evaluators (animalcharityevaluators.org; see also Bockman, this volume) and Faunalytics (faunalytics.org) are acquiring more expertise and bringing out more and more studies. Regarding the ideas outlined in this article, here are some suggestions for further research:

* How does the impact of moral and non-moral (e.g., food) advocacy compare?
* Is consistency or flexibility more attractive to outsiders?
* Regarding vegan products and dishes, which proportion is sold to vegans, which to reducers?
* Are people who are already eating plant-based diets for health reasons more open to animal rights arguments?
* Which are the best messages to advocate for reduction?

Notes

1 The focus in this article is on farmed animals (raised for food products mainly), but obviously animals are also used for clothing, research, and entertainment.
2 Possible starting points could be the publication of *Animal Liberation* by Peter Singer (1985), or – going further back – the foundation of the Vegan Society in the UK (1944).
3 Animal rights activists and vegans or vegan organizations may also use health and taste (as well as environmental arguments), which can be considered non-moral arguments.
4 Of course, in a way, *all* vegans, vegetarians, animal rights activists, are idealists. In this article, however, I uses the term idealistic to point to a certain *approach* or strategy, and distinguish that from a pragmatic approach (which is also used by people who are basically idealists in the sense that they do have an ideal and are working towards it).
5 See, for instance, Charles Duhigg (2012) on how change usually occurs in steps.
6 This is easy to see when looking at statistics of vegans versus meat reducers. While the former still hardly even show up on the radar, reaching at most 2 percent of the population in some countries (but usually less), people who reduce consumption of animal products may number 20 percent or more (depending on how they are defined). To see how the number of vegetarians and vegans has been quite stable over the last few decades, see Simcikas (2018).
7 To illustrate this with numbers: in the fluid milk category, plant-based milks now take a global marketshare of about 12 percent (Bailey 2018). It is obviously not the (at most) 1 percent vegans who are consuming all these dairy alternatives.
8 For an argument against using health and environmental arguments, see Reese (2017).
9 I'm talking about what is sometimes referred to as chattel slavery or *de jure* slavery, and am not denying the existence of modern-day *de facto* slavery.
10 This model was further popularized by Chip and Dan Heath in their book *Switch: How to Change Things when Change is Hard* (2011).
11 In Far East Asia, we can safely assume that the commercial selling of tofu and other plant-based staples has been around for many centuries.

References

Bailey, T. (2018) Dare not to dairy. What the rise of dairy-free means for dairy... and how the industry can respond (Rabobank). Retrieved from https://services.rabobank.com/publicationservice/download/publication/token/tRJL6VMr6xTTT5eFd6m0
Bockman, J. (this volume). Accomplishing the most good for animals. In K. Dhont and G. Hodson (Eds.), *Why We Love and Exploit Animals: Bridging Insights from Academia and Advocacy*. Abingdon: Routledge.
Duhigg, C. (2012). *The Power of Habit: Why We Do What We Do and How to Change*. New York: Random House.
Festinger, L. (1957). *A Theory of Cognitive Dissonance*. Stanford, CA: Stanford University Press.
Fhaner, G., & Hane, M. (1979). Seat belts: Opinion effects of law-induced use. *Journal of Applied Psychology*, 64, 205–212.
Going global: Beyond Meat set to launch product in 50 countries. (2018, May 11). Food Ingredient List. Retrieved from www.foodingredientsfirst.com/news/going-global-beyond-meat-set-to-launch-flagship-product-in-50-countries-worldwide.html.
Haidt, J. (2006). *The Happiness Hypothesis: Putting Ancient Wisdom to the Test of Modern Science*. New York: Basic Books.

Health and Environmental Implications of US Meat Consumption & Production (n.d). *Johns Hopkins Bloomberg School of Public Health*. Retrieved from www.jhsph.edu/research/centers-and-institutes/johns-hopkins-center-for-a-livable-future/projects/meatless_monday/resources/meat_consumption.html.

Heath, C. & Heath, D. (2011). *Switch: How to Change Things When Change Is Hard*. London: Random House.

Hosie, R. (2017, Dec. 28). McDonald's mcvegan burger launches today as excited customers say it's 'so good'. *Independent*. Retrieved from www.independent.co.uk/life-style/food-and-drink/mcdonalds-mcvegan-vegan-burger-launch-finland-sweden-vegetarian-excited-plantbased-soy-a8131766.html

Lampert, E. (2016, Jan. 28). If you're on a plant based diet, stop calling yourself vegan! *Ecorazzi*. Retrieved from www.ecorazzi.com/2016/01/28/if-you-are-on-a-plant-based-diet-stop-calling-yourself-vegan/.

Leenaert, T. (2017). *How to Create a Vegan World: A Pragmatic Approach*. New York: Lantern Books.

Loughnan, S., & Davies, T. (this volume). The meat paradox. In K. Dhont and G. Hodson (Eds.), *Why We Love and Exploit Animals: Bridging Insights from Academia and Advocacy*. Abingdon: Routledge.

Loughnan S., Haslam N., & Bastian B. (2010). "The role of meat consumption in the denial of moral status and mind to meat animals," *Appetite, 55,* 156–159.

McCarthy, N. (2017, Jan. 17). The number of Americans going gluten-free has tripled since 2009. *Forbes*. Retrieved from www.forbes.com/sites/niallmccarthy/2017/01/17/the-number-of-americans-going-gluten-free-has-tripled-since-2009-infographic/#1a24dc2f32f2.

Minson, J. A., & Monin, B. (2012). Do-gooder derogation: Disparaging morally motivated minorities to defuse anticipated reproach. *Social and Personality Psychological Science, 3,* 200–207. doi: 10.1177/1948550611415695

Norwood, B. (2018, Jan. 18). Food demand survey, Oklahoma State University, Department of Agricultural Economics, 5. http://agecon.okstate.edu/files/January%202018.pdf

Pellman-Rowland, M. (2017, Dec. 10). Tyson Foods injects more money into plant-based meat. *Forbes*. Retrieved from www.forbes.com/sites/michaelpellmanrowland/2017/12/10/tyson-foods-plant-based-meat/#4674d0c67efa

Reese, J. (2017). Survey of US Attitudes Towards Animal Farming and Animal-Free Food. www.sentienceinstitute.org/animal-farming-attitudes-survey-2017.

Rothgerber, H. (2010). "Efforts to overcome vegetarian-induced dissonance among meat eaters," *Appetite, 79,* 32–41.

Simcikas, S. (2018, Aug. 16). Is the percentage of vegetarians and vegans in the U.S. increasing? *Animal Charity Evaluators*. Retrieved from https://animalcharityevaluators.org/blog/is-the-percentage-of-vegetarians-and-vegans-in-the-u-s-increasing/

Traynor, I. (2009, May 14). Day of the lentil burghers: Ghent goes veggie to lose weight and save planet. *The Guardian*. Retrieved from www.theguardian.com/environment/2009/may/13/ghent-belgium-vegetarian-day

Zane D., Irwin, J., & Walker Reczek, R. (2016). Do less ethical consumers denigrate more ethical consumers? The effect of willful ignorance on judgments of others. *Journal of Consumer Psychology, 26,* 337–349.

Zaraska, M. (2016). *Meathooked: The History and Science of Our 2.5-million-year Obsession with Meat*. New York: Basic Books.

16

ANIMALS AS SOCIAL GROUPS

An intergroup relations analysis of human-animal conflicts

Verónica Sevillano and Susan T. Fiske

Abstract

Just as people seem to come in types (middle-class, rich, old, homeless), so do animals (pets, predators, prey, pests). The societal images of animal species reflect socially shared beliefs about different animals – *social stereotypes*. Similar to the case of human social groups, animal stereotypes could predict human emotions and behaviors toward different animals. Accordingly, in parallel to human-human intergroup responses, this chapter develops a new human-animal intergroup framework linking humans' beliefs (*stereotypes*), emotions (*prejudice*), and behaviors (*discrimination*) toward animal species, drawing on the Stereotype Content Model and Behaviors from Intergroup Affect and Stereotypes Map. Expanding the human-animal intergroup framework, the types of human intergroup conflicts and human-animal conflicts are compared. An intergroup relations approach can contribute to the study of human-animal relations by taking into account different kinds of animal species' stereotypes, as well as similarities between human-human and human-animal conflicts.

The *Colloquy of the Dogs* (1613/2013), Cervantes' most original work after *Don Quixote*, illustrates weaknesses and wiles of human beings through a dialogue between two talking dogs. Before discussing (negative) human characteristics, the dogs, Scipio and Berganza, introduce human social beliefs about certain animals' characteristics, including themselves. Scipio says, "What I have heard highly extolled is our strong memory, our gratitude, and great fidelity; so that it is usual to depict us as symbols of friendship [...]." Berganza replies, "[...] I know, likewise, that next to the elephant, the dog holds the first place in the way of appearing to possess understanding, then the horse, and last the ape" (p. 91).

What Cervantes shows here are the social stereotypes – socially shared beliefs held in seventeenth-century Spain – about dogs, elephants, horses, and apes. As we will see, these animals are similarly perceived in the seventeenth and twenty-first century. This chapter highlights the benefits of attending to stereotypes of animal species in the same way that social psychology attends to stereotypes of human social groups. Stereotypes generate consequential biases, prejudices, and discriminatory behavior toward human social groups and, as we argue, toward animal species. Later, we describe our approach, applying a human-to-human intergroup relations framework to human-animal relations, in the larger context of guiding integrated theory in this area. We will refer mainly to conservation biology, a multidisciplinary discipline dealing with the loss of biodiversity, due to its focus on human-animal conflicts.

Differentiating animals' images: All animals are not the same

This chapter aims to show the importance of images referring to animals not just as animals, but as exemplars of different species (in the media, and surveys). Several problematic implications follow from referring to animals generically, instead of referring to specific animal species. First, mentioning animals in general leads to methodological difficulties due to the very ambiguity/generality of the term. Some people may agree to using animals for experimentation if by "animals" we mean rats, but they may not agree to their use if the animal in question is a chimpanzee. This nuance is sufficiently evident. However, opinion surveys maintain the generic use of the term "animals" in issues of social relevance (genetic modification of organisms or species), which leads ostensibly to ambiguous data. Making a comparison with humans, it would be less informative to ask in a survey whether "people" are entitled to health care (universal in Spain). It would be more meaningful, as it in fact does occur, to ask if specific social groups are entitled to health care (e.g., "do immigrants have the right to health care?").

Second, the term "animals" introduces a perceptual bias. Using the category *animals* may induce thinking of companion animals, which are seen generally as positive (friendly and intelligent), instead of wildlife animals, which are seen ambivalently (unfriendly but intelligent; Sevillano & Fiske, 2016a). Alternatively, the generic *animals* may also bias cognitive associations toward pests, which elicit disgust, rather than companion animals that are more positively viewed (Sevillano & Fiske, 2016a). The term "animals" creates individual variability in estimating the psychological distance between human and non-human animals. Even without identifying the term with pets or dangerous animals, the use of the term "animal" is generally equivalent to mammals, excluding other groups of animals. Asking people to list names of animals leads them to thinking mostly of mammals, but not exemplars of insects, amphibians, or fishes (e.g., Henley, 1969).

A way to illustrate the ambiguity and bias effect of the generic term, animal, is by using the Eurobarometer about *Attitudes of Europeans toward Animal Welfare*. The European Commission consulted European citizens about their understanding of the term "animal welfare" (European Union, 2015). Animal welfare refers to the welfare of animals for human consumption; however, it means different things to different people. Generally, European citizens are divided between those (46%) who think that "animal welfare" refers to "the duty to respect all animals" and those (40%) who refer to animal welfare as "concern for the way farmed animals are treated, providing them with a better quality of life." When attending to differences between countries, some member states (especially Sweden and Denmark, 61%) consider animal welfare as limited to farm animals, whereas others (Netherlands, 70%; Luxemburg, 62%) generalize to all animals. This plurality of meanings complicates communication between different societal groups.

The ambiguity of the terms "animals" and "animal welfare" shows the functioning of different categories of animals in people's minds. That is, all animals are not equal, and some animals are perceived as more animal-like than others.

An intergroup relations approach to human-animal conflicts

An intergroup relations perspective provides a new approach to human-animal conflict. From psychology, the extant theoretical approaches to human-animal relations, used in the biological conservation area, have been mainly those from expectancy-value models (Vaske & Manfredo, 2012) and from the social perception of risk (Bruskotter & Wilson, 2014). These models do not consider specific differences between animal species as an intergroup approach does. Within expectancy-value models, the Theory of Planned Behavior (TPB; Azjen, 1991) has been the most common framework in conservation biology, due to the precision of measurement and the explanatory power of the attitude concept (Dressel, Sandström, & Ericsson, 2015). But it does not facilitate distinguishing different species.

Conceptually close to expectancy-value models is the Cognitive Hierarchy Theory (Vaske & Manfredo, 2012). This theory was specifically developed to account for the relations between human beings and one kind of animal, wildlife. The theory proposes that general value orientations regarding wildlife will affect attitudes and norms toward specific objects, and those attitudes and norms, in turn, influence behavior. The construct of *wildlife value orientations*, novel in this theory, refers to the application of general values to a particular object, in this case, to wildlife. Two *wildlife value orientations* are proposed: "mutualism" – mutual relations between other living beings and human beings – and "domination" or "utilitarianism" – relationships of superiority of humans over other living beings (Jacobs, Vaske, Teel, & Manfredo, 2013). But this approach neglects other categories of animals.

From risk perception analysis (Carter, Shawn, Riley, & Liu, 2012; Slagle, Zajac, Bruskotter, Wilson, & Prange, 2013), animal species are seen as a source of "risk" (e.g., tigers will attack pets) and "benefit" for human's wellbeing (e.g., tigers keep the forest healthy). This approach identifies the relevant factors associated with animal species, which increase the perception of risk among people, and determines the acceptability of that risk. For example, characteristics such as unpredictability of animals' behavior and uncontrollability of human-animal interactions increase humans' fear toward animal species (Johansson & Karlsson, 2011; Johansson, Karlsson, Pedersen, & Flykt, 2012). The role of emotion in risk analysis is also relevant for explaining species' acceptance (Bruskotter & Wilson, 2014). For example, positive affect toward wolves was related to perceived benefits and greater support for wolf recovery (Slagle, Bruskotter, & Wilson, 2012). But the theory does not create dimensions for the content of animal images.

These three theoretical approaches to human-animal conflicts (TPB, Cognitive Hierarchy Theory, and risk analysis), mostly from social psychology, vary in the weight given to different explanatory terms. From TPB, the importance is given to *beliefs*, *attitudes*, and *social norms*, in this case, associated with species. The cognitive hierarchy theory focuses on *value orientations*, in this case directed toward wildlife, whereas risk analysis emphasizes animal species as *hazards*, with *benefits* and *affect* associated. None of them focus on stereotype content about animals for predicting attitude, behaviors, or perceived risks.

In this chapter, we present an intergroup relations framework, a well-established area in social psychology, and then show it can organize most of the constructs found relevant using other approaches to human-animal relations. This intergroup framework includes the construct of social stereotype applied to different animal species, the emotional reactions and behaviors associated with the stereotypes, and the types of intergroup conflict applied to human-animal relations. Several models account for intergroup relations constructs. We draw on the Stereotype Content Model (SCM: Fiske, Cuddy, Glick, & Xu, 2002) and the Behaviors from Intergroup Affect and Stereotypes Map (BIAS Map; Cuddy, Fiske, & Glick, 2007).

Animal species as social groups

In the case of animals, species may function as social groups (Hodson, MacInnis, & Costello, 2014; Sevillano & Fiske, 2016a). This statement is more challenging for social psychology than for conservation biology because it involves an assimilation between the construct of psychological social groups (a main object of study in social psychology) and animal species. Considering animal species as social groups implies attributing to animal species the distinctive qualities of human social groups: entitativity and essence. For human social groups, perceived *entitativity* refers to the apprehension of people in groups as forming a new organization of bonded people (i.e., an entity, instead of a loose collection). Thus, people perceive social groups (e.g., immigrants, Jews, rich people) as unitary and coherent, as a social entity, and not as an aggregate of individuals (Fiske, 2010). This quality implies that group members are seen as very similar, oriented to a common fate, and proximate to each other. Applied to animals, species members are also seen as highly similar, with the same goals, and close to each other.

Being a social entity makes people believe in its *essence*, a core something, a quality (often referred as "blood" or genetics) that individual members have, which makes them into group members. Similarly, people would perceive animal species as entities emphasizing the biological essence (e.g., genetics) underlying group membership (Medin & Atran, 2004). Emphasizing an animal species as an entity – a reified thing – imputes a robust biological essence to members of that species, cementing them together as a clear-cut category. The notion of animal species as real social groups leads humans to hold a variety of systematic stereotypes, socially held judgments about a social group; prejudices, affective responses to a group's members; behaviors, discrimination; and perceived threats associated with animals.

Social stereotypes → emotions → behaviors

In the popular imagination, wolves and bears *are* intelligent, strong, and aggressive; dogs and horses *are* well intentioned and skillful; cows and rabbits *are* unaggressive but unintelligent; and rats and chickens *are* neither friendly nor intelligent. Leaving experts apart, lay people may see these assertions regarding different species as quite evident. What it is not so obvious is the stereotypical content used when describing those animal species. Whereas the recognition of differences between how we perceive and use distinct animal species has been the focus of several well-known publications (Arluke & Sanders, 1996; Herzog, 2010; Plous, 2003), invoking stereotype constructs for describing animal species is uncommon.

In the case of human beings, *stereotypes* are key concepts for understanding intergroup relations. Defined as shared beliefs about social groups or categories, stereotypes portray the social image of groups of people (e.g., "women," "black people," "immigrants"). For example, the stereotype of immigrant people portrays them as untrustworthy and unskilled (Fiske, et al., 2002). Stereotypes operate as shortcuts for thinking about people, and they entail consequential biases, prejudices, and discriminatory behavior toward social groups (for a review see Yzerbyt & Demoulin, 2010).

Similarly, we propose that people hold stereotypes about different animal species. For example, the stereotype of wolves describes them as cruel and cunning (Skogen, 2001). Therefore, stereotypes, whether about immigrants or wolves, will affect people's emotions (prejudices), attitudes, and behaviors (discrimination).

Animals' stereotype content parallels humans' stereotype content

Whereas beliefs about human groups might seem very diverse, research has shown that some common characteristics underlie the content of social stereotypes. These dimensions are basic for social perception of groups or persons and identified by the stereotype content model (SCM; Fiske et al., 2002): first, warmth (What is the intention − good/bad − of another person/group?), and second, competence (What resources − abilities, power − does the person/group have to achieve their goal?). These two dimensions, Warmth (Sociality/Morality) and Competence (Capability/Assertiveness) define stereotypical content (Fiske et al., 2002; Wojciszke, Bazinska, & Jaworski, 1998). Warmth entails perceived intentionality, positive or negative; and Competence refers to perceived ability, high or low. When people think about social groups, they focus on how friendly or unfriendly their intentions are − Warmth − and how capable or incapable they are to achieve their goals − Competence.

Turning to animals, warmth and competence dimensions also seem relevant. Different animal species' behavior toward humans differs (they adopt, ignore, attack, or avoid us), reflecting diverse intentions toward humans. Animal species also present diverse capacities to enact their intentions (e.g., intelligence, strength, natural weapons). Those characteristics frequently associated with animal species, among other factors, may serve as the content base for the formation of animal stereotypes by humans, in terms of positive or negative intention and species' abilities. Let us take an example. The social image of dogs relates to being tamed/harmless/friendly but also an intelligent animal. That is, like most other *pets*, dogs are animals with a positive inclination toward humans and high skills. Not so well intentioned but similarly capable are bears, allegedly competent but not warm *predators*. Other animals are *prey*, such as sheep, who are seen as inoffensive toward humans but not especially skillful. Animals with the worst social reputation are *pests*: insects or lizards, perceived as unfriendly and clearly not gifted. Terms such as tame, inoffensive, well intentioned, unfriendly or harmful refer to the warmth dimension. Separately, attributes such intelligence or skillfulness refer to the competence dimension.

In humans, the two dimensions jointly map social groups' relative positions in a warmth-by-competence space. The best position in society is for reference groups, such as the ingroup and allies (e.g., the middle class), which are perceived as high in both warmth and competence. The worst position is for groups with no positive function in society, such as homeless people, who are perceived as low in both warmth and competence. Other groups elicit mixed positive and negative beliefs, which define an ambivalent stereotype. Groups high in warmth but low in competence – non-threatening goals but without any capacity – include older people. Those low in warmth but high in competence are groups such as rich people – groups with threatening goals plus the capacity to attain them.

Animals also have intentions and capabilities, so perhaps they also array on the warmth x competence map. Intergroup contexts (e.g., culture) are relevant in determining collective beliefs (stereotypes) about nonhuman species. Sevillano and Fiske (2016b), drawing on the SCM, argue for at least four stereotypes that capture the beliefs associated with different animal species: *Subordination* or *prey* (e.g., cows), harmless animals with common abilities; *threat-awe* or *predators* (e.g., lions), harmful animals with uncommon abilities; *contemptible* or *pests* (e.g., rats), harmful animals with common abilities; and *protective targets* or *pets* (e.g., dogs), harmless animals with uncommon abilities. To support these proposed shared stereotypes necessitated finding groups of animals all similarly perceived. We found precisely that. Through statistical analysis of data collected in North American samples, 25 animal species widely known by the general population were grouped in four clusters, in accord with the four proposed stereotypes (see Figure 16.1).

When individuals judge the degree of good intentionality (Warmth) and intelligence (Competence) of several animal species, they tend to see some animals as more similar than others. Animals such as tigers, bears, whales, leopards, and lions all were seen as highly intelligent but also as bad-intentioned animals. We generically label this group of animals as "predators" indicating their high skillfulness but

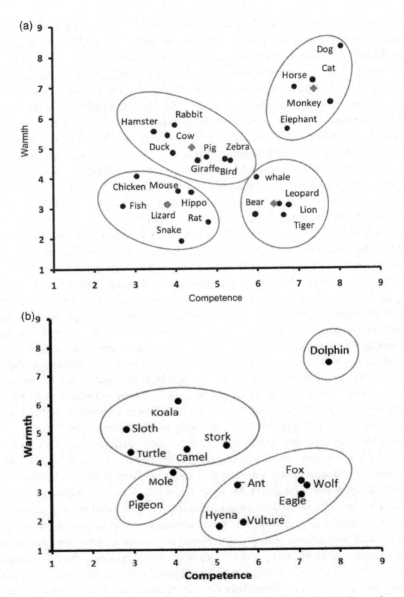

FIGURE 16.1 a) Four-cluster solution for 25 animals (US sample); b) Four-cluster solution for 14 animals (Spanish sample). *Note.* Warmth (*warm, well intentioned, friendly*), and competence (*competent, skillful, intelligent*) dimensions are rated using 9-point Likert scales, 1 (*not at all*) to 9 (*extremely*), on how the animals are viewed by society.

Source: a) Warmth and competence in animals. Sevillano, V., & Fiske, S. T. *Journal of Applied Social Psychology*, 46. Copyright © 2016 Wiley-Blackwell; and b) Sevillano & Fiske, unpublished data (*N* = 29, preliminary data).

low friendliness. Animals such as dogs, monkeys, elephants, horses, and cats were seen as high-intelligent but also as good-intentional animals. We generically label this group as "pets," "companions," or "protected targets." A large group of animals comprises ducks, cows, rabbits, hamsters, zebras, giraffes, birds, and pigs. Those animals obtained ratings moderate in warmth but low in capacity. We refer to them as "prey" or "subordination targets"; they are farm and some exotic animals. Finally, we use "pests" for the cluster of animals comprising lizards, rats, chickens, snakes, mice, hippopotamuses, and fishes. Those contemptible targets were seen as low in warmth and competence. Apparently, different animal species are perceived differentially in terms of warmth (degree of positive intent) and competence (degree of skill or intelligence) (Sevillano & Fiske, 2016b).

The SCM proposes that the different stereotypes of social groups imply different emotions toward those groups (Fiske at al., 2002). Social groups elicit different emotions based on comparison processes (upward and downward in status, competence) and outcome attributions (intent, control). For example, if someone with a disability seems incompetent and comparatively worse off, but has good intent (not to compete, exploit, or burden; that is, sharing resources, not taking them), people feel pity. But if the unfortunate person has selfish intent (is demanding), people feel contempt. Turning to people with good outcomes, the other half of the SCM elicits either pride (upward comparison, good intent) or envy (upward comparison, but bad intent). When our own tribe member succeeds, we feel pride, but when a rival does, we feel envy.

Stereotypes also imply types of discriminatory behavior (BIAS Map, Cuddy, et al., 2007). People's behavior toward others may be beneficial or harmful, that is good or bad for them. Protecting and associating generally benefit others; attacking and neglecting generally harm them. Positive behavior follows positive emotions (pride, pity) and high-warmth stereotypes. Negative behavior follows negative emotions (disgust, envy) and low-warmth stereotypes. As the behavior examples imply, regardless of benefit or harm, behavior can also be more intense, direct, and active (protecting or attacking require effort); behavior can alternatively be more indirect and passive (merely associating or neglecting require less effort).

Notably, behaviors directed toward different kinds of targets cluster by jointly considering their warmth and competence. In other words, high-warmth/high-competence targets will receive active facilitation *and* passive facilitation; we both protect and associate with the middle class, but also our companion pets. High-warmth/low-competence targets will receive active facilitation *but* passive harm; we protect but ignore old people and farm animals. Low-warmth/high-competence targets will receive active harm *and* passive facilitation; we associate with rich people and predators but also attack them when we can. Finally, low-warmth/low-competence targets will receive active *and* passive harm; we both avoid and sometimes attack homeless people and pests such as rats.

The empirical adaptation of the BIAS Map to animals (Sevillano & Fiske, 2016b; see Figure 16.2) indeed links the hypothesized emotions and behaviors. Typical

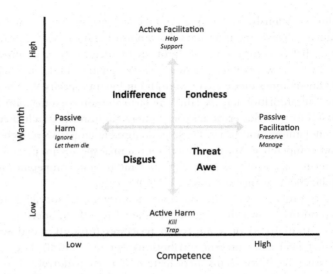

FIGURE 16.2 Four forms of prejudice and behavioral tendencies as a function of perceived warmth and competence.

Source: Updated from Sevillano & Fiske (2016b).

emotions elicited by high-warmth/low-competence animals (prey) were indifference and neutrality, whereas low-warmth/high-competence animals (predators) elicited awe and threat. These two clusters show stereotype ambivalence (mixed valence). In contrast, people show clear positive and negative feelings respectively toward high-warmth/high-competence animals (pets) which elicited fondness, and low-warmth/low-competence animals (pests) which elicited disgust. These two clusters leave no room for ambiguity.

As for the case of human groups, we proposed that the two stereotype dimensions predict different aspects of behavior. The warmth dimension – animals' intentions – will determine active behaviors, facilitative or harmful. So friendly animals (high warmth; e.g., dogs, cows) will elicit protection (active facilitation). Unfriendly animals (low warmth; e.g., lions, rats) will elicit hunting and killing (active harm). Regarding Competence, animals' ability, will determine passive behaviors that vary in valence, facilitative or harmful. So skillful animals (high competence; e.g., lions, dogs) will involve preservation (passive facilitation). Unskillful targets (low competence; e.g., rats, cows) will involve ignoring (passive harm).

Empirically, animals, based on their warmth and competence clusters, did elicit distinctive prejudices and behaviors. Such a chain of beliefs-prejudices-behaviors legitimates human practices regarding animals: *unfriendly* animals make us feel threatened and disgusted, so extreme practices such extermination or hunting are understandable. Alternatively, changing beliefs should affect prejudices and behaviors: increasing the perceived abilities of *incompetent* animals may lead us to feel admiration and awe, so positive practices such protection become relevant.

Some limitations appear in our research so far. Contrary to expectations, participants in our studies did not perceive the category of friendly but unintelligent animals as very friendly. Cows, rabbits, ducks, giraffes were seem as moderate in Warmth, whereas we predicted them to be high in Warmth. Notably, this category included very diverse animals, which may explain the moderate perceived Warmth. Some of them were typical farm animals, such as cows and rabbits, who were warmer than other, more exotic-wild animals, such as giraffes, and zebras. Some indeterminate others, such as hamsters, were warmer than birds (see Figure 16.1). These differences in warmth ratings may be indicating the influence of animal's tame-wild dimension. Future studies applying the SCM will test the influence of such a dimension and the stability of this category.

Currently, evidence regarding SCM and BIAS Map is correlational; that is, we asked people to judge real animal species on warmth and competence scales, emotions, and behaviors. We did not probe experimentally whether the Warmth and Competence dimensions per se elicit different emotions and behavioral tendencies.

Representations → emotions → behaviors

So far, we have described in detail the SCM and BIAS Map and their focus on animal stereotypes, emotional responses to them, and behavior toward them. Nonetheless, other models have addressed the links between how we perceive social targets and how we feel and behave toward them. The Value x Threat Model of Humanness Representations (Hodson et al., 2014) proposes different emotions and behaviors associated with a broad array of targets, depending on perceived value and threat. The model deals with nonhuman animals, humans, machines, and deities (gods, demons) because of the focus on perceived humanness. The perceived *value* dimension separates both humans (dictators, in-group) and deities from both nonhuman animals (predators, vermin) and machines; relative values are based on importance, which is close in meaning to targets' social position and intellectual abilities. The *threat* dimension presents targets according to the degree of perceived harm they cause, leading to a division between harmful human targets such as terrorists and sociopaths and animals such as primates and pets. The threat dimension is similar to perceived intent.

The two models have several points in common. The meaning of the main dimensions underlying targets' representations overlaps somewhat: Value is similar to competence plus social status; and threat is similar to warmth. In addition, stereotypes, the main construct for SCM, and representations are similar constructs, but at different levels. Whereas stereotypes apply to human groups and animals, as we argue, representation is a broad concept dealing with general social perception of targets, including animals.

Despite these similarities, the models also differ. The adapted SCM and BIAS Map is limited to animal species, whereas the Value x Threat Model of Humanness representation applies to a broad array of targets. This focus leads to some differences in the images of animals. For example, Value x Threat Model of Humanness considers predators as low in value and moderate in harm. The rationale behind this relies on two themes: the central factor assigned to human life (value) and the target's ability to affect the ingroup (relevance). In this sense, predators are low in value because they are not human and they do not really affect ingroup outcomes (people do not expect a personally directed attack). The framework of the model compares predators to humans and deities, which comparatively degrades predators in value and relevance. However, if the framework is limited to animal species, predators' value and relevance is high (e.g., beauty, symbolism). Correspondingly, in the SCM predators are low in warmth and high in competence. Regarding emotions and behaviors, there are also differences. The Values x Threat Model of Humanness emphasizes feeling disgust toward predators and exterminating or avoiding them, whereas SCM accentuates the feelings of threat/ awe and behaviors such extermination and preservation. As a result, SCM considers predators as targets to avoid and persecute, but also to revere and protect.

Structural variables predicting stereotype content

Where do stereotypes come from? The SCM proposes two structural variables that determine the content of stereotypes: the status of human social groups in society (high or low status) determines their perceived competence, and their perceived competition (or lack of cooperation) determines their perceived warmth. For example, those groups identified as elites or upper class are perceived as intelligent or skillful. Competitive groups are perceived as unfriendly.

In the case of animals, the different positions (status differences) that animals hold and their degree of conflict with humans (cooperation differences) relate to the contexts in which animals are present within the human social structure (e.g., food industry, therapeutic settings, protected areas, household settings, entertainment). With the aim to test the relevance of status and cooperation for animals, in a preliminary study we asked participants ($N = 198$ Spanish college students) to rate 15 animal species in perceived warmth, competence, cooperation, and status, among other variables (Sevillano & Fiske, 2017). Perceived warmth of animals was positively related to the perceived degree of cooperation of these animal species (question wording: "How cooperative toward human beings are several animal species?"). In other words, those animals cooperating with human society, helping society, are benevolently perceived. Beliefs about their competence were positively related to the degree of perceived status (question wording: "To what extent is the status of several animal species in society low or high?"). Those animals perceived as in higher social positions are perceived as competent. However, we also found a small unpredicted positive association between warmth and status, meaning that friendly and trustworthy animals are also elevated to the human sphere (high status), independently of competence.

Why do we love and exploit animals?

With regard to the aim of this book, why we love and exploit animals, our answer concerns individuals' use of stereotypical images about animal species. Granted, the differences between dogs, cats, monkeys, elephants, and horses are huge, given that they are different species. However, the social image of these animals in terms of stereotype content is similar: They are mostly friendly and skillful, so we generally love and care for them. Diametrically opposite, we hold stereotypes about animals such as rats, lizards, mice, hippopotami, fishes, chickens, and snakes – all having a negative image of unfriendly and unskillful animals. We demean and harm them. Our argument is that stereotype content partially explains different emotions and behaviors elicited by different animal species (also see Hodson et al., 2014).

Previous well-known publications (Arluke & Sanders, 1996; Herzog, 2010) have stressed the challenge of studying human–animal relations due to their extraordinary variety. This variety is described referring to the categories of animals operating at a societal level: farm animals, companion animals, etc. Instead of attending to general categories of animals, we descend to the level of animal species and explore associated stereotype content, prejudices, and behaviors. Our analysis borrows from parallels to the study of stereotypes and intergroup relations for human social groups; we hope it is potentially useful.

An example of application: Implications for conservation biology

Animal stereotypes

Conservation biology could benefit by including the construct of social stereotype in its research programs. Research usually focuses on shared beliefs about animal species, but the term stereotype is not used referring to those beliefs. In the case of wolves, their *social image* – *shared belief* or associated *characterization* – depicts animals that are cruel, recreational killers, intelligent, aggressive, dangerous, and bold (Johansson et al., 2012; Skogen, 2001; Skogen et al., 2008). For example, Bath (2000) found that around 20% of residents in Des Alpes Maritimes and Savoie believe that "wolves kill sheep out of cruelty, not for nourishment." From our perspective, these beliefs make up the stereotype of *threatening-awe predators*: seeing wolves as unfriendly but skillful animals. Despite the fact that wolves cause less damage to property and attack fewer people than bears, people express greater fear and anger toward wolves (in North America, see Dressel et al., 2015). Generally, people perceive large carnivores as a significant threat even when their attacks are uncommon (Treves & Palmquist, 2007). The ambivalent predator stereotype could explain these findings.

Conservation biology can use this perspective on social stereotypes toward the aim of improving social stereotypes of animal species, which may benefit conservation practices. Social stereotypes operate as heuristics to evaluate exemplars of a category, implying perceptual biases. This construct and its biases warn us, for example, about the labels used by ecology to classify different types of urban wildlife and its effect on how people perceive those types of wildlife and the stereotypes they seem to confirm. One of the most widely used taxonomies of urban ecology (Soulsbury & White, 2015) distinguishes between "exploiters," "adapters," or "avoiders" (McKiney, 2006). Another more recent one differentiates between "avoiders," "utilizers," and "dwellers" (Fischer, Schneider, Ahlers, & Miller, 2015). The very labels used to refer to the species (in this case, quite negative), and the terms used to describe different animal characteristics (e.g., diet, sociality, personality), will affect the social perception of the animal species because those terms carry a specific image. As an example, urban pigeons are classified into the "exploiters" category. This category reflects animal response to urbanization: "Exploiter" depends on urban resources for survival. Socially, this category associates with a competitive intent regarding food. The use of these labels may promote negative stereotypes about various species.

Social stereotypes perpetuate the negative image of a social group in society. Similarly, in risk perception, the *stigmatization* of hazards is a mechanism of risk amplification (Kasperson et al., 1988); stigma increases or attenuates the consequences of risks through communicative processes. That is, a hazard is associated with a negative image (Breakwell, 2007) as in the expression "Franken-food" for genetically modified food (see Hodson, Dhont, & Earle, this volume, for a discussion of the Frankenstein metaphor and its cultural interpretation).

The content of the animal stereotypes in conflicting contexts may relate to how lay people understand the ecology of predators, which could be called the *folk ecology* of predators. The scientific ecology of predators is explained in terms of loss or fragmentation of their habitats, predators' density, or changes in their preys' grazing practices. However, lay people may explain the behavior of predators living in city parks (e.g., foxes) as the result of predators' inherently bad behavior (stereotype of a malicious animal), instead of caused by changes in its habitat. Likewise, lay people may take the behavior of a conflictive exemplar as an example of the behavior of the whole category because both belong to the same species. Sillero-Zubiri, Sukumar, and Treves (2007) cited the case of cattle producers in Ecuador who killed nine spectacled bears because of one cattle-killing bear. Those cattle producers may perceive every bear as a killing bear because of their stereotypic perception of bears. We believe that these issues can be best understood by applying the construct of social stereotyping (and its implications) for the different animal species, so we propose its use by conservation biology.

Types of behaviors and emotions

The behavioral tendencies toward animals proposed by psychology (adapting the BIAS map to animals) are theoretically close to conservation biology's proposals, even though they are underdeveloped. For example, the human actions to mitigate human–animal conflict (i.e., with large carnivores) can be *direct* or *indirect* actions (Treves, Wallace, Naughton-Treves, & Morales, 2006).

Direct measures prevent some of the animal's actions, and they vary in the *degree of damage* inflicted on the animal. They range from preventing its passing through or foraging in an area (barriers, repellents) to physical manipulation of the particular animal (sterilization, lethal control). Indirect measures refer to actions that "allow" the action of the animal (1) in return for compensation to the people affected (incentives, compensations); or (2) by increasing the promotion of coexistence (education campaigns, participatory management). According to the BIAS map, those animal groups low in warmth and high in competence (e.g., "large carnivores" refers to bears, big cats, or wolves) will sometimes receive active harm (direct measures: be killed) and sometimes passive facilitation (indirect measures: be conserved), which is consistent with the proposal of Treves et al. (2006).

Similarly, a conceptual distinction among types of behavior regarding wildlife uses the active/passive, and positive/negative characterization: *intolerance* (negative-active), *tolerance* (no action), and *stewardship* (positive-active) (Bruskotter & Fulton, 2012). A broader perspective of wildlife management specifies types of animal species (mosquitoes/rats; bears/jaguars; deer/whales; and the rest of species), the type of impact on humans (negative or positive), and the type of management action (lethal control, protection, no management) (Marchini, 2014).

The characteristics of positive/negative and active/passive behaviors, or degree of damage and direct/indirect measures, from various conservation models (Bruskotter & Fulton, 2012; Marchini, 2014; Treves et al., 2006), resemble the behavioral tendencies advanced by psychological models previously reviewed: the BIAS Map and Value x Threat Model (Hodson et al., 2014; Sevillano & Fiske, 2016b). Behavioral dimensions of valence (help, harm) and intensity (active, passive) proposed by the BIAS Map and types of behaviors (avoid, exterminate, ignore, exploit, respect) proposed by the Value x Threat model are highly similar. The convergence of behavioral conceptualizations coming from different disciplines, psychology and conservation biology, encourages communication among them and generates new research questions.

For example, in the case of urban foxes, which carry a zoonotic infection to humans, citizens in a suburb of Munich reported how they expected people, from a list of potential behaviors, to treat foxes (König, 2008). Applying the behavioral tendencies identified by BIAS map, the list of behaviors may organize tentatively in terms of *active facilitation*, "people must help sick animals," "People should [de-] worm the foxes"; *active harm*, "the fox population in the community should be reduced dramatically"; and *passive facilitation*, "people should not attract foxes with food," "people should adapt," and "the fox population can increase." Relevant variables such as value orientations, social norms, membership in conservation groups, emotions toward animals, may predict the resulting typology of behavioral tendencies varying in valence and intensity.

Regarding emotions, SCM and BIAS Map deal with a variety of emotional reactions depending on the stereotype content. From previously used theoretical models (TPB, Cognitive Hierarchy Theory, risk analysis), only risk analysis considers affect centering on fear toward animals. This parallels SCM, which considers threat (an antecedent of fear) limited to animals perceived high in competence and low in warmth, mostly predators.

Types of conflict

This section describes, from a perspective of intergroup relations in social psychology, the most frequent human-animal conflicts analyzed in conservation biology. The relevance of applying an intergroup approach is based on two parallels. First, conservation biology explicitly conceptualizes human-animal conflicts as conflicts due to competition for resources; and social psychology conceptualizes human-human conflict in the same way. Second is the relevance of psychological and cultural variables in the understanding of these conflicts. Issues such as competition for resources, as well as considering the benefit that human beings should obtain, compared to the benefit that other living organisms should obtain, are a subject of debate in conservation biology (Miller, Soule, & Terborgh, 2014). Are natural resources (including animals) primarily for human beings' benefit? This debate is revealed in the different trends in conservation strategies (*conservation framings*): from a more utilitarian tendency toward the benefit for humans ("nature for people"), to a trend more focused on the intrinsic value of nature ("nature for itself") (Mace, 2014).

On the other hand, psychological and cultural factors also affect the perceived conflict. The management of animal species in conflict areas is frequently referred to as the management of human beings (Sillero-Zubiri et al., 2007) because individual human variables such as emotions, values, norms, attitudes, risk perception, and cultural values come into play (Manfredo & Dayer, 2004; Sjölander-Lindqvist, Johansson, & Sandström, 2015). As one example, social factors such as social norms, perceived benefits of wildlife (tourism, healthy environment, animals' cultural value), and community identity are essential for accepting or rejecting the presence of large carnivores (for a review, see Treves & Bruskotter, 2014). Therefore, this area of conservation biology is frequently called the human dimensions of wildlife management.

Given the similar base of conflicts for human-animals and human-human groups, we now review the different types of conflicts by applying the traditional categorization of conflicts in intergroup relations. Conflict among human groups is categorized into realistic and symbolic conflict (Stephan & Stephan, 2000; Yzerbyt & Demoulin, 2010). *Realistic conflicts* threaten the group's available resources or their well-being, their physical integrity. As we shall see, both realistic threats (resources and well-being) are present in the biological literature on human-animal conflicts. *Symbolic conflicts* threaten the worldview, values, and beliefs of the group to which individuals belong (e.g., North Americans value human autonomy). Human-animal conflicts may be a symbolic threat to human groups' identity (e.g., cultural practices of communities), and to the perceived higher status of human beings. In human-human intergroup relations, realistic and symbolic threats increase out-group derogation (Riek, Mania, & Gaertner, 2006).

Although conservation biology attends to both types of conflict, references are scattered unsystematically in the literature. Here we advance an integrative view of distinct conflicts using intergroup theory. In the next sections, we will use the term "realistic conflict" referring to threatening resources and human life. We will use the term "symbolic conflict" referring to threatening lifestyle, social values, and group identity.

Realistic conflict: Competition for resources

Protected areas and conflict areas reveal an essential dilemma between concerns about biodiversity conservation and ensuring human lifestyle (Treves, 2008). The literature defines human-animal conflicts as conflicts over resources, with charismatic fauna (i.e., large carnivores and herbivores) being the more studied species (European Commission, 2009; Marchini, 2014; Naughton-Treves, Grossberg, & Treves, 2003). The contested resources broadly include private property (livestock, plantations), human leisure possibilities, and human security – threat to life and health (wildlife as carriers of diseases) (Sillero-Zubiri et al., 2007; Treves, 2008). Our discussion of realistic conflict separates the threat to resources (private property, prey) from the threat to human life (physical threat); later, we will cover the threat to leisure possibilities in the symbolic threat section, because it is better framed as limiting lifestyle.

In line with competition for resources, farmers and hunters – who are negatively affected by the presence of wolves and bears (loss of cattle or prey) – show more negative attitudes toward these animals than does the general public (Dressel et al., 2015). Animals themselves are considered a resource; and their (illegal) hunting benefits people living in poverty (Duffy, John, Büscher, & Brockington, 2016). That is, human-animal conflicts are mainly defined as a conflict caused by a *realistic threat to economic resources*, in intergroup terms.

Realistic conflict: Physical threat

Realistic threat also has to do with the threat to human life or health. Fear of animal attacks increases support for lethal treatments, for limiting interaction with animals, and for supporting conservation measures (Sjölander-Lindqvist et al., 2015). The presence of wolves in an area is negatively related to agreeing to pay for the conservation of other related species, like the wolverine, toward whom the fear is extended (Ericsson, Kindberg, & Bostedt, 2007). Time living in proximity also plays against the large carnivores. In the case of wolves, but not of bears, attitudes are more negative, the longer people have lived in close proximity with them (Dressel et al., 2015). To explain this result, observe the negative effect of media coverage of wolf attacks (Fernández-Gil et al., 2016), emphasizing the aggressive stereotype of the species.

The perception of danger or physical threat is greater in affected communities or in communities that coexist with large carnivores (Johansson & Karlsson, 2011), and attitudes are more negative for those who live near protected areas or areas where large carnivores live (Heberlein, 2012). From the perspective of risk analysis, in these areas, the likelihood of damage is greater, and the control of threat is lower. From the SCM, threat is primarily elicited by high competence-low warmth species.

The secondary impact of realistic conflict may also amplify the symbolic threat. The experience of a physical threat (e.g., death of a family member) or a resource threat (e.g., loss of crops, loss of livestock) has effects, as a secondary impact, on family dynamics (e.g., assumption of responsibilities by other members), on the available food, or on the family's future economic capacity. This hidden impact of attacks or lost resources is not foreseen by the authorities and may increase not only the perceived realistic threat in the affected communities (Barua, Bhagwat, & Jadhav, 2013) but also the symbolic threat due to the threat to lifestyle.

Symbolic conflict: Lifestyle, values, and culture

Conservation biology recognizes that human–animal conflicts are not just competition for space, food, or life, but also involve different humans having different values associated with nature (Treves, 2008): utilitarianism versus conservation, domination versus subjugation, and values associated with lifestyles (rural vs. urban). Some refer to the conflicts of values as the true essence of the human–wildlife conflicts (Marchini, 2014). Apart from social values, another source of symbolic conflict is the threat to group identity. For example, the ban of traditional practices detrimental for animals (e.g., fox hunting) may be perceived as a symbolic threat to group identity. In intergroup relations, symbolic threats (e.g., perceived imposition of outgroup's cultural practices) are more relevant than realistic threats (Fiske, 2010).

Research on values has shown that people with a domination value orientation toward living beings – humans' superiority to other species (Jacobs et al., 2013) – tend to hunt and favor management practices that benefit humans. Relatedly, in the social arena, differences in human supremacy beliefs are linked to moral exclusion of animals, meat consumption, and animal exploitation (see also Dhont, Hodson, Leite, & Salmen, this volume). People believing in human superiority over animals show less moral obligation to have concern regarding all types of animals, high frequency of meat consumption, and more acceptance of animal exploitation (Dhont & Hodson, 2014; Leite, Dhont, & Hodson, 2019). In human–animal conflicts, people who hold values of domination agree with more severe responses to wildlife (e.g., trapping; Manfredo & Dayer, 2004). On the other hand, people with a mutualistic orientation – reciprocal relations between human beings and other animals (Jacobs et al., 2013) – perform more activities related to the observation of species and disagree with management practices that are detrimental to animals or that only benefit humans. Socially, materialistic cultural values – those related to physical and economic security – are associated with a more utilitarian view of the wildlife, whereas post-materialist values – values of self-expression and quality of life – are associated with the protection of nature (Manfredo & Dayer, 2004).

Given different values about animals, conservation practices seeking to benefit an animal species can be interpreted as detrimental to human beings, even more so when such practices are implemented by elites (experts, urbanites, researchers or policy-makers; Naughton-Treves et al., 2003; Sillero-Zubiri et al., 2007), with all the associated stereotypes. For example, people of rural origin perceive the practices of reintroducing the wolf as imposing the city's environmental values on traditional rural values (Blekesaune & Rønningen, 2010). Something similar occurs with the designation of protected areas or natural parks, which is perceived as an imposition against the interests of the affected population (West & Brockington, 2006).

The perceived threat of animals via conservation practices relates not only to individual values, but also to community values. Community values may promote or impede species conservation. The members of a community may consider actions such as trapping or killing a predator as part of their group identity (e.g., Marchini & MacDonald, 2012). Alternatively, the symbolic value of an animal species in a specific context can also increase acceptance and positive images of the animal (Carter et al., 2012). The focus of human-wildlife conflicts in charismatic fauna may be explained by their capacity to elicit strong emotional reactions in society – polarizing the views about their management (Marchini, 2014). Therefore, focusing on these animals feeds the human-human symbolic conflict about values.

Management actions thus involve a series of symbolic threats: They challenge human superiority to use animals, imply urban domination of rural population, and question group identity. Therefore, human-animal conflict can also incorporate a *symbolic threat*, as in intergroup relations.

Recap

The social relevance of animals (Amiot & Bastian, 2015) and the increasing human-animal interactions in urban contexts (Soulsbury & White, 2015) suggest increasing human-animal conflicts (Marchini, 2014). These conflicts may be realistic (resources, physical wellbeing) or symbolic (value and identity driven). Although always through human beings' behaviors, the realistic conflict mainly involves the human versus the animal group, whereas the symbolic conflict creates confrontations among human groups (farmers, rural, urban citizens, scientific experts, and conservation organization members), fighting over how to resolve human-animal conflict. These scenarios call for an intergroup relations approach to conservation biology.

From an intergroup framework, developing optimal conditions for intergroup contact and enhancing cooperation is key for reducing intergroup conflicts (Brewer, 1997). Notably, enhancing (perceived) cooperation between humans and animals is suitable to animal conservation policies. Some recent research in conservation biology may be framed as intergroup research on cooperative versus competitive relations. For example, several correlational and experimental studies found that perceived benefits (for humans) of animal species facilitate species acceptance (Carter et al., 2012; Slagle et al., 2013). Examples of perceived human benefits are the increase in tourism, a healthy environment, and animals' cultural value for the population. In the light of the intergroup relations approach, the perception of human benefits associated with animal species is quite similar to the operationalization of perceived benefits of certain groups for society in intergroup research (Caprariello, Cuddy, & Fiske, 2009; Falomir-Pichastor, Munoz-Rojas, Invernizzi, & Mugny, 2004; Sevillano & Fiske, 2016c; Tsukamoto & Fiske, 2018).

Perceived benefits of certain human groups for society are linked to group's perceived cooperation. For example, immigrants may be portrayed as culturally enriching a host society, doing jobs that natives refuse, growing the economy, and generally "helping" host society. In this way, immigrants are perceived as "beneficial" or cooperators with society, thus less threatening in realistic and symbolic terms. Alternatively, immigrants portrayed as violating cultural and national values of host society, using government benefits, and taking jobs from natives can lead to a negative image of immigrants as harmful or competitive toward society. Consequently, in intergroup terms, many animal species *cooperate* (unintentionally) with humans giving certain benefits (realistic and symbolic). In keeping with different stereotypic images of different species, cooperation would take different forms in different clusters of species.

Conclusion

Research and management of human–animal conflicts may benefit by adopting an intergroup relations approach applied to different animal species. The characterizations of the types of conflicts, stereotypes, emotional reactions, and behaviors associated with different animal species can explain human dimensions of wildlife management, based on abundant evidence in the case of human social groups. We summarize the contribution of the intergroup approach to existing theoretical approaches in four aspects:

1. **The relevance of social psychology constructs previously uncovered.** On the one hand, an intergroup approach highlights constructs considered classic in social psychology, but not yet applied to conservation biology: distinct stereotypes, behavioral tendencies, and conflicts. On the other hand, the intergroup approach expands the scope of affect and behavior. The SCM and BIAS Map consider a diverse set of affective reactions (disgust, contempt, admiration, threat, pity), and behavioral tendencies associated with stereotypes (passive/active help or harm). However, influenced by risk analysis, the role of emotion in human–animal conflicts is currently limited to fear (regarding an attack). Similarly, types of behaviors are not distinguished in the conservation biology literature. Different affective reactions associated with animal species play a role in predicting attitudes toward them, and behavioral tendencies vary as a function of specific species' stereotypes.

2. **Clarify the conceptualization of attitudes.** The lack of consistency in measurement and conceptualization of attitudes toward wildlife has been criticized (Bruskotter & Wilson, 2014; Dressel et al., 2015). In some cases, under the label of "attitudes and opinions," beliefs and behaviors associated

with the different species are evaluated together (e.g., König, 2008). Using the concept of stereotype to refer to the specific beliefs associated with different animal species, as well as using attitude for evaluative judgments (liking vs disliking) of the species, could help to integrate research results, as well as to develop explanatory models.

3. **Organize different theoretical constructs in an integrative view.** The assumption of an intergroup approach may help to integrate empirical findings that are sometimes difficult to organize theoretically. As we reviewed, the SCM and the BIAS Map link stereotypes, affect, and behaviors in a causal manner. In relation to the various types of human-animal conflicts, an intergroup approach categorizes them in terms of realistic and symbolic conflicts. This categorization presents the conservation biology's research on the loss of resources, physical threat, value orientations, group identity, and cultural values in a more comprehensive way. In addition, this organization can answer new research questions for conservation biology, although not for social psychology. For example, are realistic threats more important than symbolic threat? Is stereotype content of species related to distinct types of threats? Answering these questions would facilitate interdisciplinary communication between social psychology and conservation biology and also, help guide intervention programs.

In addition, conservation biology grants much importance to the threat to (economic) resources and to physical well-being (life, health) (Bruskotter & Wilson, 2014), but somehow, relegates symbolic threat to what has generically been called the "social perception of conflict" (Sillero-Zubiri et al., 2007). From an intergroup approach, the "social perception of conflict" is the conflict itself: how individuals understand the interdependent situation is the key to the conflict.

4. **Facilitate interdisciplinary communication.** As a consequence from the previous points, the adoption of an intergroup approach facilitates the transference of knowledge between social psychology and conservation biology.

The objective of this chapter is to provide a new theoretical proposal that can enrich the area of human-animal relationships and promote interdisciplinary efforts between social sciences and ecology. Ultimately, the utility of such a proposal should be evaluated with applied research.

Acknowledgment

The authors gratefully acknowledge Dr. Fernando Talayero and Dr. Beatriz Cortés for their comments on an earlier version of this chapter.

References

Ajzen, I. (1991). The theory of planned behavior. *Organizational Behavior and Human Decision Processes, 50*, 179–211.

Amiot, C. E., & Bastian, B. (2015). Toward a psychology of human–animal relations. *Psychological Bulletin, 141,* 6–47.

Arluke, A., & Sanders, C. R. (1996). *Regarding Animals.* Philadelphia, PA: Temple University Press.

Barua, M., Bhagwat, S. A., & Jadhav, S. (2013). The hidden dimensions of human–wildlife conflict: Health impacts, opportunity and transaction costs. *Biological Conservation, 157,* 309–316. http://dx.doi.org/10.1016/j.biocon.2012.07.014

Bath, A. J. (2000). *Human Dimensions in Wolf Management in Savoie and Des Alpes Maritimes, France.* France LIFE-Nature Project, Le retour du loup dans les Alpes Françaises, & the Large Carnivore Initiative for Europe (LCIE). Retrieved from: www.kora.ch/malme/05_library/5_1_publications/B/Bath_2000_Human_Dimensions_in_Wolf_Management_in_France.pdf

Blekesaune, A., & Rønningen, K. (2010). Bears and fears: cultural capital, geography and attitudes towards large carnivores in Norway. *Norwegian Journal of Geography, 64,* 185–198. doi: 10.1080/00291951.2010.528225

Breakwell, G. M. (2007). *The Psychology of Risk.* Cambridge: Cambridge University Press.

Brewer, M. B. (1997). The social psychology of intergroup relations: Can research inform practice? *Journal of Social Issues, 53,* 1, 197–211.

Bruskotter, J. T., & Fulton, D. C. (2012). Will hunters steward wolves? A comment on Treves and Martin. *Society & Natural Resources, 25,* 97–102.

Bruskotter, J. T., & Wilson, R. S. (2014). Determining where the wild things will be: using psychological theory to find tolerance for large carnivores. *Conservation Letters, 7,* 158–165. doi: 10.1111/conl.12072

Caprariello, P. A., Cuddy, A. J. C., & Fiske, S. T. (2009). Social structure shapes cultural stereotypes and emotions: a causal test of the stereotype content model. *Group Process & Intergroup Relations, 12,* 2, 147–155. doi:10.1177/1368430208101053

Carter, N. H., Riley, S. J., & Liu, J. (2012). Utility of a psychological framework for carnivore conservation. *Oryx, 46,* 525–535.

Cervantes, M. (1613/2013). *The Exemplary Novels.* Trad. Walter K. Kelly. Tübingen, Germany: Belle Époque press.

Cuddy, A. J. C., Fiske, S. T., & Glick, P. (2007). The BIAS map: Behaviors from intergroup affect and stereotypes. *Journal of Personality & Social Psychology, 92,* 631–648.

Dhont, K., & Hodson, G. (2014). Why do right-wing adherents engage in more animal exploitation and meat consumption? *Personality and Individual differences, 64,* 12–17. doi: 10.1016/j.paid.2014.02.002

Dhont, K, Hodson, K., Leite, A. C., & Salmen, A. (this volume). The psychology of speciesism. In K. Dhont & G. Hodson (Eds.), *Why We Love and Exploit Animals: Bridging Insights from Academia and Advocacy.* Abingdon: Routledge.

Dressel, S., Sandström, C., & Ericsson, G. (2015). A meta-analysis of studies on attitudes toward bears and wolves across Europe 1976–2012. *Conservation Biology, 29,* 565–574.

Duffy, R., St. John, F. A. V., Büscher, B., & Brockington, D. (2016). Toward a new understanding of the links between poverty and illegal wildlife hunting. *Conservation Biology, 30,* 14–22. doi: 10.1111/cobi.12622

Ericsson, G., Kindberg, J., & Bostedt, G. (2007). Willingness to pay (WTP) for wolverine Gulo gulo conservation. *Wildlife Biology, 13*(sp2), 2–13.

European Commission (2009). *Coexistence with large carnivores: The challenge and the opportunity.* Retrieved from: http://ec.europa.eu/environment/nature/conservation/species/carnivores/pdf/coexisting_with_large_carnivores_catalogue_s.pdf

European Union (2015). *Special Eurobarometer 442: Attitudes of Europeans towards animal welfare.* doi: 10.2875/884639

Falomir-Pichastor, J. M., Munoz-Rojas, D., Invernizzi, F., & Mugny, G. (2004). Perceived in-group threat as a factor moderating the influence of in-group norms on discrimination against foreigners. *European Journal of Social Psychology, 34*, 135–153.

Fernández-Gil, A., Naves, J., Ordiz, A., Quevedo, M., Revilla, E., & Delibes, M. (2016). Conflict misleads large carnivore management and conservation: brown bears and wolves in Spain. *PLoS ONE, 11*, 1–13. doi: 10.1371/journal.pone.0151541

Fischer, J. D., Schneider, S. C, Ahlers, A. A., & Miller, J. R. (2015). Categorizing wildlife responses to urbanization and conservation implications of terminology. *Conservation Biology, 29*, 1246–1248. doi: 10.1111/cobi.12451

Fiske, S. T. (2010). *Social Beings. Core motives in Social Psychology* (2nd edition). New York: Wiley.

Fiske, S. T., Cuddy, A. J. C., Glick, P., & Xu, J. (2002). A model of (often mixed) stereotype content: competence and warmth respectively follow from perceived status and competition. *Journal of Personality & Social Psychology, 82*, 878–902.

Heberlein, T. A. (2012). *Navigating Environmental Attitudes.* New York: Oxford University Press.

Henley, N. M. (1969). A psychological study of the semantics of animal terms. *Journal of Verbal Learning & Verbal Behavior, 8*, 176–184.

Herzog, H. A. (2010). *Some We Love, Some We Hate, Some We Eat: Why It's So Hard to Think Straight About Animals.* New York: Harper.

Hodson, G., Dhont, K, & Earle, M. (this volume). Devaluing animals, "animalistic" humans, and people who protect animals. In K. Dhont & G. Hodson (Eds.), *Why We Love and Exploit Animals: Bridging Insights from Academia and Advocacy.* Abingdon: Routledge.

Hodson, G., MacInnis, C. C., & Costello, K. (2014). (Over)Valuing humanness as an aggravator of intergroup prejudices and discrimination. In P.G. Bain, J.Vaes, & J.-Ph. Leyens (Eds.), *Humanness and Dehumanization* (pp. 86–110). London: Psychology Press.

Jacobs, M. H., Vaske, J. J., Teel, T. L., & Manfredo, M. J. (2013). Human dimension of wild-life. In L. Steg, A. Van den Berg, & J. I. M. Groot (Eds.), *Environmental Psychology: An Introduction* (pp. 77–86). Oxford: British Psychological Society & John Wiley & Sons.

Johansson, M., & Karlsson, J. (2011). Subjective experience of fear and the cognitive interpretation of large carnivores. *Human Dimensions of Wildlife, 16*, 15–29.

Johansson, M., Karlsson, J., Pedersen, E., & Flykt, A. (2012). Factors governing human fear of brown bear and wolf. *Human Dimensions of Wildlife, 17*, 58–74.

Kasperson, R., E., Renn, O., Slovic, P., Brown, H. S., Emel, J., Goble, …Ratick, S. (1988). The social amplification of risk: A conceptual framework. *Risk Analysis, 8*, 177–187.

König, A. (2008). Fears, attitudes and opinions of suburban residents with regards to their urban foxes. *European Journal of Wildlife Research, 54*, 101–109. doi: 10.1007/s10344-007-0117-z

Leite, A. C., Dhont, K., & Hodson, G. (2019). Longitudinal effects of supremacy beliefs and vegetarianism threat on moral exclusion (vs. inclusion) of animals. *European Journal of Social Psychology, 49*, 179–189. doi: 10.1002/ejsp.2497

Mace, G. (2014). Whose conservation? Changes in the perception and goals of nature conservation require a solid scientific basis. *Science, 345*, 1558–1560. doi: 10.1126/science.1254704

Manfredo, M. J., & Dayer, A. A. (2004). Concepts for exploring the social aspects of human–wildlife conflict in a global context. *Human Dimensions of Wildlife, 9*, 1–20. doi: 10.1080/10871200490505765

Marchini, S. (2014). Who's in conflict with whom? Human applied ecology and human dimensions in biological conservation, in L. M.Verdade, L. M. Lyra-Jorge, & C. I. Piña

(Eds.), *Applied Ecology and Human Dimensions in Biological Conservation* (pp. 189–209). doi: 10.1007/978-3-642-54751-5_13

Marchini, S., & Macdonald, D. W. (2012). Predicting ranchers' intention to kill jaguars: case studies in Amazonia and Pantanal. *Biological Conservation, 147*, 213–221. doi: 10.1016/j.biocon.2012.01.002

McKinney, M. L. (2006). Urbanization as a major cause of biotic homogenization. *Biological Conservation, 127*, 247–260. doi: 10.1016/j.biocon.2005.09.005

Medin, D. L., & Atran, S. (2004). The native mind: biological categorization and reasoning in development and across cultures. *Psychological Review, 111*, 4, 960–983.

Miller, B., Soule, M. E., & Terborgh, J. (2014). "New conservation" or surrender to development? *Animal Conservation, 17*, 509–515.

Naughton-Treves, L., Grossberg, R., & Treves, A. (2003). Paying for tolerance: rural citizens' attitudes toward wolf depredation and compensation. *Conservation Biology, 17*, 1500–1511.

Plous, S. (2003). Is there such a thing as prejudice towards animals? In S. Plous (Ed.), *Understanding Prejudice and Discrimination* (pp. 509–528). New York: McGraw-Hill.

Riek, B. M., Mania, E. R., & Gaertner, S. L. (2006). Intergroup threat and outgroup attitudes: a meta-analytic review. *Personality & Social Psychology Review, 10*, 336–353.

Sevillano, V., & Fiske, S. T. (2016a). Animals as social objects: groups, stereotypes, and intergroup threats. *European Psychologist, 21*, 206–217. doi: 10.1027/1016–9040/a000268

Sevillano, V., & Fiske, S. T. (2016b). Warmth and competence in animals. *Journal of Applied Social Psychology, 46*, 276–293. doi: 10.1111/jasp.1236

Sevillano, V., & Fiske, S. T. (2016c). Fantasia: Being emotionally involved with a stereotyped target changes stereotype warmth. *International Journal of Intercultural Relations, 54*, 1–14. doi: http://dx.doi.org/10.1016/j.ijintrel.2016.06.001

Sevillano, V., & Fiske, S. T. (2017, July). *Animals as social objects: status, cooperation, and threat towards human beings.* 18th General Meeting of the European Association of Social Psychology. Granada (Spain).

Sillero-Zubiri, C., Sukumar, R., & Treves, A. (2007). Living with wildlife: the roots of conflict and the solutions. In D. Macdonald, & K. Service (Eds.), *Key Topics in Conservation Biology* (pp. 255–272). Oxford: Blackwell.

Sjölander-Lindqvist, A., Johansson, M., & Sandström, C. (2015). Individual and collective responses to large carnivore management: the roles of trust, representation, knowledge spheres, communication and leadership. *Wildlife Biology, 21*, 175–185. doi: 10.2981/wlb.00065

Skogen, K. (2001). Who's is afraid of the big, bad wolf? Young people's responses to the conflicts over large carnivores in Eastern Norway. *Rural Sociology, 6*, 203–226.

Skogen, K., Mauz, I., & Krange, O. (2008). Cry wolf! Narratives of wolf recovery in France and Norway. *Rural Sociology, 73*, 105–133.

Slagle, K. M., Bruskotter, J. T., & Wilson, R. S. (2012). The role of affect in public support and opposition to wolf management. *Human Dimensions of Wildlife, 17*, 44–57.

Slagle, K. M., Zajac, R., Bruskotter, J. T., Wilson, R. S., & Prange, J. (2013). Building tolerance for bears: a communications experiment. *The Journal of Wildlife Management, 77*, 863–869.

Soulsbury, C. D., & White, P. C. L. (2016). Human–wildlife interactions in urban areas: A review of conflicts, benefits and opportunities. *Wildlife Research, 42*, 541–553.

Stephan, W. G., & Stephan, C. W. (2000). An integrated threat theory of prejudice. In S. Oskamp (Ed.), *Reducing Prejudice and Discrimination: The Claremont Symposium on Applied Psychology* (pp. 23–45). Mahwah, NJ: Erlbaum.

Treves, A. (2008). The human dimensions of conflicts with wildlife around protected areas. In M. J. Manfredo, J. J. Vaske, P. J. Brown, D. J. Decker, & E. A. Duke (Eds.), *Wildlife and Society: The Science of Human Dimensions* (pp. 214–228). Washington, DC: Island Press.

Treves, A., & Bruskotter, J. T. (2014). Tolerance for predatory wildlife. *Science, 344,* 476–477. doi: 10.1126/science.1252690

Treves, A., & Palmqvist, P. (2007). Reconstructing hominin interactions with mammalian carnivores (6.0 - 1.8 Ma). In K. A. I. Nekaris, & S. L. Gursky (Eds.), *Primates and their Predators* (pp. 355–381). New York: Springer.

Treves, A., Wallace, R. B., Naughton-Treves, L., & Morales, A. (2006). Co-managing human-wildlife conflicts: A review. *Human Dimensions of Wildlife, 11,* 1–14.

Tsukamoto, S., & Fiske, S. T. (2018). Perceived threat to national values in evaluating stereotyped immigrants. *The Journal of Social Psychology, 158,* 2, 157–172. doi: 10.1080/00224545.2017.1317231

Vaske, J. J., & Manfredo, M. J. (2012). Social psychological considerations in wildlife management. In D. J. Decker, S. Riley, & W. F. Siemer (Eds.), *Human Dimensions of Wildlife Management* (pp. 43–57). Baltimore, MD: Johns Hopkins University Press.

West, P., & Brockington, D. (2006). An anthropological perspective on some unexpected consequences of protected areas. *Conservation Biology, 20,* 609–616.

Wojciszke, B., Bazinska, R., & Jaworski, M. (1998). On the dominance of moral categories in impression formation. *Personality and Social Psychology Bulletin, 24,* 1245–1257. doi: 10.1177/01461672982412001

Yzerbyt, V. Y., & Demoulin, S. (2010). Intergroup relations. In S. T. Fiske, D. T. Gilbert, & G. Lindsay (Eds.), *The Handbook of Social Psychology* (5th ed., pp. 1024–1083). New York: Wiley.

ASKHAM BRYAN
COLLEGE
LEARNING RESOURCES

17

THE MORAL MARCH TO MEATLESS MEALS

The scripted Hebrew meat prohibitions versus the unscripted path to becoming vegetarian or vegan

Paul Rozin and Matthew B. Ruby

Abstract

The path to vegetarianism or veganism is typically gradual and unscripted. The usual moral (compassionate or environmental) and health motivations set goals to cut back or eliminate meat consumption, but there are no formal or informal instructions for how to achieve these goals. Instead, individuals make up their own evolving set of rules. In contrast, the rejection of pork (and other prohibitions) are precisely specified in the Orthodox Jewish tradition, which considers and resolves situations where there would otherwise be indecision. The contrast is clear for the issue of meat contamination, where the sympathetic magical "law" of contagion comes into play. This chapter, largely constituted by a debate between two fictional Orthodox Jewish rabbis, one a vegan and the other an omnivore, illustrates how some aspects of a path to vegetarianism could be scripted, with the aim of reducing but not eliminating conflict situations. The Hebrew tradition raises and attempts to solve some of the practical and conceptual problems of enacting prohibitions. A discussion following the dialogue deals with the similarities and differences between the Kosher dietary prohibitions and the Hindu beef prohibition, and the path to vegetarianism.

The following is primarily a set of conversations between two (fictional) Orthodox Jewish rabbis. Rabbi (abbreviated to Reb) Ruby is a vegan, while Reb Rozin is an omnivore, except that he observes the laws of Kashrut (including prohibition on meat of pigs and shellfish, as well as mixtures of meat and dairy in the same meal, and other elaborations of these prohibitions). The subject is the process of becoming a moral vegan or vegetarian, and the ethical and logistical dilemmas associated with dividing foods into edible and inedible in light of issues like trace contamination. They discuss the Hebrew dietary prohibitions, particularly against consuming pork, and contrast the explicit rules of Kashrut about what foods are acceptable with the unscripted, informal, and individualized ways in which moral vegetarians decide what foods are acceptable. In these conversations, they debate many of the common arguments for and against eating animals.

[*The rabbis are sitting outside together. Reb Rozin is consuming what looks like a cheeseburger.*]

RUBY: Reb Rozin, I have to say, I am disturbed to see you eating what appears to be meat and dairy together.

ROZIN: Reb Ruby, be assured, I would never do such. The hamburger is Kosher, and what looks to you like a slice of cheese is actually non-dairy vegan cheese.

RUBY: Well, the Jewish laws, the Halacha based in the Talmud, dictate that one cannot do something that looks forbidden to others, so as not to set a bad example or to have them suspect you of sin.

ROZIN: Yes, you are correct, and that is precisely why I brought the vegan cheese wrapping and left it here on the bench so that you might notice on your own that it was non-dairy cheese, but I guess you didn't notice it. The Halacha permits it if there is a visual indication that it is not a transgression.

RUBY: Still, I find just the appearance of meat and dairy together to be upsetting.

ROZIN: Well, I do not, and as we both know, what I am eating is Kosher. Some observant Jews would do what I am doing, and some would not. I don't need to assure you that the hamburger meat comes from a kosher butcher, from a cow that was ritually slaughtered according to the Torah, with all blood drained out from the meat.

RUBY: I know the Torah permits both the non-dairy cheese, and the properly slaughtered cow, but that is not the ideal, and I reject all animal products.

ROZIN: That seems rather extreme, to me. The Torah explicitly allows us to eat meat. Genesis 9:3: "Every moving thing that lives shall be yours to eat; like the green vegetation, I have given you everything."

RUBY: Yes, but even what you cite indicates that there was a prior phase of veganism in the Garden of Eden. Genesis 1:29–30: "Behold, I have given you every seed bearing herb, which is upon the surface of the entire earth, and every tree that has seed bearing fruit, it will be yours as food." This means that the ideal state is vegan!

ROZIN: Maybe it was ideal in the Garden of Eden, but after the flood, G-d[1] saw fit to allow humans to eat meat. Genesis 9:2: "The fear and the dread of you shall be upon all the beasts of the earth and upon all the birds of the sky – everything with which the earth is astir – and upon all the fish of the sea, they are given into your hand."

RUBY: Yes, but this was only as a response to the corruption and weakness of humans. Why should we not honor His original design, and live righteously, as vegans?

ROZIN: G-d meant to let humans eat meat. Not only is this said explicitly in the Torah, but G-d explicitly forbids the consumption of blood: Genesis 9:4: "You must not, however, eat flesh with its life-blood in it." There would not be an explicit prohibition on blood, stated a number of times, unless eating the rest of the meat was permitted. Blood is the life essence. Blood is the symbol of life and is only for G-d. It wouldn't be forbidden unless there was a real tendency to like it. And G-d made us like it, so He is teaching us partial restraint, by letting us eat the bloodless meat.

RUBY: You can't really get rid of all the blood, if you could, the meat wouldn't be red. So G-d is really saying, don't eat meat.

ROZIN: I think G-d generally says what is on His mind. He could have just said, "Don't eat meat." G-d made me, and he made me love the taste of meat. I'm sure it tastes better with the blood, but I willingly make the compromise of eating more or less bloodless meat.

RUBY: G-d works in strange ways. Surely, He does not approve of the way, now-adays, most animals raised for food are treated, and eventually killed. That wasn't so in biblical times. To give all animals and plants into the hand of humans does not imply the right to mistreat and slaughter the animals with that hand. Rather, the Lord meant for us to *care* for the animals. That's why the Lord had Noah load a pair of each animal on to the ark. Didn't you once tell me that "Judaism is not about our rights, it's about our responsibilities"?

ROZIN: The Bible considers G-d at the top, then humans, and then all of the other animals. Where in the Bible does it say humans should be nice to animals, and not slaughter them? In the Bible, animals are like plants, not like humans. G-d gave explicit permission to eat meat He made us, and He made us love meat. And G-d loves meat. Look at all those animal sacrifices He asked for!

RUBY: The animal sacrifices don't mean He loves meat. After all, Judaism doesn't believe in a corporeal G-d who eats anything. In fact He doesn't have a body, so how can He eat at all? G-d has dominion over everything, and the sacrifice of animals is an explicit recognition of that fact. It is ridiculous to say that G-d loves and eats meat.

ROZIN: G-d is omnipotent. He could eat without having a body.

RUBY: I don't think we can be sure about what or whether G-d eats. But it is important to remember that after the flood, G-d permits meat eating with restrictions, but it is a concession. Some modern rabbis believe that the intent in the Bible is a diet free of animal products, as in the Garden of Eden. Some rabbis, like David Rosen, claim that one can infer from parts of the Bible that one should not cause unnecessary cruelty to animals. So, it is possible to derive from the Bible and the Talmud that the ideal state for Jews is to be vegan. And it is clearly the case that to be vegan is to be kosher. Virtually all the food restrictions in the Bible have to do with meat, and the mixing of meat and dairy. Orthodox Jews who must keep separate dishes for dairy and meat need not do so if they become vegan. It may be easier, in terms of daily life, to be a vegan than to eat meat and dairy. And doing so will respect a higher morality, a kindness to animals and to the earth. But back to our main concern. How do you feel about killing animals so you can have a tasty meal? Is your pleasure so much more important than their lives?

ROZIN: It isn't just my pleasure. Hundreds of people can get pleasure and nutrition from one killed cow. So far as I know, a full adult cow yields about 500 pounds of meat. So if a person eats a quarter pound of beef each day, 2000 people can get meat pleasure for one day from eating a cow, or one person will eat a cow in about 6 years. So a 60-year-old, assuming beef is the only meat they eat, eating a quarter pound of beef a day, will eat only nine or so cows in their life.

RUBY: Yes, nine dead cows, for the pleasure of chewing on some beef. Doesn't it bother you that each person will kill nine cows in their life, just for some pleasant sensations in their mouth?

ROZIN: But we have to eat meat to flourish– it is a matter of life and death for us. Just as it is for the lions who eat zebra. How else can we get enough protein?

RUBY: Don't you know any healthy vegetarians? I haven't eaten any animals for 10 years, and I'm doing very well.

ROZIN: But they say meat is necessary for life, and everyone, well almost everyone, eats it.

RUBY: Just think, Adam and Eve, and Abraham and Sarah, Noah and so many others, ate no meat, dairy, or eggs. And most of these early Hebrews lived many hundreds of years, and they were healthy. It's not about health. It's killing versus pleasure. Seriously, don't you feel bad about that?

ROZIN: Yes, so meat may not be necessary, though it tastes wonderful. Why did G-d make animal flesh taste so good? To teach us restraint? Most scholars believe that thousands of years ago, humans were eating animals. Our teeth and our guts suggest that we ate both plants and animals.

RUBY: Sure, since the flood, and that's a long time, humans have been eating meat. But originally, perhaps like gorillas or orangutans, humans did not eat meat.

ROZIN: But all of those are wild animals. We *bred*, we *made* domesticated cows, so they are our creation, not G-d's.

RUBY: So you're saying we're allowed to eat things that are our own creation. So why not eat domesticated dogs and cats? By that logic, we could eat our *children, after all, we made them!*

ROZIN: Good point. Really good point. But not eating meat is so revolutionary. Almost everyone eats meat.

RUBY: Yes, and almost everyone eats pork, and the 1.3 billion people who live in China are the biggest pork eaters of the world, maybe half of the world's pork. So what? You represent a small but special minority. Anyway, you were on the way to being a vegetarian before we got started. You weren't eating pork and shellfish and that wasn't so hard. You're just expanding what you already do.

ROZIN: It's not the same. Pork is polluting, as the Torah says, while these other meats, like beef, are okay. Your motivation for eating them is very different– it's thinking about the animal's rights, not our moral purity. It's like the world's billion Hindus, who won't eat beef or kill cows. For them, cows are sacred, and it is because of the animal's soul that they don't eat beef. For them pollution comes from consuming/destroying something sacred. For us, it is not reverence for pigs, but rather it is G-d's word, and we find pigs disgusting, and consuming pork pollutes our soul.

RUBY: But you will note that many Hindus are vegetarians. Hundreds of millions of them, about half of the vegetarians in the world, aren't eating any animals at all.

ROZIN: Speaking of animals, you have a pet cat, yes?

RUBY: Yes, I do.

ROZIN: I bet your cat eats only kosher, no meat and dairy at the same time, no pork.

RUBY: That's right, my cat eats kosher.

ROZIN: But I bet your cat eats meat, like beef and chicken.

RUBY: Yes, she does. She needs to eat meat to be healthy.

ROZIN: Well, if the cat eats meat, why can't we?

RUBY: Because we are omnivores. We can eat a broad range of plants, and we like many of them. We also have more advanced mental abilities than cats. We may love the taste of meat, but we know it means killing animals, so we can refrain from eating meat and not kill animals. As stated in the Bible, we are higher creatures than other animals.

ROZIN: Okay, I don't eat pork, and I agree, cows are sort of nice, so maybe I will be able to eat less beef. I will sacrifice some pleasure, and I guess it won't be a problem. If I eat 50 percent less beef, then I will save the lives of half of the cows that I would have eaten.

RUBY: Good start.

A month later

RUBY: How are you doing, eating half of the beef you used to?

ROZIN: It is okay. I sort of feel a little noble about this. Maybe I can cut down to one quarter of the meat I used to eat.

RUBY: Are you cutting back on other types of meat, or just beef? I know you've always eaten a lot of chicken.

ROZIN: Well, cows are more like humans, and they are probably smarter than chickens. I just don't think that a chicken life is worth as much as a cow life.

RUBY: Okay, so you are trying to spare the most meaningful lives. We already calculated that you might eat about nine cows in a full life time. Well, you get about 3 pounds of meat from an adult chicken, as opposed to about 500 pounds from an adult cow. That means you have to kill about 170 chickens to get the same amount of meat as the meat from one cow. Do you think 170 chicken lives are worth less than one cow life?

ROZIN: I never thought of that…I guess 170 chickens are worth more than one cow.

RUBY: And given that the average American eats about 40 pounds of chicken a year, or about 13 chickens a year, that equals about 800 chickens over a 60-year span. That's a *lot* of lives. So are you going to cut back on eating chicken as well?

ROZIN: It's the logical thing to do. I will cut back on both of them, 50 percent down in each. Actually, it won't be as hard because I like beef more than chicken, and I've already cut back on that.

A month later

ROZIN: I have done it; I am 50 percent down on chicken and beef. It actually wasn't so hard, but I do crave beef.

RUBY: Do you also crave pork?

ROZIN: I've never eaten pork, but I am curious to know what it tastes like, and I sort of like the smell of bacon.

RUBY: Interesting. I have always found pork and bacon disgusting, and recently, in my tenth year as a vegetarian, I came to find beef disgusting, so I am not tempted to eat it.

ROZIN: I guess you are lucky, you don't have temptation.

RUBY: I did at the beginning, but it passed with time. Have you considered becoming a partial vegetarian, by cutting out all mammals and birds from your diet?

ROZIN: That's asking for a lot.

RUBY: You know that killing half the number of chickens, maybe 400 in a lifetime instead of 800 is an improvement, but you are still killing animals. It is not as bad to kill one person than two, but only a little better.

ROZIN: You know, it is really hard to eat *no* mammals or birds. Eating is a lot less pleasurable, and it's socially difficult. Most other people eat meat, and they would have to make special food for me.

RUBY: Come on, you are used to that, when you eat with people who aren't kosher.

ROZIN: True, but this is a bigger imposition.

RUBY: But isn't it worth the slight imposition to not kill animals? And by eating delicious plant-based food, maybe your friends will consider cutting back on their meat consumption, like you.

ROZIN: Yes, but it is practically impossible to eliminate *all* meat. For example, some vegetable soups are made with chicken broth. Or, let's say I am eating an entirely vegetarian meal, but I am sitting next to someone who is eating a hamburger. Some of those beef molecules from the sizzling hamburger could get in the air, and I could breathe them in. I don't know if we could ever fully avoid taking in meat.

RUBY: Well, you know that Jewish scholars have devoted a lot of time and thought to this, because we worry that the whole world is contaminated. Do you know, according to a friend of mine who is a geologist, if someone in Europe drinks a glass of water, they consume at least a few water molecules that went through Adolf Hitler!

ROZIN: Wow, that is disturbing, but you have to have water. But as you note, we Jews have solved the contamination problem, getting around all those tiny amounts. We have the 1/60th rule. If a contaminant falls into an otherwise kosher food, by accident (or you don't know about it), and the volume of the contaminant is less than 1/60th of the total, it remains kosher.

RUBY: Yes, weren't our ancestors clever, thorough, and creative? But it's even better than that. Suppose you're at a cookout, and some bacon, 1/80th, accidentally falls into your kosher food. It's still kosher, right?

ROZIN: Right.

RUBY: But now, suppose in that same food, some shrimp, 1/90th of the volume, also falls in by accident. Now 1/80 + 1/90 is more than 1/60. And if you remember your calculus, lots of tiny amounts of non-kosher foods could easily accumulate to more than 1/60.

ROZIN: And our Talmudic ancestors figured out a way to deal with that too, as I remember. Only the principal (highest concentration) contaminant matters, so one doesn't have to worry about all those epsilons.

RUBY: Right, so we can use our kosher laws to get around the problem of meat micro-contamination for vegetarians. The 1/60th rule for all meat.

ROZIN: Nothing is simple. Of course blood is a big prohibition, but you can't get all the blood out of the muscle before you eat it, and you probably can't get enough out so it is less than 1/60th, so that rule doesn't hold for blood. But generally, the 1/60th rule solves many problems. So I could go on a mammal- and bird-free diet. But if I *know* that some gelatin, derived from pigs, is a micro-contaminant (less than 1/60th) in an otherwise kosher food, I'm not sure I can eat it.

RUBY: That is a problem. It has to do with what it means to be "by accident." You didn't put it in, but you know someone else, by definition intentionally, put it in the food. By the way, although I know the 1/60th rule, I wouldn't eat anything that I knew had any pork in it. I would just find it disgusting.

ROZIN: Well I wouldn't find it disgusting, and because it is acceptable according to the laws of Kashrut, I would eat it. In a way I should get more credit than you, because I have to overcome temptation, and you don't. Anyway, okay, I don't want to kill nice animals. I will try it. But wait. Sometimes you can eat meat and not be a party to killing animals.

RUBY: What?

ROZIN: Well, I will give you an example: roadkill deer. They are already dead, and you are not participating in killing if you eat them.

RUBY: I suppose that's true, if you don't find it disgusting.

ROZIN: I don't. I might even wait near a place where road kills frequently happen and get some morally-okay meat that way. But I'll give you a more common example. Suppose you are eating with a friend in a kosher restaurant, and she is eating a kosher hamburger. She can't finish it. The server will take her partly eaten hamburger and throw it out in the kitchen. Now I could eat that leftover, destined to be waste, without being involved in killing. What do you think of that?

RUBY: You could, if you don't find someone else's food disgusting.

ROZIN: It is challenging, sort of Talmudic, to consider ways of eating meat without participating, at least indirectly, in killing. So here's another one. Now I couldn't eat pig eyes, but I could eat cow eyes and be a moral vegetarian.

RUBY: What?

ROZIN: The cow eye, for us, is a waste product. *No one* kills a cow for the eyes. So if I eat the eyes, I am morally okay.

RUBY: That's really disgusting, but fair enough. That's a big difference between Kashrut and vegetarianism. For us, pigs and shellfish are inherently polluting. But for a moral vegetarian, meat eating is only immoral for a consequential reason, either participation in killing, or as you know, another moral reason— participation in a food system which is not sustainable.

ROZIN: Well, it's hypothetical. I too find cow eyes disgusting. But there is a case where I could be involved. I'm pretty sure moral vegetarians are most upset about eating baby animals.

RUBY: Right, because they are cute, and because they have so much more life to live.

ROZIN: Right, so veal is a big no no. But I love calf's liver, and in the USA (not France), calf's liver is not popular, so it is essentially a waste product. No one here kills a calf for the liver.

RUBY: That's true, but by eating that, you are contributing to the economic value of killing the calf.

ROZIN: True, I guess, but the calf will be killed anyway. I think I can eat road kill deer, unfinished cow meat, and calf's liver, at least in the USA.

RUBY: By that logic, you could also argue that you could eat beef tongue. No one kills a cow for the tongue.

ROZIN: Wow, that's right, and I love tongue. Thanks.

RUBY: But with most of your exceptions, except the road kill, you are contributing to the income of the people who kill animals. And you still don't get around the issues of how the animals are treated before they are killed. But okay for the road kill, and maybe the unfinished food, if you don't find it disgusting.

A month later

RUBY: How are you doing?

ROZIN: Fine. I've cut way back on mammals and birds. I do still crave them, and once in a while, at a good restaurant, I will eat something really good and meaty. And then there is, of course, the waste meat I eat. And I don't worry about traces. So I'm pretty good, morally.

RUBY: That's a lot of progress! Of course, you don't eat shellfish, but you're still eating fish, no?

ROZIN: Yes. Fish spend most of their lives in the ocean and have a normal life until they are caught, and I imagine, sometimes they are killed rapidly, like frozen.

RUBY: Some fish are farmed under crowded conditions. Don't you think that fish feel pain? Even freezing is a slow and painful death, and besides, they would likely suffocate before they freeze.

ROZIN: Who can tell?

RUBY: Well, you seem to think cows can feel pain.

ROZIN: Yes, I do.

RUBY: So you *think* cows feel pain, and you *think* fish don't. Are you going to make a major moral decision on such flimsy evidence?

ROZIN: Yes. It's all I have. Maybe asparagus feels pain?

RUBY: Asparagus doesn't have a nervous system. But okay, I can't prove fish feel pain, but if you ever saw a fish struggling after a fisherman reels it in and it is on the ground, gasping, you have to admit that it's upsetting. But okay, so you *think* cows feel pain?

ROZIN: Yes.

RUBY: So why do you not eat beef, but you do drink milk?

ROZIN: You don't kill cows to get milk.

RUBY: Well, actually you do. What do you think they do to the male offspring of dairy cows? They don't produce milk. They either kill them, or fatten them up some, and then kill them. And regardless of the calf's sex, they're usually taken away from their mother within a day of being born.

ROZIN: But that's not directly about the milk-producing cows.

RUBY: No, but it's a significant part of the process. And what do you think they do with a dairy cow after she can't produce milk anymore?

ROZIN: I can guess what happens. But you know, cows have to die.

RUBY: You would think after producing milk for years for humans, the cow could retire in a grassy field for the rest of her life.

ROZIN: That would be nice.

RUBY: But you agree that cows feel pain. So humans have bred cows to have big mammary glands, and to produce lots of milk, which causes pressure... that probably hurts.

ROZIN: When they get milked then, it is a great relief.

RUBY: Yes, like when you wear really tight shoes, so when you take them off, you will feel good. And what do you think happens to the baby cows that are supposed to get the milk?

ROZIN: I hope they can nurse, with the excess milk going to humans.

RUBY: A very small number of farms do that, but in many farms, they are taken away from their mother within a day of being born, so all the milk can go to humans. Is that fair, is it kind to the cow mother to take her offspring away?

ROZIN: I guess not.

RUBY: And I'm sure you know that in some modern dairies, cows are kept in small stalls, so they can't move around much.

ROZIN: I don't like that. But they are kept at a good temperature and they get quality food. Maybe that's a perfect life. Maybe they have a rich inner life? Maybe it would be better if they had a television, or could go outside.

RUBY: I don't know about TV... but walking in a grassy field would be humane.

ROZIN: In some dairies, the cows are outside, and just come in to be milked. But you know, it's not obvious to me that, in a stall where they could turn around, get good medical care and have comfortable temperatures, they aren't in cow heaven. After all, I bet there are many people who would consider it great to sit on a sofa all day and watch television, with pizza and soda served up a few times a day. Maybe, if they were sitting on a padded and comfortable chair with a toilet installed under it, they wouldn't have to get up at all.

RUBY: Nonetheless, considering the discarding of the newborn, and the dairy cows' fate to be killed when production drops, out of compassion, you should stop eating dairy products. So that means no more milk, or cheese, or yogurt. Dairy should be prohibited. You already practice a dairy prohibition. You can't eat dairy with meat by the laws of Kashrut, and you can't even eat dairy for up to 6 hours after eating meat.

ROZIN: I guess it all depends on how the cows are treated. If they could nurse their young and move around more, it could be fine. Humans could tell us how bad it would be for them to stay in a room with just a television, pizza, and soda, but cows can't tell us. I guess it all depends on what cows like. Maybe we could give them a choice of a stall with hay, and an open field with grass. I have a friend in the dairying business, and he says he would rather be a cow in a modern dairy, with some access to the outside, than on a traditional small farm. Better food, and much better medical care.

RUBY: But you will admit that some dairy cows, we don't know what percent, are probably treated in an inhumane manner.

ROZIN: Yes.

RUBY: So unless you can know in detail how they are treated, it is better to give up dairy products.

ROZIN: Well, the same holds for meat. If cows are raised in an open field, and killed humanely, one might feel it is morally okay to eat them. Of course, we don't care about that for pigs, because we will never eat them, but I feel bad for pigs that are mistreated.

RUBY: Well, I just don't like killing animals so we can eat them. I don't like cutting their lives short, but I can understand that some would feel it is okay to eat cows if they were treated well before being killed. But back to milk, you don't know how dairy cows are treated, unless it is your cow, or it is from a farm you know, so why not just stop eating dairy products?

ROZIN: Maybe. At this rate, there may soon be nothing left for me to eat. You kill a plant when you eat it, except for fruit. Eating, by its nature, involves killing living things.

RUBY: There are some people who worry about that, but I don't. Fruitarians, of which there aren't many, will only eat the fruit, seeds, nuts and things like that because that doesn't involve killing the plant.

ROZIN: So they think it is OK to "sterilize" a plant, but not to kill it. That's a questionable distinction.

RUBY: It is always difficult to make distinctions like this. That's why we have the Talmud to think it all through over the centuries. And we vegans and vegetarians can profit from that.

ROZIN: OK. Look, I will cut back on dairy products, especially because of the separation of the mother cow from her calf. But I will eat eggs. I know you are going to tell me about the crowding and cruel treatment of chickens, but some of them are probably treated well, and since I can't tell, I will just consume the eggs. And I think they are not fertilized, so I am not going to kill future chickens. Maybe commercial chickens are happy with their food, medical care, and when they are prevented from pecking and hurting each other.

RUBY: But they may be deprived of natural social interactions. There is probably more potential pecking damage because they are kept in such crowded conditions, where they often don't even have enough space to turn around. That's another long argument. Very long. We don't have Talmudic wisdom to tell us what to do about eggs or milk, so we have to figure it out on other grounds. But not harming animals is a basic principle.

ROZIN: I personally think that not harming animals is important, but I don't think you can find anything in the Hebrew Bible that says humans should be kind to animals, that they should treat animals better than plants.

RUBY: That may be true. There is some disagreement about that by biblical scholars, and there are some references to being kind to animals in the Talmud.

The old texts may not say anything about being kind to animals, but not all of our morality comes from the Bible.

ROZIN: What about insects? There are a lot of different cultures where people enjoy eating them.

RUBY: I haven't eaten insects, and can't say I've ever given it much thought.

ROZIN: Well, they are technically animals, and maybe they feel pain. That's a big issue for you.

RUBY: Yes, that would be an issue. Because they're so small, if they do feel pain, eating half a pound of them instead of half a pound of beef or chicken might cause even more suffering.

ROZIN: But the Torah says it is okay to eat a few kinds of insects. Leviticus 11:21–23: "Yet these may ye eat of every flying creeping thing that goeth upon all four, which have legs above their feet, to leap withal upon the earth. Even these of them ye may eat; the locust after his kind, and the bald locust after his kind, and the beetle after his kind, and the grasshopper after his kind. But all other flying creeping things, which have four feet, shall be an abomination unto you."

RUBY: Yes, I guess I knew that, though of course insects have six legs. But from the language in Leviticus, we don't know precisely what the species were. And of course, these species were in the Middle East, not the United States.

ROZIN: That's pretty technical, but it is in the Torah. I think I would eat locusts, grasshoppers and beetles, at least out of curiosity, to see if they tasted good.

RUBY: There are about a million species of insects. You think that because the Torah allows the consumption of four species, you can eat any of them? And then there is the pain issue for insects, but I will admit, that in terms of the environmental costs, insects are a lot better than cows or chickens.

ROZIN: So here's what I'm going to do. I don't want to hurt mammals or birds. Certainly, I don't want to kill them. So I won't eat them, with the protection of the 1/60th rule, so I can still survive, and don't have to worry about contamination. But I will eat road kill and waste meat because that doesn't involve my participation in killing. And as a reward to myself, I will allow myself to have some really good beef or chicken once in a while, like at a very good restaurant. And maybe fewer dairy products... but I keep thinking of Hindus. They revere cows and don't kill or eat them, but they consume a *lot* of dairy. Indians eat more yogurt than any other country. Maybe they treat their dairy cows well, I guess they must, because they are sacred. So I think I could eat yogurt in India, and in some cases, dairy in the USA.

RUBY: You may be right.

ROZIN: But you know, if everyone starts drinking skim milk, then cream will become a waste product, so I would be able to eat whipped cream and some ice cream. That would be great.

RUBY: Only if you could call cream a waste product. You see how complicated this is– can I eat this, can I eat that? We don't have this problem with the laws of Kashrut. You know, as far as observing Kashrut, if you follow the laws, your life is straightforward, and you are not always worrying if it is okay to eat something. As my friend and colleague, Reb Levi Haskelevich has told me, it is easier to be 100% orthodox than 90% orthodox. If 90%, you are always worrying about which 10% you will allow. And I would say that the two stages of rejecting animal foods that are most stable, most free of constant decision problems, are vegetarian and vegan.

ROZIN: I get it. And you are going to say, for being a moral vegetarian, it is better to go 100% and be a vegan. Then it is just no animals or animal products, with some rule about unintentional traces. It's easy to make food decisions. It is easier to be 100% vegan than a 90% vegan. Why worry about what you can eat?

RUBY: You know a small group of Rabbis now believe the intent of the Bible is to eliminate animal product eating, and that's why there are so many complicated meat and dairy rules. The simplest way to keep kosher is just not to eat dairy or meat. Vegan food is kosher, and vegan restaurants are kosher. There is now a growing movement for religious vegan Jews in Israel, making up about 5% of the population, the largest vegan incidence in any country. 100% vegan is 100% kosher, and you don't have to worry about mixing meat and milk, or have separate dishes for meat and milk.

ROZIN: But I still feel okay eating waste meat, some insects, eggs, and a little dairy. Just as a special treat, once in a while, I might eat a delicious piece of meat or some ice cream... of course, not at the same time. And there won't be too many hard decisions, because it is easy to distinguish animals from plants.

RUBY: Okay, your choice. But that treat does involve killing an animal. And, every time this opportunity arises, you have to *think* about whether it is okay to treat yourself and kill.

ROZIN: Yes, it is a weakness, but I eat less than 1/60th of the meat I used to eat!

RUBY: Hah! I have been making you feel bad every time you come by, but I want you to feel good about what you have done. You are doing even more to make the world a better place, over and above saving animal lives and avoiding animal mistreatment. You know that recently, there has been a lot of attention to the idea of sustainability. The earth's population is growing, and we will need food to feed everyone. That means resources and land. And meat is a very expensive way to feed people. It takes about 8 pounds of corn to get one pound of beef. Now beef meets all of our needs for a range of amino acids, the components of protein. Corn does not meet those needs, nor do beans. But corn and beans together provide a satisfactory "profile" of required amino acids, and they are the staples of traditional Mexican cuisine.

ROZIN: But I, like most people, maybe even Mexicans, like beef better than corn and beans. So once again, there is the pleasure issue.

RUBY: Yes, but if you are a vegetarian or vegan for reasons of compassion, a positive side effect is that you are helping the environment.

ROZIN: It's like a free add-on. A positive side effect. I'm for the environment. I like the idea of a positive side effect.

RUBY: Right, and more and more, the sacrifice of pleasure is less and less. First of all, many vegetarians and vegans get to dislike the taste of meat. Second, modern society is getting better and better at making plant foods that taste like meat. Now that doesn't work for me, because I find imitation meats disgusting, but you really love meat. They make stuff from plant proteins that really looks and smells like meat. It has the texture of meat, and much of the flavor. And now, they can get bacteria to make heme, a blood component that imparts a meat flavor. There are some entirely plant-based "burgers" you can buy now that taste just like meat. And this is just the beginning. It's going to get easier and easier to be a vegetarian. If you're kosher, you will be able to have something that looks and tastes like a grilled cheese with bacon and lobster sandwich, but is entirely made from plants!

ROZIN: What about cultured meat? No animal is killed, just a sampling of some muscle cells.

RUBY: That's a real challenge for vegans. If they find meat disgusting or bad tasting, cultured meat would be of no use. But for compassion and the environment, it should be fine. Cultured pork muscle might be an attractive option for some vegans, but for kosher Jews, it is a definite no. Their issue is not compassion or the environment, but that pork is spiritually and inherently bad. Same with beef for the Hindus. There are many motivations for meat avoidance. Meat eating is really psychologically complicated.

ROZIN: Okay, I have made some substantial changes in my diet, motivated by compassion for animals and concern for the environment. But I want to push back a little, moving on from my example of cultured meat. A big question is whether the aim of vegans is reducing animal suffering and avoiding killing, and improving the environment, or just not eating any animal products. Cultured meat would save animal lives. Consider a predator animal that is killing many herbivores. Killing it would save animal lives, and then eating it should be okay. Is it even possible to be a pure vegan, given all the problems of contamination? You could make vegan contamination rules, as in the kosher tradition, but why should you? The Hebrew view is that pork is inherently polluting, but that is not the vegan view for animal products. Trace contaminants have no significant impact on animal lives or suffering or on the environment. So cutting out 99% of animal products is almost the same as cutting out 100%, and it is a lot harder to cut out 100% than 99%. Purity is really hard to achieve, with minimal gain and great cost.

RUBY: These are good points that I frankly hadn't considered.

ROZIN: And then there is the question of "what is meat?" Insects are animals, related to crabs, lobster, and shrimp. Eating insects, and farming them, could save many other animal lives, and be good for the environment.

RUBY: Well, insects may have some virtues as a human food. They are often seen as disgusting, but disgust can be overcome, and if they don't suffer, this is an arguable point from the vegetarian perspective.

ROZIN: And finally, you point out that if Orthodox Jews became vegan, they could abandon all the rituals involving animal products, like not mixing meat and dairy, and even having separate dishes for meat and dairy. But these rituals have meaning, they connect us to our ancestors, and are expressions of our member-ship in our tradition. Food is a lot more than nutrition and pleasure, it is about identity, social interaction, moral values, and for many people, spiritual beliefs. My Judaism is very important to me, and I do not want to abandon the rituals that connect me to it.

RUBY: Okay, I get it. Meat is really complicated. This is just a first chapter in our discussions. We will have to continue to get to the meat of the matter.

General discussion

The above dialogue raises many issues about meat consumption and avoidance. We discuss them here out of the dialog context, to identify some fundamental issue and conflicts, and address some strategies to improve animal welfare and promote environmental sustainability. Across the world, meat is one of the most preferred and one of the most prohibited foods. This is true in both traditional and modern Western cultures (e.g., Fessler & Navarrete, 2003; Rozin, 2004). The chemo-sensory attraction of meat is similar across most cultures, and at least in Western cultures, across millennia. But the problematic aspects, which motivate the prohibitions, have changed over the millennia, especially in the last 50 years. The compassionate component of meat avoidance, the discomfort about animal suffering and the killing of animals, has almost certainly been present for millennia. The health concerns are modern, rising to prominence in about the last 50 years. The articulated environmental concerns are even more recent, and appear to be rapidly spreading (e.g., Fox & Ward, 2008; Rosenfeld, 2018; Ruby, 2012). However, the spiritual concerns were predominant in biblical times, and are present in many traditional cultures. In the rabbi dialog we have presented, we see the contrast between an ancient, specific set of spiritually-based meat prohibitions, and the modern, morally-motivated movement to reduce or eliminate meat intake. The compassion and more abstract world-oriented environmental concerns are not present in the Hebrew scriptures. It is the word of God that establishes the Hebrew meat prohibitions, and consumption of the prohibited animals acts against the soul, spiritually polluting the self.

We highlight in the rabbi dialog two problems with most prohibitions. One is defining the category of prohibited foods. For example, are insects meat? Is there a sense in which a plant-based imitation meat that looks and tastes like meat belongs in the "meat" category? The second is the universal problem of micro-contamination. The Hebrew Bible sets up rules about this, but vegetarians and vegans have to invent, perhaps with advice from others, what the operative rules are. To what extent do vegetarians and vegans actually make up their own, perhaps idiosyncratic contamination rules, and to what extent do they just decide in the moment when they face a contamination issue? Alternatively, how do personal links between vegetarians and vegans, or organizations or communities of vegetarians or vegans, locally construct rules that apply to their group?

The dialogue illustrates how the two laws of sympathetic magic bear directly on Kosher practices, and also on the path to vegetarianism (Rozin & Nemeroff, 2002). The opening scene of the dialog has to do with whether the appearance of violating the kosher laws (non-dairy cheese on meat) is itself a violation. The law of similarity holds that "the image equals the object – if it *looks* like a tiger, it *is* a tiger."

Although there is a suggestion in Jewish law that this holds, in fact, Orthodox Jews differ on whether it is acceptable, for example, to consume vegetarian bacon bits (Nemeroff & Rozin, 1992). Similarly, it is our observation that vegetarians differ on the acceptability of imitation meat ("looks like meat, is meat"). Of course, the law of similarity is highly adaptive in the ancestral environment, and operates in animals as well as humans. It is only in the modern era, with an abundance of images, that the operation of the law is problematic. The last ten tigers you saw were almost certainly images of tigers.

The law of contagion holds that once in contact, always in contact. When two objects come into physical contact, there is a transfer of essence from one to the other. So far as we know, this is a uniquely human cognition, but it is present in hunter-gatherers and subsistence agriculturalists (Apicella, Rozin, Busch, Watson-Jones, & Legare, 2018). Because of issues of micro-contamination, the enactment of this law is problematic, since molecular levels of contamination are ubiquitous (Nemeroff & Rozin, 1994). Here, the Kosher laws address the problem directly with the 1/60th rule, whereas the budding vegetarian has to improvise, as we have discussed. An interesting question here is whether Orthodox Jewish vegetarians "import" the kosher dietary rules to help them deal with issues like contamination. Informal interviews with a few such individuals suggest to us that in general, they do not import these rules, but more systematic study is warranted.

The avoidance of meat, for spiritual, health, compassionate, or environmental reasons, is opposed primarily by the pleasure most people get from consuming meat. At one level, giving up meat is like giving up cigarettes: there are positive hedonic forces on one side, and less proximal forces on the other side. In both cases, there is a long-term health concern, which is better documented and more extreme for smoking. But the liking for cigarettes must overcome an initial aversion, whereas for meat, there is probably some innate appeal. Most critically, the forces acting against liking for meat are much more diverse than those against cigarette smoking, and include at least three types of moral forces (compassion, spiritual/divine, and environmental; see Ruby, 2012, and Rosenfeld, 2018 for an overview).

There are two major reasons to study the development of vegetarianism in individuals. One has to do with understanding moral decision making, and a second, more practical reason, is to promote more prosocial behavior with respect to animal welfare and the environment. First, becoming a vegetarian is a very apt arena in which to investigate the conflict between pleasure and other motives (health and/or moral). It is particularly interesting because one's own health can be a powerful motivation, but is qualitatively different from the moral concerns about divine commandments, animal welfare, or environmental sustainability. One would expect that the nature of the opposition to pleasure would have implications for how the vegetarian pathway is negotiated. Clearly, a vegetarian motivated primarily by health concerns would experience a major change in diet if research uncovered that eating meat was actually very healthy, but this discovery would likely have minimal influence on a vegetarian primarily motivated by moral concerns. Attitudes to meat and meat preferences may manifest in different ways: for example, morally motivated vegetarians are more likely to find meat disgusting, and hence internalize their meat avoidance, than are health motivated vegetarians (Rozin, Markwith, & Stoess, 1997). Of course, many vegetarians have multiple motives (e.g., Amato & Partridge, 1989; Rosenfeld, 2018; Rozin et al., 1997; Ruby, 2012).

Most critically, on the pathway to moral vegetarianism or veganism, what are the more common strategies, and what is the role of social forces and informal norms? We suspect, because vegetarianism and veganism minimize eating conflicts, that they are the most common "stopping" points on the moral meat rejection trajectory, but this is, of course, an empirical question (see Amato & Partridge, 1989; see also Piazza et al., 2015, for some supporting evidence). The moral meat trajectory offers moral psychology a very dynamic arena, with many instances of morally-laden choices made on a daily basis, and in the absence of an agreed set of laws or rules.

Every possible bite of food can have moral consequences. People have to find ways to negotiate this so they are not confronting moral dilemmas so many times a day. The Hebrew dietary laws, motivated by divine commandments, provide an articulated, scripted base for meat prohibitions. The problem of contamination is explicitly articulated, and boundary rules are established. There is variation, of course, in degree of orthodoxy, but the dietary laws are established early in life. The situation for the beef avoidance of approximately a billion Hindus is similar, though the contamination rules are less articulated. It is not a matter of compassion or health, but rather a divine conception of life that includes reincarnation. Some of the same doctrines that support beef avoidance are generalized, for hundreds of millions of Hindus (about half of the world's vegetarians) to rejection of all meat (Caplan, 2008; Preece, 2008; Spencer, 1993).

Psychologically, the more interesting situation is the often gradual avoidance of meat, as illustrated in Reb Rozin's journey toward vegetarianism. There are moral dilemmas all along the path, often starting with a rejection of meat of (baby) mammals (see also Piazza, this volume). At each "stage," there are boundary issues (what is a baby, what is a mammal, what about levels of contamination?). Amato and Partridge (1989) began to sketch out the stages of becoming a morally motivated vegetarian, but we do not yet know in any detail how the mental conflicts and boundaries are negotiated at each stage in the vegetarian journey. Our recent work (Ruby, Rozin, Gendelman, Li, & Peelish, in prep; Ruby, Quai Hoi, & Rozin, in prep) suggests that a first stage is what we call *conflicted omnivores*. These are people who feel bad about eating meat (for moral and/or health reasons) but continue to consume it. This may be followed by reduction in amount consumed (see Reb Rozin), or in the categorical rejection of certain types of animal food.

The second reason for studying the development of vegetarianism is to facilitate discovering strategies for decreasing meat intake, a practical result which will benefit animals and the planet. There are "how to" books that present a stage of actions that will lead to the goal of vegetarianism or veganism (e.g., Fraser, 2007; Greger, 2015). We hope they are more successful than diet books oriented toward weight loss. A growing numbers of authors are also making the case for incremental solutions, such as substantial reductions in meat intake, and have drawn attention to the difficulties of truly being a "100% pure" vegan (e.g., Kateman, 2017; McWilliams, 2015; see also Leenaert, this volume). The Hebrew dietary laws, the Muslim avoidance of pork, and the Hindu prohibition of beef, along with the fact that roughly half of the world's vegetarians are Hindu, offer one mode of decreasing meat intake, but each of these depends on religious beliefs. For the less religious or secular world, ways of reducing meat intake are less likely to be divinely inspired, but can be, and often are, deeply moral (Rosenfeld, 2018; Ruby, 2012). Raising consciousness about the moral costs of eating meat provides one avenue, along with providing more and more sensory appealing meat substitutes. Of course, as we have noted for orthodox Jews, plant-based imitations of pork may only be effective for those who do not behave in accord with the law of similarity. Furthermore, we should never underestimate the power of social pressure in influencing food choice, which is emerging in some groups of Westerners, and is already present in much of Hindu India (e.g., Appadurai, 1981; Caplan, 2008; Nandy, 2004).

We consider the documentation of, and the understanding of, the process of becoming vegetarian or vegan as a major project, an important way of understanding moral change in moral psychology, a fundamental aspect of understanding the relationship of humans to their food, and an important contribution to the creation of a world that is kind to all animal life and is sustainable. We will need to learn from many sources and disciplines; this chapter represents the blending of information from ancient biblical sources and scholarly work on the modern dilemmas concerning meat consumption. We are just beginning to approach the meat of this very important problem.

Note

1 In some Jewish traditions, it is a convention to refer to the higher power as "G-d" in written form, because of an interpretation of Deuteronomy 12:3, in which Jews were instructed not to erase or deface the Name of God. By only writing the name in an incomplete form, this precludes anyone else from later defacing the name.

Bibliography

Adams, C. J. (2003). *Living Among Meat Eaters: The Vegetarian's Survival Handbook*. New York: Bloomsbury Publishing.

Amato, P. R., & Partridge, S. A. (1989). *The New Vegetarians: Promoting Health and Protecting Life*. New York: Plenum Press.

Apicella, C., Rozin, P., Busch, J., Watson-Jones, R. & Legare, C.H. (2018). Evidence from hunter-gatherer and subsistence agricultural populations for the universality of contagion sensitivity. *Evolution and Human Behavior, 39*, 355–363.

Appadurai, A. (1981). Gastro-politics in Hindu South Asia. *American Ethnologist, 8*, 494–511.

Arbit, N., Ruby, M., & Rozin, P. (2017). Development and Validation of the Meaning of Food in Life Questionnaire (MFLQ): Evidence for a New Construct. *Food Quality and Preference, 59*, 35–45.

Bastian, B., & Loughnan, S. (2017). Resolving the meat-paradox: A motivational account of morally troublesome behavior and its maintenance. *Personality and Social Psychology Review, 21*, 278–299.

Caplan, P. (2008). Crossing the veg/non-veg divide. Commensality and sociality among the middle classes in Madras/Chennai. South Asia. *Journal of South Asian Studies, 31*, 118–142.

Fallon, A. E., & Rozin, P. (1983). The psychological bases of food rejections by humans. *Ecology of Food and Nutrition, 13*, 15–26.

Fessler, D. M., & Navarrete, C. D. (2003). Meat is good to taboo: Dietary proscriptions as a product of the interaction of psychological mechanisms and social processes. *Journal of Cognition and Culture, 3*, 1–40.

Fiddes, N. (1991). *Meat: A Natural Symbol.* New York: Routledge.

Foer, J. S. (2009). *Eating Animals.* London, NY: Little, Brown and Company.

Fox, N., & Ward, K. (2008). Health, ethics and environment: A qualitative study of vegetarian motivations. *Appetite, 50*, 422–429.

Fraser, B. (2007). *Thrive: The Vegan Nutrition Guide to Optimal Performance in Sports and Life.* Toronto, ON: Penguin Books.

Graça, J., Calheiros, M. M., & Oliveira, A. (2015). Attached to meat? (Un) Willingness and intentions to adopt a more plant-based diet. *Appetite, 95*, 113–125.

Greger, M. (2015). *How Not to Die: Discover the Foods Scientifically Proven to Prevent and Reverse Disease.* New York: Flatiron Books.

Grunfeld, D. I. (1972). *The Hebrew Dietary Laws* (two volumes). London/Jerusalem/New York: Soncino Press.

Heinrich-Böll-Stiftung, Chemnitz, C., & Becheva, S. (2014). *Meat Atlas: Facts and Figures About the Animals We Eat.* Berlin: Heinrich Böll Foundation.

Holmes, O. (2018, March 17). "There is no kosher meat": The Israelis full of zeal for going vegan. Retrieved from: www.theguardian.com/world/2018/mar/17/there-is-no-kosher-meat-the-israelis-full-of-zeal-for-going-vegan

Hynes, M. (Host). (2018). *Tapestry* [Radio series]. Toronto: CBC. Available at: www.cbc.ca/player/play/1237137475982

Joy, M. (2011). *Why We Love Dogs, Eat Pigs, and Wear Cows: An Introduction to Carnism.* San Francisco, CA: Conari Press.

Kateman, B. (2017). *The Reducetarian Solution: How the Surprisingly Simple Act of Reducing the Amount of Meat in Your Diet Can Transform Your Health and The Planet.* New York: TarcherPerigree.

Leenart, T. (this volume). Steakholders: How pragmatic strategies can make the animal protection movement more effective. In K. Dhont and G. Hodson (Eds.), *Why We Love and Exploit Animals: Bridging Insights from Academia and Advocacy.* Abingdon: Routledge.

McWilliams, J. (2015). *The Modern Savage: Our Unthinking Decision to Eat Animals.* New York: St. Martin's Press.

Nandy, A. (2004). The changing popular culture of Indian food: Preliminary notes. *South Asia Research, 24*, 9–19.

Nemeroff, C., & Rozin, P. (1992). Sympathetic magical beliefs and kosher dietary practice: The interaction of rules and feelings. *Ethos, 20*, 96–115.

Nemeroff, C., & Rozin, P. (1994). The contagion concept in adult thinking in the United States: Transmission of germs and interpersonal influence. Ethos, *22*, 158–186.

Piazza, J. (this volume). Why people love animals yet continue to eat them. In K. Dhont & G. Hodson (Eds.), *Why We Love and Exploit Animals: Bridging Insights from Academia and Advocacy.* Abingdon: Routledge.

Piazza, J., Ruby, M.B., Watkins, H., Luong, M., Kulik, J., & Loughnan, S. (2015). Rationalizing meat consumption: The 4 Ns. *Appetite, 91*, 114–128.

Povey, R., Wellens, B., & Conner, M. (2001). Attitudes towards following meat, vegetarian and vegan diets: An examination of the role of ambivalence. *Appetite, 37*, 15–26.

Preece, R. (2008). *Sins of the Flesh. A History of Ethical Vegetarian Thought.* Vancouver: UBC Press.

Reese, J. (2018). *The end of animal farming: How scientists, entrepreneurs, and activists are building an animal-free food system.* Boston, MA: Beacon Press.

Rosenfeld, D. L. (2018). The psychology of vegetarianism: Recent advances and future directions. *Appetite, 131*, 125–138.

Rothgerber, H. (2014). Efforts to overcome vegetarian-induced dissonance among meat eaters. *Appetite, 79*, 32–41.

Rothgerber, H. (2015). Can you have your meat and eat it too? Conscientious omnivores, vegetarians, and adherence to diet. *Appetite, 84*, 196–203.

Rozin, P. (2004). Meat. Entry in *Encyclopedia of Food* (S. Katz, Ed.). (pp. 666–671). New York: Scribner.

Rozin, P., Ashmore, M., & Markwith, M. (1996). Lay American conceptions of nutrition: Dose insensitivity, categorical thinking, contagion, and the monotonic mind. *Health Psychology, 15*, 438.

Rozin, P., & Fallon, A. (1980). The psychological categorization of foods and non-foods: A preliminary taxonomy of food rejections. *Appetite, 1*, 193–201.

Rozin, P., Haddad, B., Nemeroff, C., & Slovic, P. (2015). Psychological aspects of the rejection of recycled water: Contamination, purification and disgust. *Judgment and Decision Making, 10*, 50–63.

Rozin, P., Markwith, M., & Stoess, C. (1997). Moralization and becoming a vegetarian: The transformation of preferences into values and the recruitment of disgust. *Psychological Science, 8*, 67–73.

Rozin, P., & Nemeroff, C. (2002). Sympathetic magical thinking: The contagion and similarity "heuristics." In T. Gilovich, D. Griffin, & D. Kahneman (Eds.), *Heuristics and Biases: The Psychology of Intuitive Judgment* (pp. 201–216). Cambridge: Cambridge University Press.

Rozin, P., Ruby, M. B., & Cohen, A. B. (2019). Food and eating. In D. Cohen & S. Kitayama (Eds.), *Handbook of Cultural Psychology* (2nd Ed., pp. 447–477). New York: Guilford.

Ruby, M. B. (2012). Vegetarianism: A blossoming field of study. *Appetite, 58*, 141–150.

Ruby, M. B., Alvarenga, M. S., Rozin, P., Kirby, T., Richer, E., & Rutsztein, G. (2016). Attitudes toward beef and vegetarians in Argentina, Brazil, France, and the USA. *Appetite, 96*, 546–554.

Ruby, M. B., Rozin, P., Gendelman, B., Li, Y. J., & Peelish, N. (in prep). *Conflicted Omnivores: Incidence, Associations, and Implications for Behavioral Change.* Manuscript in preparation.

Ruby, M. B., & Heine, S. J. (2011). Meat, morals, and masculinity. *Appetite, 56*, 447–450.

Ruby, M. B., & Heine, S. J. (2012). Too close to home: Factors predicting meat avoidance. *Appetite, 59*, 47–52.

Ruby, M. B., Heine, S. J, Kamble, S., Cheng, T. K., & Waddar, M. (2013). Compassion and contamination: Cultural differences in vegetarianism. *Appetite, 71*, 340–348.

Ruby, M. B., Quai Hoi, N. S., & Rozin, P. (in prep). Dissonance reduction strategies among omnivores in the USA and Australia. Manuscript in preparation.

Singer, P. (1995). *Animal Liberation.* New York: Random House.

Singer, P., & Mason, J. (2007). *The Ethics of What We Eat: Why Our Food Choices Matter.* Emmaus, PA: Rodale Books.

Spencer, C. (1993). *The Heretic's Feast. A History of Vegetarianism.* London: Fourth Estate.

Twigg, J. (1979). Food for thought. Purity and vegetarianism. *Religion, 9*, 13–35.

18

THE GROUND OF ANIMAL ETHICS

Carol J. Adams and Matthew Calarco

Abstract

The following conversation between Carol J. Adams and Matthew Calarco is part of an ongoing dialogue aimed at trying to discern points of commonality and divergence between a feminist care ethics approach to animal issues and the approach to animal issues found in the writings of French philosopher Jacques Derrida. Here, Adams and Calarco discuss the "ground of animal ethics," by which they mean the root or foundation for our ethical relations with animals. In standard philosophical treatments of animal ethics (found in the writings of influential authors like Peter Singer and Tom Regan), it is often argued that *reason* is what compels us to extend ethical treatment to animals. On this account, whether we have any genuine feelings of affection or concern for the well-being of animals is largely irrelevant; what matters for these authors is that we intellectually grasp the idea that extending ethics to animals is required by reason. Adams and Calarco suggest that the feminist care ethics tradition and the Derridean-inspired approach to the question of the animal proceed from a very different orientation, with a focus on such themes as care, relation, embodiment, suffering, and passion. Although the feminist care ethics tradition and Derrida's work are characterized by considerable differences in philosophical commitments and heritage, on several points the authors uncover surprising and productive intersections that they believe hold promise for innovative work on animal ethics in the future. The inspiration for this conversation and their larger work is their commitment to activism, and the understanding that activism benefits from engaged theory.

Introduction

Welcome to our conversation! In the pages that follow you will find us engaging with the issue of animal ethics. We begin by considering the writings of a French philosopher who discusses being seen naked by his cat (and its implication for his sense of self and philosophy), and then move toward a discussion of suffering, mortality and what a good death would mean for us all. For several years, we – Matt and Carol – have been exploring, through the framework of an extended dialogue, the commonalities and divergences of two approaches to theorizing about human relationships with animals (e.g., Adams & Calarco, 2016). We do so because of an understanding that activism is informed by theory, and if theories that we believe are important and relevant remain little known and misunderstood, they cannot inform activism.

We use the term "ground of animal ethics" to mean the root or foundation for our ethical relations with animals. Philosophical conversations about animals often presume that there is one main approach to animal ethics, derived from Anglo-American thought and characterized by writings such as Peter Singer's *Animal Liberation* and Tom Regan's *The Case for Animal Rights*. The French philosopher we consider most closely, Jacques Derrida, was very influential in the development of an alternative branch of philosophy called "Continental" or "modern European" philosophy. In his book, *Zoographies: The Question of the Animal from Heidegger to Derrida*, Matt explores these writings more closely (Calarco, 2008). Carol has been involved for several decades in developing and expanding feminist ethics of care philosophy to incorporate the-more-than-human world. The anthology she co-edited with Josephine Donovan on this subject, *The Feminist Care Tradition in Animal Ethics: A Reader* provides the history of this approach, and offers many of the foundational essays on the subject (Donovan & Adams, 2007).

The robust theoretical approaches to addressing animals' status found in Continental philosophy and feminist ethics of care philosophy offer an alternative to the traditional approaches, and widen our understanding of how theory happens – that theory takes into consideration the body, emotions, and the socio-political context in which it is being offered. This is the engaged theory of which we talk and to which we aspire. In this sense, there is a further grounding: grounding activism in theory – though, we want to note, we do not want to fetishize theory as something that is immutable.

This is our second dialogue. In our first dialogue (Adams & Calarco, 2016) we addressed what sorts of intersections (or not) might be found between Jacques Derrida's work and Carol's *The Sexual Politics of Meat* (Adams, 1990). Our goal is to make accessible philosophical issues, not so much by comparing and contrasting but through dialogue and exploration.

Carol

You exit the shower, walk down the hall naked, and encounter the gaze of the little kitty who lives with you. This is an ordinary event for many who live with companion animals. But one towel-less man's experience became epochal: Jacques Derrida's. He then used his encounter with the little kitty to begin a discussion about animals in philosophy; a discussion that inherently challenges what philosophy (and not just philosophy) has done to animals.

Derrida insists to us that the cat is neither allegorical nor representational nor fictional. That cat, the cat whose gaze meets Derrida's naked body, was an actual cat, who stretched and padded down the hall; who licked milk from a bowl (probably), and spent a part of each day grooming herself. Here is the first thing I notice when I reread Derrida's (2008) late work, *The Animal that Therefore I Am*: Derrida spends a lot of time insisting on the reality of this cat, this cat whose gaze does not avert itself when a naked Derrida enters her range of vision. Seeing how Derrida must insist on the reality of this specific cat suggests something about his audience; that he is talking to people who think allegorically and representationally – that is, theoretically – rather than literally, as activists might, say, those who trap, neuter and release feral cats, or those who offer foster care for cats awaiting adoption from a shelter. He erects a boundary around this cat: she is not Alice's kitty from *Alice in Wonderland*, nor Montaigne's cat. She is not a cat from "Kafka's vast zoopoetics." Derrida's genealogy of cats that he introduces here is interesting because he is offering a negative lineage of fictional cats. The cat who lives with him, he is saying, is not fictional, is not a part of this fictional group of cats because *she is a real cat*.

I notice this because if I were beginning to talk to a group of say, animal people – animal activists, animal shelter providers, animal sanctuary workers, animal foster families – and I mentioned a cat in the hallway looking at me, they would, without a second thought, assume I was speaking of a literal cat. But Derrida has a different audience; one accustomed to fictional cats who represent something else, rather than themselves.

One of the difficulties for animal activists is speaking literally about literal animals to all the different audiences that are comfortable with the allegorical and representational and fantastical. Zoopoetics and zootheologies and zoomythologies – people inhabit these zoo-spaces perhaps unaware and unquestioning of the functioning of allegory as a distraction from the "real" lives of animals.

It seems as though Derrida, by discussing the actual cat in his life, wants to resist all the allegorical distortions that can occur in addressing the lives of animals. Well, in addressing the *deaths* of animals, too. Language often moves away from the literal facts of animals' lives and deaths especially at the point at which humans are most implicated--causing their deaths in myriad ways. Veering from the literal often results in anthropocentric distortion. Derrida illustrates this point in *The Animal that Therefore I am,* as he segues from the little kitty with her gaze, to the doors of the abattoir, where we often avert our own gazes. Derrida must use words such as "torture," words such as "industrial, mechanical, chemical, hormonal, and genetic violence" to convey something about this literal knowledge (p. 26). He acknowledges "men do all they can in order to dissimulate this cruelty or to hide it from themselves" (p. 26) They hide it from themselves by avoiding the literal, the hidden lives stunted in factory farms and slaughtered through mass production. The literal little kitty leads Derrida to acknowledge and name these hidden aspects of human's oppression of the other animals.

Susan Fraiman (2012) compares Derrida's experience with his cat with my experience of my pony's death at the hands of some youth who were target practicing in the woods behind our pasture.

> Adams's paradigmatic animal encounter brings her "face to face with death." And this is true across Adams's corpus; more often than not, the animals we encounter there are neither canny companions nor prurient pets but the decaying corpses we euphemistically call meat. Made suddenly aware, the night her horse is shot, that she is feasting on dead cow, her first response is similar to Derrida's; shrinking back in shame at the "strangeness" of animals, she dramatizes the nonrecognition allowing them to be killed for human use. Like Derrida, her subsequent work proceeds in a critical mode; instead of celebrating intimacy with animals, she, too, is more interested in tracing the discursive patterns authorizing human violence against them.
>
> *p. 110*

But Fraiman makes a distinction between my "hailing" and Derrida's. She writes: "In contrast to Derrida, however, Adams does not respond to her shame by blushing to be ashamed." Fraiman continues, "Instead, her shame yields quickly to a second impulse: 'I also recognized my ability to change myself: realizing what flesh actually is, I also realized I need not be a corpse eater.' ... Exposed in her shame, she is moved not to cover but rather to examine and reimagine herself. ...A 'metaphysical shift' ...over to the side of animals [is]... a shift inextricable from its occurrence in the early 1970s, underwritten by the civil rights, antiwar, and women's liberation movements."[1]

J. Coetzee suggests that a trajectory of thought arises from an encounter with an animal (which Coetzee labels "a conversion experience" [Cavalieri, 2009, p. 89]). After the conversion experience, he argues, educated people "who place a premium on rationality" proceed to seek backing for their new perspective in the writings of thinkers and philosophers. Or, as Derrida discovers, *not* find such backing in the writings of philosophers and use that absence as the basic for reflection. Or, in my case, worrying about the idea for 15 years and drawing on feminist literary criticism to call the dissimulating of the cruelty, the hiding of what humans are doing to other animals, the "absent referent" (Adams, 2015, p. 21). As I explain in *The Sexual Politics of Meat*, behind every meal of meat is an absence: the literal death of the nonhuman animal whose place the meat takes. The absent referent is that which separates the meat eater from the other animal and that animal from the end product. The structure of the absent referent allows the animal to disappear both literally and conceptually. Humans do not regard meat eating as contact with another animal because it has been renamed as contact with *food*. Who is suffering? No one, people rationalize.

Matthew

Allow me to pick up on a couple of the themes you mentioned, specifically with regard to Derrida's emphasis on the "real" cat with which he has an encounter. You are right that much has been written about this encounter, but there are several ethical and ontological themes in this portion of his text that have received inadequate attention, and your remarks have helped to pinpoint some of those issues for us.

Like you, I think it is quite important to note the details of the scene of Derrida's address and how different that scene might be from the kinds of audiences that you or I or many of our readers might encounter or participate in as activists, educators, etc. These kinds of details concerning the scene of enunciation are, in the context of Derrida's thought, not minor ones, and in this essay he makes a determined effort to mark the place of delivery of this particular text. The delivery of the paper takes place in front of a number of long-time friends and philosophers in Cérisy, France, at a conference dedicated to his work. It is the third in a series of such conferences "*autour de travail de Jacques Derrida*" (around, or in regard to, the work of Jacques Derrida). It is interesting to note that all three conferences, which are spread out across some fifteen years, touch on themes central to his discussion of the question of the animal,[2] something Derrida himself did not notice until after he had written his remarks. He uses this coincidence to insist that the so-called question of the animal has been stalking *him* for some time and has in a certain sense been awaiting him, dragging him along a certain path. This approach to his presentation will be important for understanding the scene of his encounter with the cat as he analyzes it later in the text. In brief, we can see already that his discourse cannot serve as a foundational ground for any kind of animal ethics, for it is itself already built upon another experience that is non-foundational or, to use his words, "abyssal."

There is another, less prominent and almost invisible scene at work in Derrida's address, and that is the scene of the reception of his work, especially in the United States and other English-speaking countries where Derrida's work is quite popular but not well understood. His work is often read here as belonging to a strain of "postmodern" idealism and anti-realism that would suggest we are trapped in the "prison-house of language" (to borrow Fredric Jameson's phrase, which conjures up images of human beings caught up in the nets of language with no way of accessing reality). So, when Derrida refers to his cat as being a "real cat," this is bound to appear naive to some of his readers. After all, how could someone as sophisticated as Derrida be talking about a *real* cat, one independent of human language and cognition? Is he not aware that his thought has trapped us within language and made such reference impossible? Such would be the standard reading and reception of his work.

Derrida is cognizant of such presuppositions about his work. He knows that most readers will think he is referring to a text about or a figure of a cat but never a real, singular cat who exceeds the bounds of cognition and language. It has now been some twenty years since Derrida first delivered this piece, and fortunately this kind of misreading of his work is slowly being displaced. As has been evident all along, Derrida has never claimed reference is impossible; instead, he has sought to insist on the unavoidable need to engage in *interpretation* when referring to Others. The point of deconstruction is to mark the limit at which the Other (whether human or non-human) exceeds one's ability to achieve full mastery and full knowledge of that Other. Put simply, deconstruction takes us carefully and critically to the edges of language and cognition so that we might glimpse that which lies beyond its purview. Thus, when Derrida insists repeatedly that he is referring to a real cat, he is in fact saying something that is consonant with his previous work. If one reads his words carefully, one sees Derrida repeatedly referring to a cat who is quite real but is not for that reason somehow fully *transparent* or fully *present* to him.

In one sense, I would agree that *The Animal that Therefore I Am* and the very personal nature of that text differ profoundly from the more abstract and impersonal critical analyses of the question of the animal that one finds in his earlier writings. At the same time, though, one could also place all of this work in the same line, especially if the more abstract critical work that he did early on is seen as the necessary prerequisite for the more experimental and personal work that he carries out in the late 1990s. One could even reverse the chronological order of presentation and suggest that the abstract theoretical texts are themselves only possible *in response to* the kind of event with an animal that Derrida describes in *The Animal that Therefore I Am* – which is to say, that the philosophical question of the animal is, for Derrida, a response to something that exceeds the space of philosophy and its mode of critical questioning. In this sense, then, *The Animal that Therefore I Am* could be read as simultaneously being both one of his earliest and latest texts, and one that both makes possible and is a consequence of his writings on the question of the animal.

Carol

After Derrida discusses his thoughts after his encounter with the little kitty, he characterizes what has happened in the past two hundred years vis-à-vis animals as a "war on pity." It is a double-edged war: discouraging sympathy for what animals are experiencing and dismissing "sympathy" as a philosophical intervention.

When you and I talked about *The Animal that Therefore I Am* at one time, we discussed Derrida's categorization of two slopes of animal discourse, the first, the traditional discourse of philosophy – men writing books – and the second, "woman's discourse." That what Derrida is saying, in a sense, is "In order for me to talk about animals, I have to leave philosophy behind, I have to take up 'woman's discourse – poetry.'"

Matt, you have said in our previous conversations that both Derrida and I place animal ethics on an alternative foundation, namely, the domain of care, emotions, and passion as opposed to reason and cognition. We who write from this perspective are concerned with *attention*. As Rita Manning (1992, p. 45) notes, caring requires "a willingness to give ... lucid attention to the needs of others." Care is a kind of ground, and relation with others is also a kind of ground in determining how I talk about and work for animals.

Together with Josephine Donovan, we assembled writings about animals based on a feminist understanding of sympathy and care. We called the book, *The Feminist Care Tradition in Animal Ethics* (Donovan & Adams, 2007). Though this approach is robust in theory and examples, it often fails to be engaged on equal grounds with analytic or Continental philosophical approaches to animals. Is it because of the socialization of men (and needless to say, most philosophers) to consider sympathy and compassion for animals as unmanly and feminine? Often analyses of the ethical approach to animals are summarized in this way: rights-based, utilitarian, or the new Continental philosophical approach. How woefully inadequate a characterization of animal ethics it is, and yet, perhaps, symptomatic of the more general derision of compassion in society at large. Sometimes you can find a nod to a religious-based notion of compassion (and such a focus, on the religious influences on "compassion" makes the feminist tradition invisible), but, in general, these surveys reflect a sort of male-oriented fetishization of reason by the gatekeepers of philosophy.

Our work, intervening in ethical theory from a sympathy tradition, draws upon Carol Gilligan's *In A Different Voice* (1982). Gilligan identified a women's "conception of morality" that is "concerned with the activity of care ... responsibility and relationships," as opposed to a men's "conception of morality as fairness," which is more concerned with "rights and rules." Feminist ethic-of-care theory derives from Gilligan and her understanding of "a morality of responsibility" in contrast to the masculine "morality of rights." Gilligan identified how the "morality of rights" emphasizes "separation rather than connection" and focuses more on the autonomy of the individual than on the context and relationship (p. 19). Though sometimes her critics say Gilligan is essentializing, that is equating women with a quality inherent to our gender, in fact, her theory does not essentialize the qualities of caring but identifies how their association with women has caused them to be unvalued.

Until the recent development of approaches to animal ethics signaled by certain aspects of Continental philosophy, US-based analytic philosophical approaches seemed to dominate the field, in particular the work of Tom Regan and Peter Singer. Their positions arise from the dominant philosophical position that devalues, suppresses, or denies emotions. This means that a major basis for the human-animal connection – love – is not engaged by their approach. Since the exclusion of the emotional response is a major reason why animal abuse and exploitation continue, it seems contradictory for animal defense advocates to also claim, as Regan (1983) and Singer (1975) have done in *The Case for Animal Rights* and *Animal Liberation*, that feelings are inappropriate guides to ethical treatment. As Donovan and I argue in *The Feminist Care Tradition in Animal Ethics,* both rights and utilitarian theory developed in the eighteenth century, the so-called Age of Reason, and reflect its rationalist ideological roots. They envisaged a society of rational, autonomous, independent agents whose territory or property is entitled to protection from external agents (other people and the government). Problematically, these theories require an assumption of similarity between humans and animals; they become both anthropocentric and anthropomorphic in their arguments. That is, they assume "human" as the model, and the human form as the example.

The ontology that underlies rights discourse presumes a society of equal autonomous agents, who require little support from others, who need only that their space be protected from others' intrusions. Animals are not equal to humans; domestic animals, in particular, are for the most part dependent for survival upon humans. We therefore have a situation of unequals and need an ethic that recognizes this fact. In general, the vision of the equal, autonomous individual (male) represented in rights theory, ignores the network of supporting persons (usually female) who enable his autonomy – that is, who raise him, feed him, clothe him, and so forth. In short, rights theory ignores the fact that most humans and animals operate within an interdependent network, and it provides no obligation to care for those who are unable to operate autonomously. New work by Sunaura Taylor (2017) in her book *Beasts of Burden: Animal and Disability Liberation* shows how limiting it is to think of any of us as autonomous subjects.

Finally, the rights approach (as well as utilitarian interest theory) tends to be abstract and formalistic, favoring rules that are universalizable or judgments that are quantifiable. Many ethical situations, however, including those involving animals, require a particularized, situational response – one that considers context and history, and is a response that may not be universalizable or quantifiable. (This is a summary of a discussion that can be found in the Introduction to *The Feminist Care Tradition in Animal Ethics*, Donovan & Adams, 2007, pp. 5–6.)

Rather than being an abstract, rule-based, universal ethics, the *feminist ethics of care* is a flexible, situational, and particularized ethic. As with feminism in general, care theory resists hierarchical dominative dualisms, which establish the powerful (humans, males, whites, able-bodied) over the subordinate (animals, women, people of color, disabled). The feminist ethic of care sees animals as individuals who have feelings, who can communicate those feelings, and to whom therefore humans have moral obligations. An ethic of care also recognizes the diversity of animals – one size does not fit all; each has a particular history. Insofar as possible, attention needs to be paid to these particularities in any ethical determination regarding them.

A feminist ethic-of-care approach to animal ethics also offers a *political* analysis that considers why animals are being abused, used, killed, and consumed. We place individual instances within a political understanding. The feminist care approach does not lose sight of the importance of each individual animal but also develops a more comprehensive analysis of her situation. As Josephine Donovan eloquently summarized it: "Contrary to Kantian rationalism, it envisages both the personal and the political. Like Buber, people exercising attentive love see the tree; but they also see the logging industry. They see the downed cow in the slaughterhouse pen; but they also see the farming and dairy industry. They see the Silver Spring monkey; but they also see the drug corporations and university collaboration" (2007, p. 192).

I am trying to figure out if attention – which engages from a position of care, and acknowledges relationships – is *exposure* in Derrida's lexicon?

Matthew

Your attempt, along with Donovan's, to rethink animal ethics by way of a feminist ethics of care is a genuinely important development and challenge to the analytic ethical discourse that has thus far dominated the discussion. To follow up on some of the points you raised: First, I would note that Gilligan's general reorientation of ethics around relationship and care is a crucial corrective to the overly abstract and decontextualized forms of ethical discourse that have long dominated analytic normative theory. Second, the effort you and Donovan make to extend feminist care ethics to animals is transformative, especially given that most of the early versions of feminist care ethics either ignored nonhuman life altogether or explicitly relegated nonhuman life to second-class status.

A third point worth mentioning, and one that also resonates with much of current ecofeminist theory and ethics, is that your approach, and that of Donovan, is grounded in *difference* rather than symmetry, reciprocity, and identity. You not only allow for difference but stress its irreducibility, and you emphasize the importance of imagination and empathy with respect to Others in ethics. These are all themes that are either downplayed or non-existent in the sort of mainstream philosophical ethics deriving from folks like Peter Singer and Tom Regan but that are consonant with ethical thinkers in minor, alternative traditions that have become important for many people working in critical animal studies.

In terms of other problems with mainstream approaches, I am not persuaded that utilitarian (Singer) or rights-based ethical theory (Regan), or any other theoretical normative approach for that matter will capture all of the salient features of animal ethics; nor do I accept the notion that the primary task of animal ethics is to determine and then refine a single, ultimate ethical theory and framework. This kind of belief in and desire for monism (that is, a sole, correct framework) in ethics is a serious problem inasmuch as it compels us to take the richness of ethical relationships with, between, and among human beings and animals and other beings and reduce it down to a single register of analysis.

As you might have guessed, then, the scenario that you describe above where deontology and consequentialism are now passé and feminist ethics goes entirely undiscussed because of the emergence of a "new" Continental approach to animal ethics (that will somehow move beyond these earlier approaches) is not something I would endorse. In my own work, I have tried to do something very different with ethics and have tried to open a space for thinking about ethics in an entirely different way. Rather than trying to get rid of consequentialism, deontology, feminist ethics (or virtue ethics, social contract theory, and so on) in the name of a more Continental-based approach to ethics, I think the more important thing to do is to reframe the way we think about what we are actually doing when we engage in ethics (both theory and practice). This is a point we can return to in more detail later.

The feminist ethical notion of grounding ethics and relation in embodiment (which is to say, in full-bodied, enfleshed creatures) constitutes, in my estimation, one of the fundamental sites of overlap with a number of important Continental ethical thinkers.

Since we are considering Derrida's ideas about ethical relationships with animals, I would add that for Derrida in particular, a thought that begins from the finitude of embodied being (understood as the exposure, nudity, and vulnerability concomitant with embodiment) goes right to the very heart of nearly all of his work on animals. His idea of finite exposure speaks less to the double sense of attentiveness you mention and more to the opening that makes this kind of ethical disposition possible. He is by no means opposed to that kind of ethical attentiveness (indeed, the general thrust of his work is consonant with it), but (as with Levinas) his focus lies more in considering its conditions of possibility in exposure. All of this is to say that Derrida's thinking starts from a reflection on the ways in which finite, embodied beings are affected by Others of various sorts.

You mention the manner in which Derrida delineates two major kinds of philosophical discourse on animals and the emphasis he gives to the location of women among these two kinds of discourse. To reconstitute that section of the text very quickly, Derrida suggests that the history of discourse on animals has been dominated by a specifically male philosophical discourse that is grounded in an objectifying gaze, a gaze that refuses to acknowledge that the knowing philosophical subject sometimes finds itself already being watched and looked at by another animal. The dominant philosophical discourse disavows this animal look and only registers its own outward gaze and subjective site. In taking his point of departure from the very animal gaze that the male philosophical tradition disavows, Derrida joins the other form of discourse on animals, among which he counts poets and prophets, men and women who have seen themselves being seen by another animal and who register that decentering glance in their discourse.

Derrida states that the very *bodies* of the signatories mark the difference in discourse here; and although this other, non-traditional mode of discourse is not inhabited solely by women, he seems to find only a small number of men (and no statutory representatives of the philosophical tradition) on this side of the discussion, only a few men who are willing to begin thinking from their embodied exposure to an animal rather than from a masterful gaze directed outward toward animals and others of various sorts. In effect, then, he is marking the space of his text as non-male, non-masculine, and non-philosophical in the traditional sense of the term. That said, even the most charitable reading of Derrida has to acknowledge that his negotiating of the highly charged, gendered, and hierarchical dimensions of embodiment, nudity, and exposure leave something to be desired; and I think it would be misleading to say that his work does full justice to the complicated set of issues that arise here. Feminist thought, along with other critical and radical discourses on embodiment, are to my mind much better situated to help us work through the kinds of concerns you raise in your remarks.

Carol

David L. Clark, an English professor who has written compellingly about the presence of animals in human atrocities (specifically the Holocaust) and is interested in how animals are at odds with philosophy – in a sense interested in animals *after* philosophy – suggested to me that Tom Regan and Peter Singer are not symptomless in their choice of philosophical approaches. They might have done something differently; they might have recognized the contingent nature of rights and utilitarianism. Regan and Singer had the possibility of articulating their position in a way that didn't make them "dated" or anachronistic, but seemed to refuse to do so. We have discussed how animal issues are tacked onto their preceding philosophical orientations which is another way of saying that the theoretical and normative commitments of most philosophers seem to precede their activist commitments. This allows their normative commitments to trump what is in front of their face. For me, after the publication of *The Sexual Politics of Meat* (and for that I drew more on feminist literary criticism than feminist philosophy), I found that the feminist ethics of care helped me articulate how we might try to talk about our relationships with the other animals. It is an activist model that creates links between communities and, I believe, care opens up possibilities.

Matthew

You and I both come to these philosophical issues from an activist orientation. This gives a pragmatic orientation to our work. We ask: "Given that we have a problem in regard to animals, what kinds of frameworks best help us think about our critical strategies and responsibilities?" It is the understanding that there are real problems with animals that prompts us to search out normative frameworks and critical theories to help us respond. As such, if we ran into a wall with a given framework, we would not hold onto to it at all costs. Reconfiguring our relationships with and responsibilities toward animals might require experimentation with multiple kinds of frameworks.

The normative theories we have been discussing (consequentialism, deontology, etc.) are typically understood as mutually exclusive, stand-alone theoretical frameworks that provide an ultimate account of moral consideration and moral status. In other words, they are supposed to provide a final answer to the questions of the essential nature and scope of ethics, of who counts ethically and how much, of where ethical consideration is supposed to begin and end. In opposition to this approach, I would suggest that such ethical questions need to remain structurally open, and that the answers offered to these questions should be seen as permanently revisable. Along related lines, I would also suggest that ethical theories should be understood positively as a set of pragmatic tools that might help us become attentive to registers of ethical reality that we might otherwise miss and that might assist us in contesting existing ethical dogmas and prejudices. Thus, rather than viewing any of the dominant ethical theories as capturing the ultimate truth about ethics in terms of scope and content, those theories could be rethought as helping us to attend to such things as sentience, subjectivity, relation, compassion, love, and so on across myriad registers and modes of relation.

If we approach ethics outside of the language game of analytic normative theory (with its emphasis on logical coherence, monism, rigor, etc.), some of the central ideas of the dominant approaches to animal ethics can be given a different life and allowed to circulate differently alongside your approach. Attending to animal sentience (as central to the utilitarian approach) or the subjective lives of animals (as central to the rights tradition) need not lead directly into a full-fledged consequentialist or deontological approach to animal ethics. Such dispositions and practices can be delinked from that language game and deployed in other contexts, beyond morality and normative theory. They are not the exclusive province of Singer, Regan, or the traditional normative theories within which they develop their frameworks.

So, from the perspective that I am giving voice to here, feminist ethics would not be something that can or should be overcome by a new or traditional alternative ethical theory. Instead, it would be seen as an irreducibly important approach to ethics inasmuch as it urges us to consider the importance of networks of relation and responsivity to singularity over and against a culture and way of life that have closed us off from these aspects of ethical life.

Of course, I would also want to rehabilitate certain strains of rights-based and utilitarian thinking in this same vein, taking the theories away from the philosophers and activists who wish to turn them into ultimate theories and bringing them back into the space of thought and life as challenges to our tendency to overlook the bodies and singular lives of animals, humans, and beings beyond animals and humans. In effect, then, I am suggesting that we place ethics in a space beyond good and evil, beyond morality. The aim of ethics in this extra- or non-moral space would be to dispense entirely with the search for ultimate accounts, decision-procedures, moral judgments, monistic theories, and so on, and instead to provide ways of thinking through the implications and potentialities of relation, encounter, and exposure. What possibilities for thought and for life are opened up by a given ethical theory? What registers of relation and experience might we be able to glimpse that we might have otherwise missed without the assistance of a particular framework? It would be these kinds of extra-moral questions and concerns that would guide the development of ethical discourse and practice.

Carol

I love this approach and hope that we are able to experience the openness to multiple approaches within animal activism. Animal activism has often chosen the opposite: a fetishizing of one approach and the dismissal of others. This approach of animal activism – often linked to vocal male leaders in the animal rights movement – delimits conversation, exploration, and solutions. What might be a robust encounter with philosophical thought is made instead canonical and unable to be challenged. One hierarchy, male leadership, inscribes another, the one and only approved philosophical approach. In this way, the movement and its theoretical approach always seems to move away from its ground of the diverse lives and experiences of animals. This also influences how animal suffering is discussed, so the claim that the movement is the voice of the voiceless emerges. Such a claim is a conservative, ableist, inaccurate claim. By privileging human voice, without explicitly acknowledging that the reference is the human voice, it leads to a savior mentality in activism. During the summer of 2018, Julia Feliz, Meneka Repka, Carolyn Bailey, and I addressed this (and other activist issues) in a "Consistent Bill of Anti-Oppression," writing "Shouldn't our role be to ensure we bring attention to the abuses and exploitations done by humans rather than to assume we speak for them [the other animals]? ... We could ask [instead], 'Why do we live in a society that privileges the human voice socially and politically?'" (see www.consistentantioppression.com/the-why-what/saviorism/).

I suspect the summoning of "the voiceless" is an attempt to evoke the unacknowledged suffering of the other animals. The feminist ethic of care is concerned, of course, about suffering. When one is inured to suffering one is able to rationalize the suffering of another. I suspect that this shows not the inability to care, but a fear of painful consequences of caring. The majority of people who consume dead animals do not want to know how animals die. So I want to acknowledge that caring about someone's suffering matters.

As we talked in preparation for this dialogue, you mentioned that by the time you have gotten around to talking about the suffering of animals, the animals have already gotten under your skin. They have gotten under our skin because we have paid attention and we have experienced something through that attention-paying. But the feminist ethic of care cannot be reduced to an ethic that focuses on responding to the issue of suffering. Caring about animals does not require that they suffer. I find it strange that this issue of suffering is the direction in which Derrida appears to turn. Derrida writes: "The *first* and *decisive* question would rather be to know whether animals *can suffer*. 'Can they suffer?' asks Bentham, simply yet so profoundly."

Let me say, first, how glad I am that Derrida said this. It's a grounded, applied ethics. But Derrida's turn to Bentham disconcerts me because when we make that claim that animals matter because they "suffer" we have already given up an important aspect of the debate: we accept that we have to prove about animals something we don't have to prove about the carnophallogocentric subject. This is how Derrida refers to the dominant schema whereby one is recognized as a full subject. Genuine human subjectivity within this schema is understood as involving the consumption of animal flesh (carno-), being a man (phallo), and being a competent speaker and user of language (logos). This subject – the dominant subject of Western culture – does not have to prove that they suffer to matter as an individual. Catharine MacKinnon put it this way, "But fundamentally: Why is just existing alive not enough? Why do you have to hurt? Men as such never had to hurt or to suffer to have their existence validated and harms to them be seen as real. It is because they are seen as valid and real to begin with that their suffering registers and they have rights against its harm" (2007, p. 326).

Have we already given up the debate when we make that claim that animals matter because they "suffer"? What about an animal who does not suffer – doesn't that animal still matter? Why does only the victimhood of the animal bring her into the realm of moral compassion? And why would a movement that wants to alleviate this very status of victimhood choose to take it as its symbol and rallying cry? Is this not how we end up with the claim of being the voice of the voiceless when they aren't voiceless at all?

The little kitty matters to Derrida not because she is suffering, nor, I would suggest, in that encounter is her *capacity* to suffer at question. (More likely, it is his capacity to suffer at her paws!)

Perhaps Derrida finds Bentham's question "can they suffer" *the* question to ask because it refuses logocentrism – which as we indicated above assumes that a subject possesses the ability to be a competent speaker and user of language (logos). Yet, the minute you turn to Bentham you have made this larger conservative turn that assumes we have to prove suffering for an animal to matter. This assumption accepts a fundamental difference between those animals and the logocentric individuals you are trying to disturb, since those individuals don't have to prove their suffering in order to matter. They *de facto* matter. And can you disturb someone when you accept a basic premise of their ontology? To need to see suffering seems to limit exactly the exposure Derrida is describing in terms of how we are to think about experiencing other animals. Does Derrida's invocation of Bentham prove the paucity of the philosophical tradition in terms of its consideration of animals?

Matt, I would like to hear your thoughts on why Bentham appears here in Derrida's work.

Matthew

One of the more intriguing features of *The Animal that Therefore I Am* is this invocation of Bentham's ethical question about animals. Derrida more typically prefers to read classical figures in the history of philosophy and uncover lingering dogmas and/or latent radical potentials in their work. He rarely analyzes minor figures like Bentham, and he rarely follows so closely in step with the positions of another philosopher. And yet, even though Derrida seems to be invoking Bentham uncritically, citing him as an authority, and following him along the path of a utilitarian discourse that can have very conservative consequences, I would suggest that Derrida is treating Bentham much like the other philosophers he analyzes. There is a strong *reading* of Bentham in *The Animal that Therefore I Am*, one that pushes against the grain of his text and the received interpretation we find among contemporary consequentialists.

To illustrate this point, let me take up the issue you raise of how Bentham's framework potentially limits ethical consideration to animal sentience (understood as animal suffering). To be sure, this reading of Bentham is primarily how consequentialist approaches to animal ethics have proceeded, placing almost exclusive emphasis on animal pain and pleasure and figuring both within the context of a utilitarian calculus. But we should note that this particular understanding of sentience is of little interest to Derrida and that his reading of Bentham is quite different from Singer's. For Derrida, Bentham's question *"Can they suffer?"* is read against the grain in order to move the analysis back one step further than the typical utilitarian approach would go. Bentham's question about animal suffering is, for Derrida, a question about the nature of animal bodies and bodies as such. A being who suffers (in the etymological sense of bearing or undergoing something) is first and foremost a finite and embodied being who is ex-posed (literally, placed or positioned outward), open, thrown beyond itself, touched by and touching other bodies, a being who lacks full sovereignty and is subjected to forces beyond its purview. This kind of suffering circulates among a network of affects and relations that precede articulate speech, even if those relations can subsequently be marked in language. As such, a thinking that begins from finitude and exposure contests one of the chief limits of logocentrism, and this is why Derrida suggests that Bentham's question (if read against the grain) actually testifies to an epochal break in the history of Western thought. Rather than asking if animals have one of the "propers" of man – that is, one of the classical traits (such as "logos") that are supposed to define an exclusively human nature and give humans ethical priority – Bentham asks whether they can suffer; and in so doing, he places animals and human beings on the same plane of finite exposure.

In brief, then, Derrida suggests that Bentham's question allows us to shift the ground of the Western philosophical tradition, and allows us to think the ground of animal ethics differently. The shift in the form of the question of the animal that we find in Bentham allows us to move away from a focus on the essence of the human and into the space of *passivity*. Both human beings and animals share in this passivity, according to Derrida, but it is not a "property," a trait, a capacity, or something that one has under one's power. The passivity at issue is radical, and it captures both human beings and animals within its scope. It is important to notice the various terms Derrida uses to stress the fundamental nature of this passivity and the way that it falls outside of the sovereign control of those who are captured by it: sufferance, passion, not-being-able, vulnerability, nonpower, and so on.

In this sense, suffering is understood by Derrida to be something much more basic and much more radical than sentience. Suffering is understood as the site in which relation takes place and through which affect passes. Suffering is not equivalent to the abstract concept of embodiment but rather refers to the fact *that* there are exposed, vulnerable, singular bodies which are open to and affected by otherness and relations *of all sorts*. This is why Derrida actually shares much in common with your concern that our ethical encounters with animals not be reduced to suffering and victimhood. To be embodied is to be exposed to suffering in the utilitarian sense, yes, but not just that. It also involves being exposed to overwhelming joys, beauties, and remarkable relations of many sorts. Derrida is thinking from and in view of the "ground" of the potential Others that might affect humans and animals; and this is why for him, if radical passivity is the ground of animal ethics, then animal ethics lacks a solid foundation. Finite embodiment provides us with no assurances, certainties, or foundations. It places us all – animals, humans, and all other finite beings – in a shared space of experimentation, the effects and consequences of which can neither be delimited nor mastered in advance.

Carol

What you say is so helpful in helping to free Bentham (and Derrida!) from the utilitarian lens in which I was hearing the statement about suffering. As you know, I have spent a great deal of time thinking about our shared finitude with all animals since we began our discussions several years ago. While I was caregiving for my mother, through years of Alzheimer's disease destroying certain aspects of her, and then as she was dying, my veganism moved to a deeper place. It wasn't a conscious act. Perhaps I was caught in the ungrounding of the ground of animal ethics. I don't know, but I know it kept me connected to others when I felt isolated by my care-giving. And the tenderness of our care for her as she was dying is a part of my deepened veganism. As I wrote in *Critical Inquiry*: "I had the opportunity to be a part of the completion of the mother–child relationship, but I am sadly aware that the mother–child relationship in animal agriculture is broken from birth forward. I had the full experience of a relationship, and farmed animals, especially cows and their calves, have nothing. Also, my mother had what I believe to be a good death; this made much more vivid the staggering nature of the bad deaths the other animals experience in becoming or producing food for nonvegans"[3] (2017, p. 788). My sense of the meaning of veganism opened further as my mother's life closed.

I also recognized how the nonverbal aspects of caregiving – especially when the cognitively impaired use language differently or not at all – decentered human exceptionalism. When something was wrong with my mother yet she was unable to say what it was, we used our senses to try to figure it out. These same senses can be used in our relationships with other species.

When my mother was dying, my father, my sisters and I worked very, very, hard with my mother's caregivers to give my mother a "good death" – a death at home, with words, singing, massage, hugging, touching. When I read some of the Continental philosophers I feel their depiction of death is exactly the opposite of this: heroic, solitary, masculine, something we face by ourselves. It seems as though this depiction of human death results as well in a privileging of human death.

Practically and theoretically, death is something we have in common with all animals as you say, we are all exposed vulnerable bodies. Why, for philosophy, isn't that something that connects us to animals? If we worked out of the idea of a good death, and we had deconstructed the notion of species – that often is sustained or upheld by the idea that animals lack of consciousness about their own death – why wouldn't we then want everyone to have a good death, or as "good" a death as pos-sible, including the animals? And so, here we have come back to our beginning –the finitude of the human in encountering other animals, as Derrida described after his encounter with that very real cat.

Acknowledgments

Vasile Stanescue suggested Matt and Carol collaborate on a discussion about Derrida; this is the second conversation, the first can be found in Annie Potts (2016, ed.), *Meat Culture*.

Notes

1 Fraiman is quoting from *Neither Man nor Beast*, p. 155.
2 For Derrida's previous conference contributions on these themes, see Derrida (1982, 1993).
3 I describe this and other ways veganism helped me as a caregiver in the chapter "Veganism and Caregiving," in Adams, Breitman, and Messina (2014, pp. 183–204).

References

Adams, C. J. (1990–2015). *The Sexual Politics of Meat: A Feminist-Vegetarian Critical Theory.* (Bloomsbury revelations edition). London: Bloomsbury.
Adams, C. J. (1995–2018). *Neither Man nor Beast: Feminism and the Defense of Animals.* London and New York: Bloomsbury.
Adams, C. J. (2017). Towards a philosophy of care through care. *Critical Inquiry, 43* (4), 765–789.
Adams, C. J., Breitman, P., & Messina, V. (2014). *Never Too Late to Go Vegan: The Over-50 Guide to Adopting and Thriving on a Plant-Based Diet.* New York: The Experiment.
Adams, C. J., & Calarco, M. (2016). Derrida and the sexual politics of meat. In A. Potts (Ed.), *Meat Culture* (pp. 31–51). Leiden and Boston, MA: Brill.
Calarco, M. (2008). *Zoographies: The Question of the Animal from Heidegger to Derrida.* New York: Columbia University Press.
Cavalieri, P. (2009). *The Death of the Animal: A Dialogue.* New York: Columbia University Press.
Derrida, J. (1982). The ends of man. In J. Derrida, *Margins of Philosophy* (pp. 109–36), trans. Alan Bass. Chicago, IL: University of Chicago Press.
Derrida, J. (1993). *Aporias: Dying – Awaiting (One Another At) the Limits of Truth*, trans. Thomas Dutoit. Stanford, CA: Stanford University Press.
Derrida, J. (2008). *The Animal that Therefore I Am.* In M.-L. Mallet (Ed.), trans. David Wills. New York: Fordham University Press.
Donovan, J. (2007). Attention to suffering. In J. Donovan & C. J. Adams (Eds.), *The Feminist Care Tradition in Animal Ethics: A Reader* (pp. 174–197). New York: Columbia University Press.
Donovan, J., & Adams, C. J. (2007). *The Feminist Care Tradition in Animal Ethics: A Reader.* New York: Columbia University Press.
Fraiman, S. (2012). Pussy panic versus liking animals: Tracking gender in animal studies. *Critical Inquiry, 39*(1), 89–115.
MacKinnon, C. (2007). Of mice and men: A fragment of animals rights. In J. Donovan & C. J. Adams (Eds.), *The Feminist Care Tradition in Animal Ethics: A Reader* (pp. 316–333). New York: Columbia University Press.

Manning, R. (1992). Just caring. In E. B. Cole & S. C. McQuin (Eds.), *Explorations in Feminist Ethics: Theory and Practice.* Bloomington, IN: Indiana University Press.

Potts, A. (Ed.) (2016). *Meat Culture.* Leiden and Boston, MA: Brill.

Regan, T. (1983). *The Case for Animal Rights.* Berkeley, CA: University of California Press.

Singer, P. (1975). *Animal Liberation: A New Ethics for Our Treatment of Animals.* New York: Harper Collins.

Taylor, S. (2017). *Beasts of Burden: Animal and Disability Liberation.* New York: The New Press.

19

SO WHY DO WE LOVE BUT EXPLOIT ANIMALS?

Reflections and solutions

Gordon Hodson and Kristof Dhont

Abstract

Humans prioritize and value humans over animals in virtually every way, at considerable cost to both animals and humans. In addressing why people paradoxically love but exploit animals, the present book uniquely assembled some of the most prominent voices on human–animal relations to bridge insights between academia and advocacy. This chapter synthesizes their overall discussions around three broad themes. First, we consider the *nature of the problem*, including topics such as animal welfare, undervaluing animals, competition with animals, social (dis)identification with animals, cultural influences, ideology and politics, and the intersection of speciesism with anti-human prejudices (e.g., racism). Second, we explore *how people live with the paradox*, involving discussions of biases in human thinking in general (and with regard to animals in particular), complications inherent in the notion of "moral" thinking, and the human propensity to rationalize the status quo. Third, we reflect on *solutions and remedies*, including a focus on psychological constructs (e.g., perceived human–animal divide; empathy), plus calls to redirect future goals and actions, bolster humane education, leverage prosocial aspects of human psychology to benefit animals, and engage in evidence-based advocacy.

In terms of human–animal relations, we certainly live in interesting times. Some right-wing extremist and neo-Nazi groups celebrate and ritualize the consumption of milk at their gatherings, expressing their belief that alternative "milks" made from soy and other plant-based materials are part of a conspiracy to feminize men (Frankin-Wallis, 2019). US President Donald Trump has used the term "local milk people" in addressing his political base, presumably as a dog whistle to White nationalists and alt-right groups, given that lactose tolerance is more common in countries with largely White populations (Hoffman, 2017; "Local milk people,"

2017). To some on the far right, drinking milk signals White superiority over other races (Stănescu, 2018). These rather bizarre trends nicely capture a central theme of this book: humanity's thinking about animals can be personal and idiosyncratic but it can also be group-based, motivated, ideological, and often political. We do not simply eat or wear animals for hedonistic or (supposed) health reasons. Rather our use of animals signals our sense of dominance over others and our willingness to exploit others for our own gain (see also Dhont & Hodson, 2014). Yet these basic facts run contrary to our widely-held beliefs that we are personally good, moral, and enlightened beings. To reconcile these contradictions, psychological and societal forces steer us away from thinking about the harm that we cause to animals in our daily lives. Yet our thinking about animals is critical to both the public discourse and to scientific study. Indeed, our thinking about animals is deeply symbolic, conflicted, and psychologically meaningful.

What does this internal conflict tell us about humanity in general? Before addressing this question it is important to note that people generally consider animal cruelty in terms of isolated incidents and specific perpetrators. In 2015 Cecil the lion (a protected animal in Zimbabwe) was slain by an American dentist; the public was outraged at the animal's slow death that transpired over 10–12 hours (Willingham, 2018). Consider also the recent case of a Canadian police officer caught on video repeatedly driving over an injured deer with his police truck. This incident lasted an agonizing 15 minutes, with the officer opting for this slow and painful method instead of saving the animal or using his gun to euthanize it (Southwick, 2019). Such incidents have understandably cued outrage and concern in the public. But in many ways these specific actions are manifestations of a broader cultural, and perhaps essentially human, outlook toward animals.

We hear, for instance, about the supposed necessity of killing (and eating) animals in order to *save them* from extinction (Flocken, 2018; Jones, 2018). Such conflicted and paradoxical thinking barely draws attention. People have become accustomed to such bizarre contradictions and paradoxes, often as a direct result of propaganda and "education" from those with power. Outside of the human-animal domain, consider the widely accepted assumption that we must go-to-war-for-peace. Consistent with such paradoxes evident in human affairs, we likewise justify our oppression of animals as natural, normal, and necessary (Joy, 2010; Piazza et al., 2015). Such rationalization becomes "baked in" to the way that society views animals and their relation to humans, from the way we filter information, the way we turn our heads (or open our eyes), to which rights we value and defend. Being the dominant predator on the planet affords humans this privilege, yet we wield such power not with dignity, humility, and a sense of responsibility but rather with cowardice, malice, brute force, and at times, cold indifference.

Consistent with the themes discussed in this book, such thinking about animals comes at a tremendous cost to animals. But it also comes at a considerable cost to humans. Furthermore, it is becoming evident that our contemporary thinking about animals is unsustainable – our exploitation of animals exerts an impossible toll on the planet and biosphere. With the human population now exceeding 7.5 billion people there is little doubt that human lives are more valued and prioritized than other sentient life forms. We live in an age now termed the *Anthropocene*, so named to recognize that humans are shaping the planet actively and formidably. At the same time, researchers are documenting a *biological annihilation* of life on the planet, with the 6th extinction event already fully underway (see Ceballos, Ehrlich, & Dirzo, 2017). Among mammals studied, over 40% have experienced losses exceeding 80%. Consider also a recent UN paper documenting that close to 1 million species are at real risk of extinction (Watts, 2019). Such numbers speak for themselves; that we can let entire species approach or reach extinction at the same time that our own species has never been so plentiful is poignant evidence that humans are (over)valued to the detriment of other species.

Below we explore the knowledge to be gleaned from the experts recruited for this book, drawing from multiple disciplines within academia, and on insights from those advocating for animals on the front lines. We outline several recurring themes from these writings, broadly split into three categories: (a) understanding the nature of the problem; (b) analyzing how the majority of people live with the problem; and (c) insights into solutions and remedies.

The nature of our problem with animals

Animal welfare and rights

A recurring theme of this book concerns society's lack of concern for animal wellbeing, if not its contribution toward worsening animal welfare. As noted by Sorenson (this volume), the meat industry and its advertising partners go to great lengths to conceal and obfuscate the true extent of suffering inflicted on animals for human gain. These very industries ironically argue that they engage in these (harmful) activities *for* the welfare of animals being exploited. Such narratives aim to minimize the concerns of vegans and activists as radical, unreasonable, misguided, and unnecessary. Sorenson also discusses the questionable notion of "happy meat," the idea that it is acceptable to kill (and eat or wear) animals for human purposes providing that the animal is supposedly well treated. Unfortunately, what it means to treat animals "well" is a moving target bolstered by human rationalizations. Pierce (this volume) lays out many of the problems with how animal welfare is construed and implemented. She argues that the so-called "Five Freedoms" that have long governed acceptable practice for the treatment and welfare of animals ironically ignore the most essential factor: *freedom*. Rather than turning to the vast array of scientific research detailing the impressive capabilities and feelings of animals, we create formal rules to convince ourselves that we are attending to the needs of animals. As the dominant species, we determine these needs for animals, or derive tests (e.g., *preference testing*) that provide answers congruent with our own needs. Here Pierce echoes the concerns of primatologist Frans De Waal (2016), who demonstrates that human biases in thinking shape the very methods we use to explore the inner lives of animals, doing the animals a great disservice but also setting back science and enlightenment.

On a related point, Woods and Hare (this volume) lament the vast number of animals and financial resources wasted in animal research designed to cure diseases afflicting humans, all at a terrible cost to animals who live unnatural lives in conditions that deprive them of their very nature. Bockman (this volume) similarly explains the problems with our human-focused needs in medical research that disregard the needs of the animal "participants."[1] He discusses how animal welfare is particularly dire in domains such as research and farming, where we most strongly assert the need to exploit animals. In many cases, animal-based research is not even capable of solving the human problem at hand, such that their pain and elimination highlights the cheapness of animal life for animal experimenters and society at large (see Hodson, Dhont, & Earle, this volume).

The fact that humans generally ignore the needs and welfare of animals has become the central raison d'être for the establishment of Farm Sanctuary, an organization founded on the principle that it is important to care for animals, particularly those who have suffered injustices through the corporatized agricultural system prevalent today. Founder Gene Baur (this volume) lays out the alternative vision and philosophy at Farm Sanctuary, one based on compassion and respect for animals. In his words, "Interacting with animals in cruel and violent ways affects us negatively, while treating them with kindness helps improve our lives and theirs, reducing stress, while building empathy and connection." Animal welfare thus represents a fundamental cornerstone in building fair and just societies. This sentiment is echoed in the illuminating discussion between Rozin and Ruby (this volume), where they argue that making decisions about animal welfare and ethics can be complex and multifaceted but that "not harming animals is a basic principle" that needs to be at the forefront of moral decision making.

Undervaluing animals

The fact that humans care so little about animal welfare is linked to the general undervaluing of animals relative to humans. Animals are considered to have lesser capabilities to perform tasks, think, and feel emotions, and also considered to be "worth" less. As Hodson and colleagues (this volume) discuss, this undervaluing of animals (and by extension those who protect animals) plays a central role in maintaining the status quo and justifying human dominance over animals (see also Dhont, Hodson, Loughnan, & Amiot, 2019). As such, they argue, "Its removal could seriously jeopardize ideologies that justify inequality as an acceptable practice in the abstract." That is, thinking of animals as less valuable and thereby affording them fewer rights lays the foundation for social injustices more broadly. Alternatively, if we extend rights to animals and increase their value, we unpick the fabric of hierarchically-based human social lives.

In their chapter Haslam and colleagues (this volume) argue that "the human/ animal distinction plays a role in the perception of social groups." That is, how we think about animals impacts how we think about other humans. We find it difficult to argue with this basic point. Moreover, as they observe, although we can use animal labels and monikers to describe loved ones such as children ("oh, you cheeky monkey!") and lovers ("aren't you a cuddly bear?"), we overwhelmingly use animal metaphors and comparisons in the pejorative. That is, being considered animal-like is generally considered an insult, and a strong one at that, which further illustrates the point that animals are undervalued.

By undervaluing animals we tend to see animals *instrumentally* (see Pierce, this volume; Sorenson, this volume), that is, in terms of their exploitation value. Thinking of "humans first" results in human priorities that virtually always trump the concerns of animals. As observed by Holden and Herzog (this volume), the meat industry is growing increasingly comfortable with the idea of deliberately breeding blind or featherless chickens, shaping animal bodies and abilities entirely on the grounds of serving human interests. Put another way, few would advocate breeding blind or featherless chickens to release them into the wild. Our moral choices, of course, represent the point at which rubber hits the road. As Bockman (this volume) disturbingly reveals, only 1.5% of American donations go to animals, with only 1% of those going to farm animals (who are particularly exploited and suffering). We make choices to prioritize humans over animals. On this point Dhont, Hodson, Leite, and Salmen (this volume) review recent research showing that when given explicit choices that pit human charities against animal ones animals routinely come out the poorer, especially (but not only) among participants expressing speciesist views.

Animals as competitive threats to humans

The way that humans disregard the welfare of animals, and undervalue them relative to humans, is consistent with the idea that people feel in *competition* with animals. Bastian and Amiot (this volume) discuss how those in more impoverished environments are less likely to have companion animals in their lives and feel less attached to animals. The reasoning is that when times are tough people direct resources away from animals and toward themselves. This is consistent with the literature reviewed above (Bockman, this volume; Dhont et al., this volume) showing lower donations to animal (*vs.* human) causes.

As noted by Bastian and Amiot (this volume), the notion that we push back against groups with which we are in (or perceive) competition holds considerable sway in the broader intergroup literature. In their chapter, Sevillano and Fiske (this volume) provide an analysis highlighting what can be gained by applying theories from human intergroup relation research (e.g., about racism or sexism) to the domain of human-animal relations. Fiske and her colleagues (e.g., Fiske, Cuddy, Glick, & Xu, 2002) have long demonstrated that the perception of human outgroups largely falls along two dimensions: *competence* (how capable *vs.* incapable are they?) and *warmth* (how friendly *vs.* hostile are they?). When it comes to thinking about human outgroups (e.g., White people, Black people, men, women) these dimensions of perception go a long way to understanding our reactions toward outgroups. It turns out that perceived warmth and competence are also meaningful dimensions of evaluation that can predict behaviors (e.g., avoidance) toward animals rather well (see Sevillano & Fiske, 2019). Our thinking about animals is, in part, dependent on whether we see ourselves as being in competition with other animals (e.g., wolves) or in cooperation with them (e.g., dogs). Those particularly deemed threatening are those considered high in competence but low in warmth, that is, predators (e.g., sharks).

Psychologically, of course, humans have complicated relationships with animals. Not only do most people see themselves in *competition with* some animals (and animal supporters), but at some level we recognize our very *dependence on* animals. Leenaert (this volume) refers to all of us as "steakholders," each with something to gain and lose. With so much at stake, it sharpens our psychological concerns and thinking about us versus them. In his words, "We have – or at least think we have – become very dependent on the use of animals for the continuation of our society." It is little wonder, therefore, that our thinking about animals is so central to our lives, and that we make decisions that nudge the goal-posts in favor of outcomes that suit humans. In their discussion, Adams and Calarco (this volume) outline the other side of the issue: animals are very dependent on us. For this reason, the *feminist ethics of care* recognizes an inherent inequality between humans and animals and stresses our interconnectedness with animals and our duty to care for animals. This approach argues that dependence on the other should not serve as the grounds for conflict, but rather for responsible benevolence. This is an admirable and achievable goal. But given that people often see humans and animals in competitive conflict, and the promotion of animal rights can itself form the basis of human-on-human conflict, interspecies dependency will likely remain a source of conflict until humanity undergoes a serious rethink of human-animal relations. As we traverse the Sixth Mass Extinction, it is possible that only a Second Enlightenment can right this ship.

Social identification concerning animals

Many of our personal and social identities are wrapped up in animals. Intergroup or interspecies competition heightens categorization in our thinking, sharpening us-versus-them distinctions. As discussed by Hodson and colleagues (this volume), the Interspecies Model of Prejudice predicts that perceiving a larger gulf between humans and animals is a precursor to thinking about human outgroups as animal-like, which itself begets prejudice, discrimination, and the denial of rights. At some level, this process hinges on people *disidentifying* as animals, a strong human tendency (see also Woods & Hare, this volume). Animals thus serve as a key marker for identifying our own group and reinforcing its status in the hierarchy of beings.

In their chapter, Bastian and Amiot (this volume) point out that humans are naturally drawn to animals (the *biophilia hypothesis*), and that we can exhibit attachment patterns with animals mirroring that with our children. The authors then discuss the importance of social identity to social life. Whereas researchers have pursued the implications of increasingly common or shared identities (e.g., "American" as opposed to White or Black), even up to the level of identification with humanity, Bastian and Amiot have pushed these implications further, testing identification with animals via their Solidarity with Animals scale. They find that people with pets are more identified with animals generally, as are vegetarians. They also find that the more that meat-eaters consume meat, the less they identify with animals. This interesting pattern is consistent with findings by Earle and Hodson (2017), whereby greater beef consumption predicts ever greater anti-vegetarian prejudice. Together such findings show that our eating habits are systematically linked to our thinking about animals and people who protect animals. Bastian and Amiot also describe the moral implications: those who identify more with animals are less likely to prioritize human lives over other animals, less likely to feel superior to animals, and less likely to show prejudices toward other human outgroups (e.g., racism). How we think about ourselves, in relation to animals, clearly matters with regard to multiple life outcomes.

As noted by Sorenson (this volume), advertisers sell us an identity that focuses on meat (see also Bastian & Amiot, this volume). They market notions of privilege and pleasure, not to mention positive health, while masking the true costs of meat. People come to identify as "meat eaters," and some do so very strongly. This is no accident or by-product of eating meat, but rather the consequence of heavy marketing and socialization around animal consumption. Loughnan and Davies (this volume) also frame their chapter around identity, particularly with regard to animals (as the *eaten*) and consumers (as the *eaters*). Thus eating meat is not a simple exercise, based largely in nutrition, but rather reflects the roles we see entities playing in society. This is reminiscent of Carol Adam's notion of the *absent referent* – how the animal identity is removed from our plates and replaced with "meat" (see Adams & Calarco, this volume). In that process the animal is "disappeared," leaving only the meat eater as a participant and the animal psychologically replaced with "food." In describing the meat paradox, Loughnan and Davies discuss the inherent clash in our own identities – as compassionate and moral beings on the one hand, and as consumers and exploiters of animals on the other. It is little wonder that this clash of identities requires substantial psychological infrastructure to keep it afloat.

Interestingly, Leenaert (this volume) argues that we should be moving away from concerns with rigid groups and identities around meat consumption. He argues for a more pragmatic approach, recognizing that *eating less meat* is a positive step forward in the process of eliminating animal suffering caused by humans. He advises against a rigid focus on being vegan: "Although being vegan is a very useful and praiseworthy thing to do or be ... overfocusing on being vegan – and certainly on being perfectly vegan – may be misguided." Here Leenaert challenges us to think beyond group identities. In fact, he argues that the animal rights movement needs to "be more inclusive and create a bigger tent." Here the onus is on animal advocates to welcome others into the fold and be more inclusive to gain more support for animal rights and veganism. Given that meat eaters consider vegans and vegetarians as threats (Dhont et al., this volume; Hodson et al., this volume), this approach of inclusivity and de-escalation is worthy of consideration.

Culture as shaper of animal-relevant thoughts and actions

Which animals we adore versus loathe, or eat versus avoid, is largely determined by cultural norms (Bastian & Amiot, this volume; Dhont et al., this volume). As observed by Dhont and colleagues, bullfighting and whaling have strong and positive cultural significance in Spain and Norway respectively, but each country considers the practices of the other rather barbaric or unnecessary. Culture shapes thinking about animals. Along similar lines, much of the pro-meat message in society is carefully shaped by industry and advertisers (Baur, this volume; Sorenson, this volume), and culturally reinforced by norms of masculinity (Adams & Calarco, this volume; Loughnan & Davies, this volume). As recognized by Haslam and colleagues (this volume) our thinking takes shape through language and metaphor, giving it life and shared social value. And, as reflected in Rozin and Ruby (this volume), institutions such as religious organizations play a formative role in dictating what and how we eat, often in ways consistent with one's values but also in ways that conflict. In sanctifying what is edible and what is acceptable husbandry practice, culture and religion can therefore play a positive role in shaping human-animal relations in the future.

Changing the dominant culture remains one of the biggest challenges for animal advocates. A theme emphasized by Sorenson (this volume), Rothgerber (this volume), and Piazza (this volume) is that culture physically separates us from animals and from food origins. This physical separation leads to psychological separation. Part of the solution will involve psychologically reuniting humans and animals (see Loughnan & Davies, this volume). Some of this change will come naturally to us. As noted by Rothgerber, animals already play a tremendous role in how we socialize children to be kind. But we simultaneously socialize children to disregard the welfare of (many) animals and toward the consumption of meat, something that many children otherwise find aversive. It is little wonder that our thinking about eating meat is thus so conflictual.

Ideology and politics

Political ideology is increasingly central to cultural influence (see Jost, 2006). To many, topics such as meat consumption concern nutrition, or health, or the economy, but we are learning that political ideology affects how we think about animals and human-animal relations. As discussed by Hodson and colleagues (this volume; see also Dhont & Hodson, 2014), those with more right- (vs. left-) leaning ideologies are more likely to eat meat and generally exploit animals, in large part because they feel a greater sense of human superiority and dominance, and because they pushback against a supposedly threatening vegetarian lifestyle. They document additional research revealing that conservatives are more likely to return to meat consumption after quitting, not because of greater meat cravings but because they are less likely to have quit meat for social justice reasons. Dhont and colleagues (this volume) explain how the social dominance aspect of right-leaning ideologies is particularly important to understanding human–animal relations, relative to preserving traditions.[2] What is undoubtedly surprising to nutritionists, healthcare professionals, and even to political scientists is that political ideologies relevant to human social life are also relevant to human–animal dynamics. In hindsight, this should not have been surprising – ideologies are organized belief systems about how the world should operate, revolving around opinions about freedom, control, hierarchy, and resistance to change. With that in mind, how could ideologies *not* be relevant to thinking about animals?

In fact, as noted by Sorenson (this volume), speciesism itself is an ideology that drives society. As put by Loughnan and Davies (this volume), "by accepting or even celebrating hierarchy and inequality, one can help alleviate concerns about eating animals; I can love animals, and still think that they are beneath me and that their exploitation is justifiable." For his part, Baur (this volume) echoes the writings of Melanie Joy, who argues that meat eating is a "carnist" ideology, one so pervasive that we barely recognize its presence.

Of course, ideologies about eating meat and using animals go hand in hand with other ideological thinking. Rozin and Ruby (this volume) nicely illustrate not only how religious beliefs impact decisions around food consumption, but the convoluted and complex nature of these thoughts. As they observe, most religions do not simply make meat consumption permissible but communicate to followers that God (or other higher entities) *want* people to eat animals. In the Bible this is laid out in Genesis, one of the most important Christian writings detailing the supposed natural order of society (man's dominance over women, humans over animals, etc). Pierce (this volume) discusses how such ideologies can rub up against basic science, and at their extreme, encourage people to ignore science. In doing so, we fail to appreciate the characteristics and needs of animals, as Pierce argues, but we also fail to appreciate the damage that our consumption of animals inflicts on the planet (e.g., climate change). It is clear that ideological thinking can be a major obstacle to thinking clearly about our relations with animals. In dealing with this issue Leenaert (this volume) provides some very practical advice: We ought not to replace one ideology (e.g., carnism) with puritanical views relevant to eating or not eating meat (see also Rozin and Ruby). With regard to cultural veganism, for instance, Leenaert argues that "To insist on 100% purity and consistency is unnecessary, unproductive, impossible and insufficient."

The intertwining of speciesism with human-human prejudices (e.g., racism, sexism)

The work of scholars such as Carol Adams, especially with her pivotal book *The Sexual Politics of Meat*, highlight that our biases are meaningfully intertwined (see

Adams & Calarco, this volume). Bias begets bias, as violence begets violence. As the contributors to the present volume make clear, this process can take several forms. Haslam and colleagues (this volume) convincingly demonstrate how metaphors are central to the dehumanization of human outgroups. In their words, "animal metaphors may be influencing and organizing how we think, feel, and behave towards others even if animal labels are never overtly attached to them." As such, counteracting human–human biases such as racism necessitates a rethink of animals. Woods and Hare (this volume) also stress that thinking about humans as animal-like (for example, "simianization") generally brings negative outcomes to those human groups. Their observations are more nuanced and interesting, especially their discussion of the *uncanny valley* – part of the reason we dislike and/or disregard some animals is precisely because of their *similarity* (not difference) to humans. As these authors discuss, we can be fascinated and bewildered by human-like animals but this can still lead to negative consequences for those animals. A key conclusion from this line of inquiry is that better understanding the links between human–animal and human–human relations better informs both topics.

This basic idea also runs through other chapters. Hodson and colleagues (this volume) discuss the Interspecies Model of Prejudice, which suggests that some of the roots of human–human biases (e.g., racism) might find their roots in human–animal biases (e.g., thinking that humans are different from and superior to animals). A central feature of this idea is that dehumanizing an outgroup would have little or no social value or punch if not for the diminished value that societies generally attribute to animals. That is, if we did not look down on animals, likening a member from another group (e.g., a Black man) to an animal (e.g., ape) would have no social value or function. In their SD-HARM model, Dhont and colleagues (this volume) unpack why general biases like speciesism are associated positively with biases like ethnic prejudice. They ask why a person who dislikes most human outgroups also scores higher in speciesism. Much of the reason is consistent with the overall theme in this volume: humans feel and express superiority and dominance toward both targets. That is, ideologies about accepting inequality and valuing hierarchy promote common biases against other humans and against animals. The more hierarchical a society, or the more hierarchically-minded the individuals within a society, the more ethnic prejudice and speciesism will be expressed. As such, human biases such as human prejudice and speciesism are systematically (not randomly) associated with each other.

There are also negative consequences for human groups who support animals. Consider that animal rights activists are often labelled as terrorists and criminals (Sorenson, this volume). Doing so marginalizes these groups, removes their rights, and discourages society from questioning a status quo that facilitates animal exploitation. By their existence animal rights activists and advocates threaten to unpick the culturally shared rationalizations around animals and meat (see also Dhont et al., this volume; Hodson et al., this volume). Leenaert (this volume) highlights another problem with prejudice toward vegans and vegetarians – it deters people from reducing their personal meat intake. Indeed, a key reason why people *return* to meat consumption after abstaining is the perceived lack of social support (Hodson & Earle, 2018). Moreover, as pointed out by Loughnan and Davies (this volume), we experience a "human paradox," whereby we claim to care for others but nonetheless treat them negatively. They point to a whole new area of generative research – examining how various paradoxes (animal, human, environmental) relate to each other.

How we live with the problem rather than change our behavior

As reviewed above, there are ample problems with human thinking about animals. How do we, individually and collectively, live with the paradoxes and the contradictions? The contributors offer insights into this critical question.

Biases in human thinking

Gilovich and Ross (2015) nicely illustrate a fundamental human fault. As they note, it has long been believed that dogs do not perceive color. (We now know this to not be true, but dogs nonetheless perceive colors less fully than humans). Consequently, we refer to dogs as *color blind* (with the tacit contrast being *relative to us*). In contrast, dogs have a vastly more sophisticated and developed sense of smell than humans but we do not refer to ourselves as "odor blind." These authors make a fundamental point that we make comparisons relative to ourselves, and by extension, to our ingroups. Dogs are considered under-performers for color but over-performers for smell (e.g., "super smeller"), centered around the human reference point. This propensity does us a tremendous disservice, particularly when trying to understand animals (see De Waal, 2016), but also when trying to understand ourselves.

In her chapter, Pierce (this volume) explores this idea at length. How we treat animals, especially in the laboratory, reflects a human-centric focus and goal. We conduct so-called *preference testing* supposedly to determine the nature of animal needs, but we only present a narrow range of stimuli, under very artificial circumstances, and then apply the results from a narrow sample to the species as a whole (homogenizing the outgroup). This serves our own needs under the pretence of serving those of animals. As noted by Adams and Calarco (this volume), by thinking from a human perspective we "privele[ge] human voice." These authors expose an interesting implication of human-centric biases. As illustrated with the case Derrida and the cat with which he interacts, it is important to not only look *at* animals, but we need consider and explore being looked at *by* animals. This recommendation can help to disrupt human biases that prioritize humans over animals and provide greater insights into ourselves.

Holden and Herzog (this volume) highlight another problematic aspect of human thinking: the (over)reliance on heuristics or rules of thumb over more thoughtful and engaged thinking. In particular they discuss reliance on heuristics about *naturalness* when thinking about animals and making moral decisions. They raise interesting questions about animal welfare and how best to serve animals, some of which might have seemingly "unnatural" applications (e.g., breeding chickens to be blind via genetic engineering). Animal advocates will undoubtedly be torn on the value of denaturing animals to lower their stress and pain. We argue that such solutions do not deal with the central problem: exploitation. Consider the following analogy. If we were able to breed women to suffer less from the negative effects of prostitution, we would consider such action horrific. Rather, we would remove the *source* of their suffering (e.g., poverty; male dominance) not their ability to suffer when exploited. We worry that efforts to genetically modify animals for human purposes further dehumanizes animals (see Hodson et al., this volume).

Disconnected thinking about animals

When it comes to thinking about animals and meat consumption we experience a tremendous "disconnect." Adams and Calarco (this volume) remind us that the process of eating meat involves the *absent referent* – the animal is removed from equation and meat consumption becomes about food and not animals. Piazza (this volume) points out that the problem is not that humans have little or no empathy for animals but that our empathy has become disconnected from our actions. When the links between empathy and action become severed or attenuated, we no longer see the relevance of our empathy, nor allow it to direct our actions. Loughnan and Davies (this volume) similarly discuss research showing that people are motivated to sever associations between an animal and its intelligence, but typically when we seek to eat the animal. That is, when we think about food and eating we dissociate the animal (e.g., pig) from the attribute (e.g., intelligence) to enable us to love but exploit the animal. These authors also discuss how we also disconnect animals from meat; when reminded of the connection between meat/food and animals, people show more empathy for the animal and less willingness to eat the animal. Recent research demonstrates related knock-on effects; reminders of the meat-animal association induces animal empathy that lowers anti-vegan prejudice, and induces meat distress that lowers perceptions of threat by vegans (Earle, Hodson, Dhont, & MacInnis, 2019).

As framed by Rothgerber (this volume), we also need to encourage the so-called *meat epiphany* (i.e., that meat comes from animals) at an earlier age, to interfere with socialization processes that otherwise suppress and negate the epiphany.

Morass of morality

Holden and Herzog (this volume) argue that much of our moral thinking about animals is based on heuristics. This means that much of our moral thinking is not based on salient or thoughtful reasoning. Piazza's (this volume) take on morality is that we exhibit an *ethical blindspot*. In his words "we often fail to realize our actions have ethical implications or should be framed in moral terms... Most people simply fail to see eating animals as a moral issue, or if they have some doubts about its ethicality, they easily convince themselves they have good reasons for doing it." Hence moral reasoning gets circumvented when exploiting animals. Other authors have emphasized how humans are conveniently flexible with our moral concerns. Dhont and colleagues (this volume), for instance, discuss how moral concern is afforded more to some animals (e.g., companion animals) than to others (e.g., food or wild animals). It is clear that we engage in moral thinking at times, as when demanding strict laws about cruelty to "domestic" animals such as cats and dogs, but disengage moral concern for entire categories of animals that we exploit (e.g., for food, clothing, entertainment, or human-focused research) or perceive to be in competition with us for land and resources (see also Sevillano & Fiske, this volume). It is not that we cannot feel empathy for animals, or allow our empathy to guide our actions, but that we selectively engage and disengage our empathy to suit our interests.

Alternatively, several authors downplay morality in bettering animal welfare. As observed by Rozin and Ruby (this volume), although some food restrictions are based in morality, as is much of veganism, others are instead based on religion or other cultural norms. In determining what and how people eat, Baur (this volume) similarly emphasizes "norms, belief systems, and economic infrastructures" more than morals. Others, such as Leenaert (this volume) and Bockman (this volume), adopt pragmatic approaches that deemphasize morality in the specific localized instance and instead focus on doing the most good for animals overall. Bockman's central message is that we need to focus on *effective altruism*, doing the most for animals with the time and resources available, directing attention toward farm animals and away from companion animals. Leenaert goes so far as to say that "it is not *necessary* that people do the right thing for the right reason." He argues that to the animals it does not matter why people reduce or eliminate the consumption of animal products; what matters is that they eat less or no animal products. From such perspectives, focusing on the moral question drains resources and attention, and, as Leenaert discusses, risks pushing people out of the tent. Tensions over the need to increase moral concern or responsibility is evident in the discussion between Adams and Calarco (this volume). Adams advocates for a stronger ethics of care and a morality of responsibility for animals, whereas Calarco calls for a de-emphasis on morality (or at least calls for a plurality of factors to be deemed important). It strikes us that this question of morality, and its centrality to the animal question, will engage and possibly divide the field for decades to come. Even if morality and empathy are not considered essential in reducing animal exploitation, both are powerful drivers of human behavior in general. As psychologists we recommend harnessing their power to help animals but not relying solely on their impact.

The human art of rationalization

Much of what underpins our claims to love but exploit animals comes down to the human capacity to rationalize what we do as good, moral, and necessary. In many ways, the chapter by Rothgerber (this volume) is the most direct in dealing with this topic. Cognitive dissonance probably plays a stronger role in justifying meat consumption than any other human activity. A common theme that runs through the overall discussion above concerns our ability to think creatively, in ways that allow us to cause harm but not experience the stress and anxiety that otherwise accompanies such actions. Much of this book highlights the powerful systems of rationalization, at the level of the person but also at the level of society.

Both Pierce (this volume) and Sorenson (this volume) discuss social factors that rationalize animal exploitation. Pierce discusses how so-called *welfare science* purports to advocate on behalf of animals but instead helps to justify continued exploitation of animals. Society convinces us that, so long as we try to minimize overt harm, we can rob animals of what they most value (freedom, including freedom from pain). Sorenson taps a similar vein, arguing that the practice of labelling animal practices as "humane," such as the so-called "happy meat" movement, legitimizes animal exploitation. Holden and Herzog (this volume) point out that people will go out of their way to justify cruelty, as when arguing that it would be cruel to *not* allow chickens to fight to the death. These are all clever methods employed by agriculture, entertainment, and advertising agencies to distract us not only from animals but from ourselves and our own moral qualms. Humans collude to develop consensual beliefs about animals that circumvent the need for deep introspection or engage in less harmful courses of action.

In addition to psychologically divorcing meat from animals (see Hodson et al., this volume; Piazza, this volume), several authors documented some interesting evidence about the mental attributes we assign to animals, and the flexible nature of their designation. As reviewed by Bastian and Amiot (this volume), Piazza (this volume), and Loughnan and Davies (this volume), when we think about animals in terms of their food purpose we attribute them less mind and mental capacity. We thus remove basic animal nature from animals we wish to exploit (see Hodson et al., this volume). This compelling research highlights the systematic and rationalizing nature of the process. An animal to be consumed is denied of a sense of mind, but the same animal not to be consumed is spared this mental dodge (Loughnan & Davies, this volume; Piazza, this volume). Interestingly, the very attributes that we most admire in humans (i.e., mind, intelligence) we deny in others to justify their exploitation. This returns to the central theme that a quality, property, or trait matters more when it is imbued with human (*vs.* animal) essence.

The icing on the rationalization cake is the use of the "4Ns": meat consumption is deemed normal, necessary, nice, and natural (see Piazza, this volume; Loughnan & Davies, this volume; Rothgerber, this volume). These four beliefs account for up to 90% of the justifications given for meat consumption (Piazza et al., 2015); little more is needed to defend the indefensible (i.e., harming animals while loving animals). It is easy to see why – to the extent that any activity in life is commonly engaged in, needed for survival, pleasurable, and inevitable, engagement in that activity will be deemed permissible or even expected. Of course, several of these beliefs are objectively on thin ground. Whether something is normal is not a suitable justification for many behaviors. It is normal for men to be paid more than women, but this does not objectively justify the inequality. Likewise, mass shootings in the USA have become part of everyday life, but shrugging one's shoulders about its normality simply leads to further deaths. The claim that meat is necessary for humans is also dubious. The very existence of healthy vegetarians and vegans, including in competitive athletic events such as mixed martial arts, dispels the idea that meat consumption is necessary. In fact, as omnivores it is difficult for humans to make the case for meat consumption as necessary. Of course, few of these objections will have impact because people grant the 4Ns considerable weight in absolving individuals and societies from responsibility in causing suffering to other animals.

Reflections on solutions and remedies

The central problem tackled by this volume concerns how we simultaneously love and exploit animals. Much of the discussion has been admittedly pessimistic. We have demonstrated that much of the problem is rooted in basic human psychology – the need to categorize the world into us versus them, the desire to set human "needs" as priorities, and the notion that humans are superior to animals and thus entitled to exploit or disregard them. These psychological tendencies and beliefs are further shaped and reinforced by societal structures, institutions, and traditions. We have also explored how people live with the mental conflict between supposedly loving animals and being moral people on the one hand, with using and exploiting animals on the other. Much of the answer here also concerns human psychology, particularly our adept ability to reframe situations and contexts to sidestep thorny moral questions, and how we rationalize our actions as normal and even necessary, with great help from how society is organized and how our cultural practices have developed over time. Fortunately the authors of the present volume also offered considerable insights into dealing with the human-animal problem.

In tackling the issue it is useful to conceptually separate the "liking or loving animals" from the "exploiting of animals." The former is an evaluation or attitude, whereas the latter is a behavior. As argued by Hodson and colleagues (this volume), the field affords too much emphasis to the attitudes that people express toward animals. This leads us to read too much into claims about simultaneously liking and exploiting animals. It is also clear that people are selective about which animals they like or not, with clear preferences for companion animals over farm or wild animals. Focusing on attitudes or evaluations can only get us so far given the human propensity to prioritize human objectives over animal objectives. Consider the paradoxical finding that the most popular and charismatic wild animals (e.g., lions, pandas) are the *most* at risk of extinction (Courchamp et al., 2018). If simply liking animals led to their protection we would not observe such patterns. We cannot simply focus on getting people to like animals if we are to reduce animal exploitation, because liking animals is only somewhat related to willingness to exploit or disregard.

A focus on psychological constructs

One recurring theme of the volume has reflected the importance of psychological constructs in understanding human-animal relations, such as the (perceived) human-animal divide. As outlined within the Interspecies Model of Prejudice (see Hodson et al., this volume), people naturally differ in the extent to which they see humans and animals as different. Those who perceive a greater divide generally express less moral concern for animals and human outgroups, and are more prone to dehumanizing human outgroups. Fortunately, the perceived human-animal divide can be manipulated experimentally; emphasizing that animals are similar to humans (but not that humans are similar to animals) have proven to boost empathy and inclusive cognitive representations in group contexts, and to lower outgroup dehumanization. Bastian and Amiot (this volume) also illustrate the benefits of highlighting human-animal similarities, while being mindful that such approaches can induce threat. These approaches are consistent with Rothgerber's (this volume) call to "blunt dichotomization" when thinking about animals and/or the differences between humans and animals. Likewise, counteracting the general tendency to categorize certain animals as food (Loughnan & Davies, this volume), or to replace animals with the concept of food (Adams & Calarco; this volume), will play a role in successful interventions.

Another key factor concerns empathy. In addressing why we love but exploit animals, Piazza (this volume) argues that we often fail to act on our empathy, or fail to recognize how our actions and empathy are connected. This message has considerable appeal, in part because it attempts to *harness* the goodness in people rather than try to instill goodness, ethics, or morality on others. If people fail to recognize how our empathy and actions are related, education will play a considerable remedial role. It is already known that empathy is related to more frequent and more positive contact with human outgroups (see Abbott & Cameron, 2014; Hodson, 2008; see also Dhont et al., this volume). In their paper Loughnan and Davies (this volume) discuss how physical contact with farm animals has previously boosted the sense of mind reported with regard to the animal in question, along with a boost to moral concern, yet failed to reduce meat consumption intentions. We see considerable value in merging these future research streams on contact and empathy with regard to animals.

Redirecting goals

One prime goal of this volume is to bridge the interests shared by academics and advocates, two groups who spend too little time in each other's orbits despite having overlapping objectives. We consider the present volume a serious first step in such integration, but also recognize that more work is needed.

The authors of the present volume offer some very practical suggestions about redirecting our energy moving forward. Bockman (this volume), in endorsing an *effective altruism* position, suggests that we should divert more of our time and money to where the problems actually lie. As he argues, we spend much of our animal charity money on companion animals, who are currently the most protected and cared for animals, and spend very little money to deal with the horrific conditions experienced by animals on modern agribusiness farms. This is certainly a call to "think big" and radically change the status quo. Ultimately, authors such as Bockman are encouraging us to recognize that all advocacy involves choices and priorities. Leenaert (this volume) also encourages us to think big. Adopting a more practical approach, he argues for a greater focus on a reduction of animal products in animal advocacy, instead of a focus on the elimination of animal products, as a strategy to achieve the end goal of the elimination of animal products. He encourages inclusive rather than exclusive in our memberships and sharing of ideas, and abandoning concerns with purity or sacredness. Leenaert also suggests that incremental steps are acceptable and even critical, a theme also stressed by Rozin and Ruby (this volume). As psychologists, we see great value recognizing the importance of incremental change, but we also recognize Pierce's (this volume) caveat that if changes are too incremental they can solidify the status quo rather than advance rights for animals. Finding the appropriate balance represents a serious challenge for future researchers and advocates.

As psychologists we also find great value in the advice of Baur (this volume) who suggests that we need to make the transition *easy* for people to adopt. Like Leenaert (this volume), Baur astutely observes that most people want to do the right thing and not harm others, but also that people are creatures of habit and convenience. As norms change, and non-exploitative options become more available and accessible, people will be increasingly likely to change their lifestyles. A key goal for policy makers and entrepreneurs must be to develop products and purchasing methods that facilitate pro-animal choices and deter other choices. Psychologists call these *nudges*, subtle prompts that shift behavior often without the awareness or motivation of the actor.

Forging new paths will require listening to divergent viewpoints. Adams and Calarco (this volume) demonstrate the value in considering multiple ethical perspectives in order to change public thinking about animals. The debate between these two authors, as the debate between Rozin and Ruby (this volume), highlight the complexity of the problem. We will likely need to shift from thinking about a single goal to thinking about multiple goals, and to shift from thinking about one moral objective or pathway to thinking about multiple paths. Given the larger body of work by Carol Adams on the link between meat eating and sexism, a broader focus on systems of oppression will be not only more informative but may be more effective in the long-term. Indeed, given that meat eating is deeply entwined with masculinity (Loughnan & Davies, this volume; see also Rothgerber, 2013; Ruby & Heine, 2011) and symbolizes broader patriarchal values (Adams, 2015), effective solutions to human-animal relations require rethinking gender relations and challenging existing gender inequality.

Redirecting actions

One of the simplest suggestions for action redirection might also be the most difficult to implement. After laying out her case that humans generally rob animals of one of the most important factors relevant to their welfare – freedom – Pierce (this volume) argues that in the future human interests may no longer be top priority. Putting animal interests on equal footing would represent a radical change in human-animal relations. Even more radical is the consideration that humans can bear responsibility for animals in the manner that adult humans bear responsibility for their children or elderly parents, reflecting their privileged positions (see Adams & Calarco, this volume). This represents a serious challenge. People will be resistant to freeing animals because freed animals are more difficult to exploit (that is, we restrain their freedoms in large part *to* exploit animals). But even among well meaning people the freedom solution will be difficult to implement because natural habitats are being destroyed at an alarming rate, meaning that "freedom" for animals might not be attainable in the twenty-first century. Given these difficulties, and the fact that affording animals equal or greater rights than humans rubs against the grain, this simple solution will require legal instruments to constrain human behavior around animal welfare. The actions called for here therefore represent large-scale goals.

Fortunately such legal courses have proven effective in other domains. As noted by Leenaert (this volume), it required legislation to enforce the wearing of seatbelts and the banning of smoking in public places in order to change public attitudes. In most Western countries people now overwhelmingly wear seatbelts and disapprove of second-hand smoking. Sometimes it takes a change in behavior to subsequently change attitudes, a key tenet of cognitive dissonance theory. Building on this work, Rothgerber (this volume) argues that in implementing such a strategy it makes the sense to target children before the human rationalization apparatus becomes fully entrenched. Shaping children to change their behaviors around animals (e.g., meat consumption; wearing leather) can alter their beliefs and attitudes at a critical developmental stage. As noted previously, changing attitudes might not be the key factor to focus on, given that people mostly like animals already. In contrast, the type of change advocated here prioritizes the change in behavior, which subsequently brings attitudes and opinions in line with the behaviors.

Education

A common theme stressed by the contributors concerns the need for more, and better, education about animals and human-animal relations. Baur (this volume) strongly advocates for education, with implications not only for schools but for governments and institutions such as prisons. He details how giving students hands-on experiences with gardening, for example, can radically change their attitudes and behaviors regarding the nature of food. Food literacy plays an important role in such education, with government resources redirected from supporting the raising of farm animals to supporting a plant-based food ecology. The thrust of the debate between Rozin and Ruby (this volume) clearly highlights the importance of engaging in education programs and social movements that can illuminate moral issues and understanding of animals and food more generally. As they argue, "Raising consciousness about the moral costs of eating meat provides one avenue, along with providing more and more sensory appealing meat substitutes... Furthermore, we should never underestimate the power of social pressure..." From this perspective, both formal and informal routes of education can be powerful tools.

By focusing on children in particular, Rothgerber (this volume) speaks to the importance of education. The time at which children reach their "meat epiphany" (i.e., meat = animals) is a critical stage to tap into their naturally occurring concerns for animal welfare. In the words of Rothgerber, children "[lack] extensive experience eating meat and [lack] a history of justifying the behavior – thus... *children may be more amenable to behavioral change than adults.*" But such interventions, he argues, need to be thoughtful and carefully planned. He discusses how leaflet distribution exercises on the harms caused by meat are relatively ineffective, possibly because such methods induce guilt, which then trigger cognitive dissonance mechanisms to rationalize away negative emotions. One message that is consistently conveyed across authors in the present volume involves being careful to not induce guilt and reactivity or pushback, and instead to open a dialogue with opponents and be grateful for incremental changes that gradually shift society toward caring for animals.

Tapping into and leveraging human nature

Finally, to appreciate options for solutions, it is important to keep in mind that humans have a tremendous capacity for empathy and compassion, as do many other non-human animals (de Waal, 2009). The key challenge is to open dialogue with those who are not used to routinely thinking about the harms we cause to animals. An excellent demonstration of this potential is evidenced in a media exchange involving Gene Baur (Founder and President of Farm Sanctuary and author in the present book). In September 2018 he engaged in a discussion of meat consumption with a very skeptical Tucker Carlson (host) in a segment entitled *Tucker Takes on Vegan Activist.*[3] Baur quickly convinced Carlson, using reason and education, that meat consumption might cause more harm that it is worth. Carlson seemed clearly moved by the discussion, particularly given that his initial goal was to "take on" Baur and challenge pro-animal ideas. Carlson closes with: "You know, I have to say, I started this segment thinking 'this guy's probably crazy' but I think you're actually reasonable and thoughtful, and I don't agree with everything you've said but I appreciate your coming on tonight." To which Baur responded: "And I appreciate your concern about animals, and your open-mindedness to these issues." Through education and fostering discussion, Baur clearly found common ground with an opponent, drawing on the host's claim to love animals as leverage toward opening his mind to consider the harm that humans cause animals through exploitation practices such as meat consumption.

In leveraging human nature to solve the problem of animal exploitation, we need to be mindful of how to best integrate research with advocacy. As reflected in this book, human thinking about animals is deeply conflicted and rationalized, meaning that simply doing what "feels right" is not an adequate guide for improving the lot of animals. Instead, a stronger focus on evidence-based approaches to animal advocacy in developing effective interventions is critically needed. As we hope the present book makes clear, the conversation between academics and advocates will increasingly need to become a two-way communication. In the words of Eagly (2016, p. 199), "scientists should serve as honest brokers who communicate consensus scientific findings to advocates and policy makers in an effort to encourage exploration of evidence-based policy options." In turn, advocates need to communicate to academics the concrete and practical obstacles faced in the field. With many species rapidly becoming extinct, while other species are exploited in unprecedented numbers, it is incumbent upon us to speak with each other openly and with conviction to bring about societal change.

Notes

1 We use this term loosely, and put it in quotation marks, given that animals never consent to such research practices and thus by most reasonable standards cannot be considered "participants" but more as "subjects."
2 However, new research is suggesting that tradition-relevant aspects become focal when the tradition has a national flavor, such as bull fighting in Spain.
3 To watch the exchange, see https://video.foxnews.com/v/5830538163001/#sp=show-clips

References

Abbott, N., & Cameron, L. (2014). What makes a young assertive bystander? The effect of intergroup contact, empathy, cultural openness, and in-group bias on assertive bystander intervention intentions. *Journal of Social Issues, 70*, 167–182. doi: 10.1111/josi.12053

Adams, C.J. (2015). *The Sexual Politics of Meat: A Feminist-Vegetarian Critical Theory* (25th anniversary edition). New York: Bloomsbury Academic.

Brewer, N. T., Chapman, G. B., Rothman, A. J., Leask, J., & Kempe, A. (2017). Increasing vaccination: Putting psychological science into action. *Psychological Science in the Public Interest, 18*, 149–207. doi: 10.1177/1529100618760521

Carrington, D. (2019, Jan 16). New plant-focused diet would 'transform' planet's future, say scientists. *The Guardian*. www.theguardian.com/environment/2019/jan/16/new-plant-focused-diet-would-transform-planets-future-say-scientists

Ceballos, G., Ehrlich, P. R., & Dirzo, R. (2017). Biological annihilation via the ongoing sixth mass extinction signaled by vertebrate population losses and declines. *Proceedings of the National Academy of Sciences, 114*, E6089–E6096. www.pnas.org/cgi/doi/10.1073/pnas.1704949114

Courchamp, F., Jaric, I., Alber, C., Meinard, Y., Ripple, W. J., & Chapron, G. (2018). The paradoxical extinction of the most charismatic animals. *PLoS Biology, 16*, e2003997. https://doi.org/10.1371/journal.pbio.2003997

de Waal, F. (2009). *The Age of Empathy: Nature's Lessons for a Kinder Society*. New York: Harmony Books.

de Waal, F. (2016). *Are We Smart Enough to Know How Smart Animals Are?* New York: Norton.

Dhont, K., & Hodson, G. (2014). Why do right-wing adherents engage in more animal exploitation and meat consumption? *Personality and Individual Differences, 64*, 12–17. doi: http://dx.doi.org/10.1016/j.paid.2014.02.002

Dhont, K., Hodson, G., Loughnan, S., & Amiot, C. E. (2019). Rethinking human–animal relations: The critical role of social psychology. *Group Processes & Intergroup Relations, 22*(6), 769–784. doi: 10.1177/1368430219864455

Eagly, A. H. (2016). When passionate advocates meet research on diversity, does the honest broker stand a chance? *Journal of Social Issues, 72*, 199–222. doi: 10.1111/josi.12163

Earle, M., & Hodson, G. (2017). What's your beef with vegetarians? Predicting anti-vegetarian prejudice from pro-beef attitudes across cultures. *Personality and Individual Differences, 119*, 52–55. http://dx.doi.org/10.1016/j.paid.2017.06.034

Earle, M., Hodson, G., Dhont, K., & MacInnis, C. C. (2019). Eating with our eyes (closed): Effects of visually associating animals with meat on anti-vegan/vegetarian attitudes and meat consumption willingness. *Group Processes and Intergroup Relations, 22*, 818–835. doi: 10.1177/1368430219835024

Fiske, S. T., Cuddy, A. J. C., Glick, P., & Xu, J. (2002). A model of (often mixed) stereotype content: Competence and warmth respectively follow from perceived status and competition. *Journal of Personality and Social Psychology, 82*, 878–902. doi:10.1037//0022-3514.82.6.878

Flocken, J. (2018, Jan 4). Trophy hunting: 'Killing animals to save them is not conservation'. *CNN*. www.cnn.com/2015/05/19/opinions/trophy-hunting-not-conservation-flocken/index.html

Franklin-Wallis, O. (2019). White gold: The Unstoppable Rise of Alternative Milks. *The Guardian*. www.theguardian.com/news/2019/jan/29/white-gold-the-unstoppable-rise-of-alternative-milks-oat-soy-rice-coconut-plant

Gehlbach, H., Robinson, C. D., & Vriesema, C. C. (2019). Leveraging cognitive consistency to nudge conservative climate change beliefs. *Journal of Environmental Psychology, 61*, 134–137. https://doi.org/10.1016/j.jenvp.2018.12.004

Gilovich, T., & Ross, L. (2015). *The Wisest One in the Room: How You Can Benefit From Social Psychology's Most Powerful Insights*. New York: Free Press.

Hodson, G. (2008). Interracial prison contact: The pros for (socially dominant) cons. *British Journal of Social Psychology, 47*, 325–351. doi:10.1348/014466607X231109

Hodson, G., & Earle, M. (2018). Conservatism predicts lapses from vegetarian/vegan diets to meat consumption (through lower social justice concerns and social support). *Appetite, 120*, 75–81. doi: https://doi.org/10.1016/j.appet.2017.08.027

Hoffman, A. (2017, Aug 3). Donald Trump referred to "local milk people." The internet took it from there. *TIME*. http://time.com/4886389/trump-local-milk-people-meme/

Jones, L. (2018, Apr 17). Could eating rare-breed animals save them from extinction? *The Guardian*. www.theguardian.com/environment/shortcuts/2018/apr/17/could-eating-rare-breed-animals-save-them-from-extinction#img-1

Jost, J. T. (2006). The end of the end of ideology. *American Psychologist, 61*, 651–670. doi: 10.1037/0003-066X.61.7.651

Joy, M. (2010). *Why We Love Dogs, Eat Pigs, and Wear Cows: An Introduction to Carnism*. San Francisco, CA: Conari Press.

'Local milk people': Donald Trump's odd expression inspires memes (2017, Aug 4). *The Guardian*. www.theguardian.com/us-news/2017/aug/04/local-milk-people-donald-trumps-odd-expression-inspires-memes

Piazza, J., Ruby, M. B., Loughnan, S., Luong, M., Kulik, J., Watkins, H.M., & Seigerman, M. (2015). Rationalizing meat consumption: The 4Ns. *Appetite, 91*, 114–128. http://dx.doi.org/10.1016/j.appet.2015.04.011

Rothgerber, H. (2013). Real men don't eat (vegetable) quiche: Masculinity and the justification of meat consumption. *Psychology of Men and Masculinity, 14*, 363–375. doi: 10.1037/a0030379

Ruby, M. B., & Heine, S. J. (2011). Meat, morals, and masculinity. *Appetite, 56*, 447–450. doi:10.1016/j.appet.2011.01.018

Sevillano, V. & Fiske, S. T. (2019). Stereotypes, emotions, and behaviors associated with animals: A causal test of the Stereotype Content Model and BIAS Map. *Group Processes & Intergroup Relations, 22*, 879–900. DOI: 10.1177/1368430219851560

Southwick, R. (2019, Jan 9). Video of police driving over 'shrieking' deer triggers calls for criminal charges, death threats. *CBC*. www.cbc.ca/news/canada/calgary/lethbridge-deer-video-1.4972346

Stănescu, V. (2018). 'White power milk': Milk, dietary racism, and the 'alt-right'. *Animal Studies Journal, 7*, 103–128. https://ro.uow.edu.au/asj/vol7/iss2/7

Watts, J. (2019, May 3). Biodiversity crisis is about to put humanity at risk, UN scientists to warn. *The Guardian*. www.theguardian.com/environment/2019/may/03/climate-crisis-is-about-to-put-humanity-at-risk-un-scientists-warn

Willingham, A. J. (2018, Mar 7). Cecil the lion 'suffered incredible cruelty' for 10 hours before he died, a new book claims. *CNN*. www-m.cnn.com/2018/03/07/africa/cecil-lion-death-book-excerpt-trnd/index.html

INDEX

Note: italicised page references indicate illustrations; **bold** page references indicate tables; the suffix 'n' indicates a note.

ASKHAM BRYAN
COLLEGE
LEARNING RESOURCES

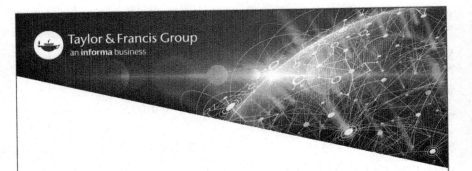

Taylor & Francis eBooks

www.taylorfrancis.com

A single destination for eBooks from Taylor & Francis
with increased functionality and an improved user
experience to meet the needs of our customers.

90,000+ eBooks of award-winning academic content in
Humanities, Social Science, Science, Technology, Engineering,
and Medical written by a global network of editors and authors.

TAYLOR & FRANCIS EBOOKS OFFERS:

A streamlined
experience for
our library
customers

A single point
of discovery
for all of our
eBook content

Improved
search and
discovery of
content at both
book and
chapter level

REQUEST A FREE TRIAL
support@taylorfrancis.com